Water Quality Modelling, Monitoring and Mitigation

Water Quality Modelling, Monitoring and Mitigation

Editors

Amit Kumar
Santosh Subhash Palmate
Rituraj Shukla

MDPI • Basel • Beijing • Wuhan • Barcelona • Belgrade • Manchester • Tokyo • Cluj • Tianjin

Editors

Amit Kumar
School of Hydrology and Water Resources
Nanjing University of Information Science and Technology
Nanjing
China

Santosh Subhash Palmate
Water Resources Division
Texas A&M AgriLife Research Center
El Paso
United States

Rituraj Shukla
School of Engineering
University of Guelph
Ontario
Canada

Editorial Office

MDPI
St. Alban-Anlage 66
4052 Basel, Switzerland

This is a reprint of articles from the Special Issue published online in the open access journal *Applied Sciences* (ISSN 2076-3417) (available at: www.mdpi.com/journal/applsci/special_issues/water_quality_modelling_monitoring_mitigation).

For citation purposes, cite each article independently as indicated on the article page online and as indicated below:

LastName, A.A.; LastName, B.B.; LastName, C.C. Article Title. *Journal Name* **Year**, *Volume Number*, Page Range.

ISBN 978-3-0365-9643-3 (Hbk)
ISBN 978-3-0365-9642-6 (PDF)

Contents

About the Editors

Amit Kumar

Dr. Kumar is a hydrologist cum environmentalist. He has worked in areas like sources, fate, and transport of carbon in water and soil, greenhouse gas emissions from reservoirs/lakes/forests, water quality, and ecological health assessment. He holds an M.Tech and Ph.D from IIT Roorkee, one of the top technical institutes in India, for which he received an MHRD fellowship. He undertook a postdoc at Hohai University's College of Hydrology and Water Resources, Nanjing, China, and now works as an Associate Professor at Nanjing University of Information Science and Technology, School of Hydrology and Water Resources, Nanjing, China. He has published more than 90 research/review articles and attended several training and conferences with support from ICAR, ETH, CAS, MOEF and DAAD. Recently, he has edited two special issues book for MDPI publication. He has edited special issues in *Land* (MDPI), *Applied Sciences* (MDPI), *Frontier in Environmental Science*, is a Topic Editor in *Water* (MDPI), is an Advisory Board Member of *Ecological Indicators*, Elsevier, as well as Associate Editor for *Applied Water Science* (Springer), and *IJRBM* (Taylor and Francis). Currently he is handling a NSFC projects and being ranked in 2% of world scientists, which was released by Stanford University in 2022.

Santosh Subhash Palmate

Dr. Palmate has worked on hydrological and climate change modeling for the last 10 years. He completed his M.Tech and Ph.D at IIT Roorkee, India. Before joining the Texas A&M AgriLife Research Center, he worked with Wetlands International South Asia, New Delhi, as a Technical Officer. Currently, he is working as a Postdoctoral Researcher at Texas A&M AgriLife Research Center, USA, and is deeply involved in research. During his academic career, he has published over 20 research articles and written 6 book chapters of international repute. Dr. Palmate is a Guest Editor in *Water* and a reviewer for various prestigious journals. He is also a professional member of EGU, AGU, IAHS, EWRA, etc.

Rituraj Shukla

Dr. Shukla has worked on hydrological and water quality modeling for the last 10 years. He completed his B.Tech and M.Tech at Indira Gandhi Agricultural University and his Ph.D at IIT Roorkee, India. Before joining the University of Guelph, he worked at the National Institute of Hydrology, Indian Institute of Technology Roorkee, etc., and handled various research projects. Currently, he is working as a Postdoctoral Researcher at the University of Guelph, Canada, and is deeply involved in research. During his academic career, he has published over 40 research articles and written 4 book chapters of international repute.

Preface to "Water Quality Modelling, Monitoring and Mitigation"

In the modern era, water quality indices and models have received global attention from environmentalists, policymakers, governments, stakeholders, water resource planners, and managers due to their ability to evaluate the water quality of freshwater bodies and groundwater aquifers. Due to their wide applicability, models are generally developed based on site-specific guidelines and are not generic; therefore, predicted/calculated values are reported to be highly uncertain. Thus, model and/or index formulation are still challenging and represent a current research hotspot in the scientific community. The inspiration for this reprint came from our desire to provide a platform for sharing results and informing young minds around the world to develop suitable models to understand water quality so that mitigation measures can be taken in advance to make water fit for drinking and for life-supporting activities.

Amit Kumar, Santosh Subhash Palmate, and Rituraj Shukla
Editors

applied sciences

Editorial

Water Quality Modelling, Monitoring, and Mitigation

Amit Kumar [1,2,*], Santosh Subhash Palmate [3] and Rituraj Shukla [4]

1 School of Hydrology and Water Resources, Nanjing University of Information Science & Technology, Ningliu Road No. 219, Nanjing 210044, China
2 China Key Laboratory of Hydro-Meteorological Disaster Mechanism and Warning, Ministry of Water Resources, Nanjing 210044, China
3 Texas A&M AgriLife Research, Texas A&M University, 1380 A&M Circle, El Paso, TX 79927, USA
4 School of Engineering, University of Guelph, Guelph, ON N1G 2W1, Canada
* Correspondence: amitkdah@nuist.edu.cn

Abstract: In the modern era, water quality indices and models have received attention from environmentalists, policymakers, governments, stakeholders, water resource planners, and managers for their ability to evaluate the water quality of freshwater bodies. Due to their wide applicability, models are generally developed based on site-specific guidelines and are not generic; therefore, predicted/calculated values are reported to be highly uncertain. Thus, model and/or index formulation are still challenging and represent a current research hotspot in the scientific community. The inspiration for this Special Issue came from our desire to provide a platform for sharing results and informing young minds around the world to develop suitable models to understand water quality so that mitigation measures can be taken in advance to make water fit for drinking and for life-supporting activities.

Keywords: water quality; monitoring; modeling; mitigations; water quality indexing

Citation: Kumar, A.; Palmate, S.S.; Shukla, R. Water Quality Modelling, Monitoring, and Mitigation. *Appl. Sci.* **2022**, *12*, 11403. https://doi.org/10.3390/app122211403

Received: 17 July 2022
Accepted: 9 November 2022
Published: 10 November 2022

1. Introduction

Due to the rapid increase in anthropogenic activity in catchments, further adverse changes in access to water resources are expected in the future [1] Under these conditions, water quality (WQ) plays an important role that determines its economic utility, including in the potable or drinking water supply, recreation, and agriculture. In the modern era, the study of and commitment to monitoring, modeling, and mitigation have become important and meaningful aspects of the environmental impact assessment process [2]. Under various circumstances, the potentially adverse impacts on ecological flora and fauna can be mitigated through the strategic design and implementation of appropriate models, tools, or techniques to diminish the severity of the effects [3,4]. Different types of nutrients, contaminants (heavy/trace metals), micropollutants, nanoparticles, microplastics, microbes, etc., disturb the ecological life in freshwater bodies [5,6]. Therefore, evidence-based pollution control is urgently needed to focus on the elementary level of water governance, known as "monitoring, modeling, and mitigation". Monitoring sets the empirical basis by providing spatio-temporal information on substance (contaminants and WQ parameters such as dissolved oxygen, biochemical oxygen demand, chemical oxygen and demand, and nutrients) loads as well as the driving boundary conditions for evaluating WQ trends and statuses and for further providing useful information to mitigate contamination and to balance ecological life [7]. Thus, modeling helps to provide long-/medium- and long-term information for times and locations where monitoring is not at all possible [8,9].

The proposed Special Issue will explore cross-disciplinary approaches, modeling, and methods and will discuss water quality risks as well as solutions for the implications for environmental sustainability and for the further conservation of ecological life. The interconnectedness of this critical problem cannot be assessed with traditional approaches; instead, inter- and trans-disciplinary approaches are urgently required worldwide to deal with water resource problems and environmental sustainability challenges.

2. Overview of Water Quality Indexing Models

In general, water quality index (WQI) models are frequently used to evaluate the WQ of freshwater bodies (e.g., lakes, rivers, and reservoirs) [10,11]. These models use aggregation techniques to convert extensive WQ datasets into a single representative value. Since the 1990s, WQI models have been extensively used to evaluate the WQ of surface water and groundwater [12] based on local criteria because they are easy to handle and free (Figure 1). The literature has reported that more than 30 WQI models have been created and introduced worldwide to evaluate the WQ of freshwater bodies [13–15]. WQI models are generally completed in four consecutive stages: (i) the selection of WQ parameters, (ii) sub-indices generation for individual parameters, (iii) the calculation of the weighting values of each parameter, and (iv) the sum of all sub-indices values to evaluate the WQI. The literature has reported a range of applications of WQI models to evaluate the WQ of freshwater systems [10,14–16]. However, most of the models that have been developed are based on site-specific guidelines and are not generic; therefore, the large uncertainty in the predictions and/or estimations made by these WQI models is coming into the picture and creating a hindrance in strategic mitigation measures for WQ control for sustainable ecological life and human use.

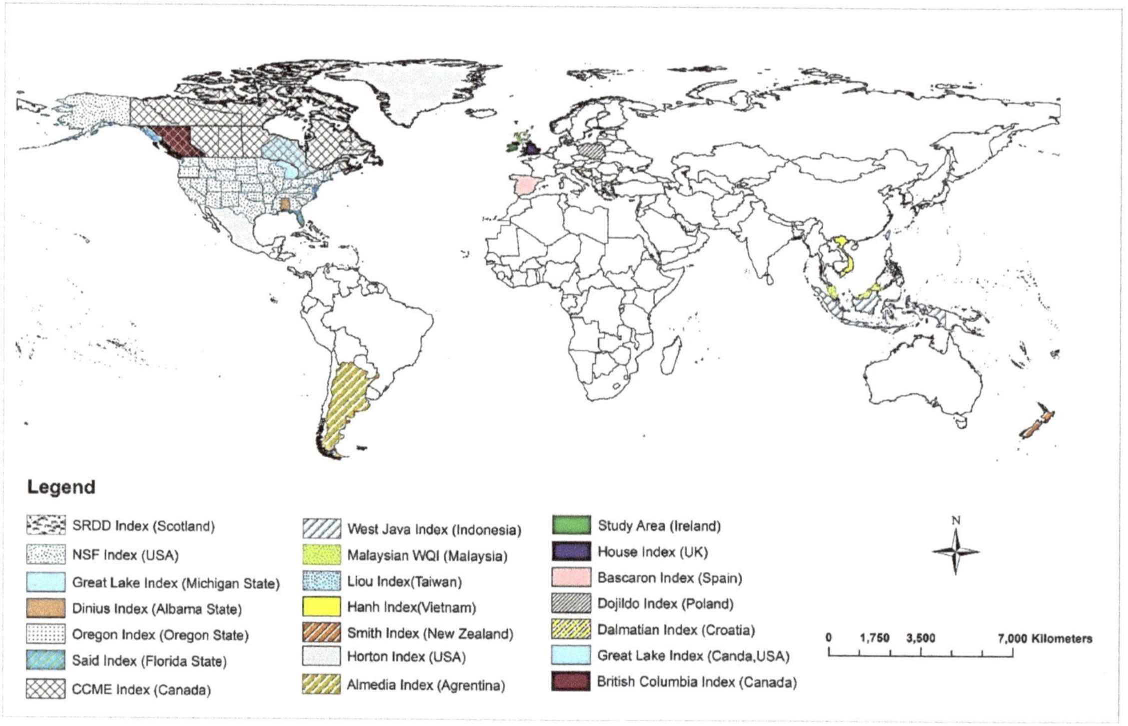

Figure 1. Commonly used WQI models worldwide [15].

3. Water Quality Models, Challenges, and Limitations

Water quality modeling (WQM) is an important tool that aids environmentalists, policymakers, water resource planners, and managers in strategic water resource management.

However, WQM represents a challenge in the scientific community due to several constraints and limitations. In general, WQ models are classified based on the types of receiving water, the complexity of the models, and the WQ parameters (e.g., nutrients, dissolved oxygen, biological oxygen demand, etc.) that the model can predict. Thus, WQM requires proper standardization, pollution hotspots, the identification of common features, and policy-relevant models. These models save labor costs, materials, and time [16] and help in effective pollution mitigation for the watershed. In the recent era, numerous models have been frequently used to simulate the water quality of freshwater bodies (streams, rivers, reservoirs, and lakes), estuaries, coastal waters, and marine ecosystems [17]. However, due to the different theories and algorithms applied in the models, their corresponding outputs are different and create huge differences in the results; thus, models could be useful and produce fruitful results when applied to solve particular environmental problems [18].

Water quality (WQ) models are generally categorized into two categories: (i) physical and (ii) mathematical models [19]. Furthermore, they can be categorized according to the complexity of model simulation, i.e., 1D, 2D, and 3D; type of approach (conceptual, physical, or empirical); data requirements; types of pollutants; area of application (groundwater, catchment, lake, river, coastal waters, etc.); nature (stochastic or deterministic); and spatial analysis [20]. In recent decades, WQ models such as ANSWERS-2000, AquaChem, MIKE SHE, AGWA, GLEAMS/CREAMS, AQUATOX, APEX, EFDC, EPD-RIV1, BASINS, HSPF, KINEROS2, LSPC, NLEAP, PRMS, QUAL2K, QUAL2E, SWMM, SWAT, WARMF, WAM, WCS, and WASP7 have been frequently used to predict WQ worldwide [21]. Because of data requirements and availability as well as types of catchment problems, the simplest reliable models are dominant over complex models [22].

WQ modeling is still challenging in the scientific domain due to the lack of expert handling of user, site, and/or regionally specific and parameter-specific information as well as inadequacies in model calibration and errors in data reporting. The uncertainty in WQM comes from various sources of errors, such as (i) parametric uncertainty, (ii) structural errors, and (iii) errors in the measurements of the input values and response uncertainty [23]. In developing countries (e.g., India and China), a uniform model standardization system has not been recognized, which limits the extensive utilization of those models for ecological and water management as a result of the lack of benchmarks and comparisons between different modeling outcomes [9,10,16]. Spatial variability is reported as a serious problem in catchment-scale WQM that generally acquires catchment behavior, representative site selection, and the integration of nonlinear biogeochemistry [24]. However, the complexity of models, the inadequate availability of data, and poor WQ data are other important limiting factors for WQM.

4. Water Quality Mitigation Measures

Water quality mitigation measures or strategies are generally intended to inform and assist communities in identifying potential alternatives to minimize the adverse impacts of pollutants on WQ and to ensure that water is safe for community use [25]. Ultimately, mitigation measures help to protect, restore, preserve, and improve the WQ of receiving water bodies. WQ protection refers to adequately treating runoff to protect downstream resources from WQ degradation [26]. Restoration comes into action if the protection strategies are not sufficient to maintain WQ standards as per the permissible limits. Stakeholders from different fields working together to achieve WQ restoration goals [27]. Water quality preservation necessitates a decision-support framework that can be used to evaluate, monitor, and optimize the effects of different drivers on WQ [28]. Furthermore, WQ improvements can be accomplished by identifying the highest priorities for WQ conditions and implementing mitigation strategies to address ongoing issues in a study area (Figure 2) [29,30]. Sometimes, the study areas do not follow the jurisdictional boundaries; therefore, several stakeholders need to work together to achieve local/regional/national WQ goals.

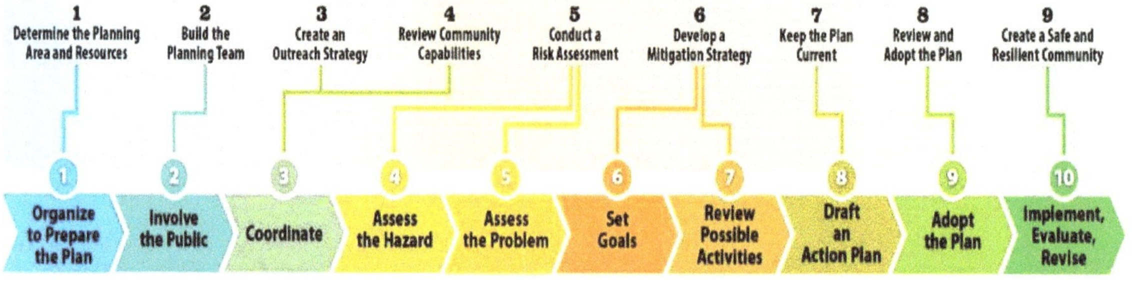

Figure 2. Implementation steps for strategies to mitigate WQ problems [2].

Water quality standards can be mitigated through regulation, remediation, and watershed management [31]. Water regulation in a specific area can control the free discharge of waste from industry or sewage treatment plants by setting standards for each pollutant released into surface waters [31]. Remediation acts, such as biological, chemical, and physical acts, help in cleaning the water contamination; (i) biological remediation is a cost-efficient method and is also called "bioremediation", which involves the use of naturally occurring organisms such as plants, bacteria, and fungi to remove or neutralize water pollutants and to breakdown hazardous substances into less toxic or nontoxic substances. Human sewage and agricultural chemicals that leach from the soil into the groundwater are generally treated by bioremediation [32,33]. (ii) Chemical remediation methods use chemicals to react with the water contaminants to remove or make them less harmful, and (iii) physical remediation includes the removal of water contamination by treating it with filtration or disposing of it. Overall, all these three remediation methods are somehow complex, expensive, and difficult to adapt.

Watershed management strategies consist of reducing the chemicals applied to land, making them more effective for nonpoint source pollution than setting pollution standards [3]. In a watershed, riparian areas promote WQ and limit pollution; therefore, their maintenance and restoration are crucial. Vegetation surrounded by a water body absorbs nutrients and provides shade to keep water cool and increase its capacity to hold dissolved oxygen (DO). Additionally, vegetation reduces runoff, promotes infiltration, and lowers soil erosion. Hence, vegetation plays a key role in the effective management of WQ through watershed management. Watershed practices that are beneficial for maintaining WQ standards include (i) regional infiltration basins; (ii) neighborhood-scale practices such as rain gardens, bioretention, and permeable pavement; (iii) stream restoration, including pooling and meandering to enhance infiltration; (iv) floodplain restoration, including floodplain benching; (v) stream (riparian) buffers; (vi) using park green space and fields to store and infiltrate water; (vii) stormwater-friendly post-construction design; and (viii) protecting and resting natural and human-made wetlands. Some important actions can be taken to get rid of polluted water before pollution ever happens and to mitigate WQ standards:

- Avoid dumping waste in and around water bodies—dumping waste in water bodies leads to water pollution over a short period time.
- Septic systems need proper maintenance and cleaning from time-to-time—leakage from septic tanks can cause groundwater pollution.
- Stakeholders and industries need to follow WQ regulations and laws—breaking local government rules can put water at a high risk of contamination due to anthropogenic activities.
- All chemicals and pesticides need to be disposed of properly—pouring these hazards into nearby drains, sinks, grass yards, or water closets can lead to them entering local water sources and water supply networks.
- Do not pour kitchen leftovers, such as cooking oils and fat substances, into the sink—pouring these substances can seriously clog sink drains as well as spread illness into utilizable water.

- Try not to use any bleaching substances for washing at home—using bleach can trigger poisoning and maybe internal burning and can eventually damage local environmental sources.
- Schedule cleanings for yards and nearby areas with your neighborhood—dirty surroundings may cause water pollution and put human health at risk.

Overall, we need to work on mitigating water quality and educating friends, family, neighborhoods, and relatives about the necessary actions for water safety.

Most WQ mitigation measures aim to prohibit illicit discharge, control erosion, reduce pollutants, and control excessive flows. Additionally, strategies consisting of implementing outreach, education, and other activities that promote infiltration, flood reduction, and stable drainage channels could be beneficial for WQ management [34]. Stormwater flow management, floodplain restoration, channel stabilization, and green infrastructure installations are the main strategies to prevent pollutant discharge into surface waters from stormwater, including wastewater. Wetland protection, rehabilitation, and restoration activities improve WQ and quantity and support the maintenance of floodplains in their natural state [35]. The protection of riparian areas and floodplains and keeping hazardous materials from source water areas can directly safeguard drinking WQ and can indirectly protect public health. Sometimes, financial resources limit the application of these mitigation strategies, so the prioritization of mitigation strategies can focus on important WQ issues that are necessary to complete in a short period of time. To overcome this, the provision of grants/funding is also essential to encourage vegetation planting and maintenance over time.

5. Conclusions

Water quality (WQ) tools and models are described and selected based on their applicability, site- or regional-specific qualities, weaknesses, strengths, and whether or not they are intended for commercial or industrial use. The outputs of models and WQ indexing are different based on the input requirements and data availability and therefore have large levels of uncertainty, are not freely available for commercial use, and require skilled model users. Model selection is a robust task in the scientific domain; therefore, when selecting suitable models for pollution control in freshwater bodies, catchments, or a specific site, there are requirements to consider, such as the availability of datasets, the complexity of the models, and the type of freshwater bodies, and the intended objectives should be modeled so that mitigation strategies can be implemented in fruitful ways.

Author Contributions: Conceptualization, A.K. and S.S.P.; methodology, A.K.; validation, A.K., S.S.P. and R.S.; formal analysis, A.K., S.S.P. and R.S.; investigation, A.K.; resources, A.K.; data curation, A.K.; writing—original draft preparation, A.K., S.S.P. and R.S.; writing—review and editing, A.K., S.S.P. and R.S.; visualization, A.K.; supervision, A.K.; project administration, A.K., S.S.P. and R.S.; funding acquisition, A.K. All authors have read and agreed to the published version of the manuscript.

Funding: This study is partially supported by National Natural Science Foundation of China (NSFC)-International young scientist Project (Grant no: 52150410400).

Acknowledgments: We are thankful to the authors for their contributions to this Special Issue, which have been integral to its success. Moreover, the excellent support from the editors (especially Section Managing Editor) and reviewers has been highly encouraging.

Conflicts of Interest: The authors declare no conflict of interest.

References

1. Xu, X.; Xu, W.; Xie, C.; Khan, M.Y.A. Prediction of the Long-Term Performance Based on the Seepage-Stress-Damage Coupling Theory: A Case in South-to-North Water Diversion Project in China. *Appl. Sci.* **2021**, *11*, 11413. [CrossRef]
2. EPA. *Integrating Water Quality and Natural Hazard Mitigation Planning;* United Stated Environmental Protection Agency: Washington, DC, USA, 2019. Available online: https://www.epa.gov/green-infrastructure/integrating-water-quality-and-natural-hazard-mitigation-planning (accessed on 11 June 2022).
3. Saby, L.; Nelson, J.D.; Band, L.E.; Goodall, J.L. Nonpoint Source Water Quality Trading outcomes: Landscape-scale patterns and integration with watershed management priorities. *J. Environ. Manag.* **2021**, *294*, 112914. [CrossRef] [PubMed]

4. Bielski, A.; Toś, C. Remote Sensing of the Water Quality Parameters for a Shallow Dam Reservoir. *Appl. Sci.* **2022**, *12*, 6734. [CrossRef]
5. Lu, H.; Yuan, Y.; Campbell, D.E.; Qin, P.; Cui, L. Integrated water quality, emergy and economic evaluation of three bioremediation treatment systems for eutrophic water. *Ecol. Eng.* **2014**, *69*, 244–254. [CrossRef]
6. Kumar, A.; Upadhyay, P.; Prajapati, S.K. Impact of Microplastics on riverine greenhouse gas emissions—A View point. *Environ. Sci. Pollut. Res.* **2022**. [CrossRef] [PubMed]
7. Kumar, A.; Taxak, A.K.; Mishra, S.; Pandey, R. Long term trend analysis and suitability of water quality of River Ganga at Himalayan hills of Uttarakhand, India. *Environ. Technol. Innov.* **2021**, *22*, 101405. [CrossRef]
8. Gupta, A.; Himanshu, S.K.; Gupta, S.; Singh, R. Evaluation of the SWAT Model for Analysing the Water Balance Components for the Upper Sabarmati Basin. In *Advances in Water Resources Engineering and Management*; Al Khaddar, R., Singh, R., Dutta, S., Kumari, M., Eds.; Springer: Singapore, 2020; Volume 39, pp. 141–151. [CrossRef]
9. Himanshu, S.K.; Fan, Y.; Ale, S.; Bordovsky, J. Simulated efficient growth-stage-based deficit irrigation strategies for maximizing cotton yield, crop water productivity and net returns. *Agric. Water Manag.* **2021**, *250*, 106840. [CrossRef]
10. Kumar, A.; Sharma, M.P.; Rai, S.P. A novel approach for river health assessment of Chambal using fuzzy modeling, India. *Desalin. Water Treat.* **2017**, *58*, 72–79. [CrossRef]
11. Lencha, S.M.; Ulsido, M.D.; Muluneh, A. Evaluation of Seasonal and Spatial Variations in Water Quality and Identification of Potential Sources of Pollution Using Multivariate Statistical Techniques for Lake Hawassa Watershed, Ethiopia. *Appl. Sci.* **2021**, *11*, 8991. [CrossRef]
12. Senoro, D.B.; de Jesus, K.L.M.; Mendoza, L.C.; Apostol, E.M.D.; Escalona, K.S.; Chan, E.B. Groundwater Quality Monitoring Using In-Situ Measurements and Hybrid Machine Learning with Empirical Bayesian Kriging Interpolation Method. *Appl. Sci.* **2022**, *12*, 132. [CrossRef]
13. Dadolahi-Sohrab, A.; Arjomand, F.; Fadaei-Nasab, M. Water quality index as a simple indicator of watersheds pollution in south-western part of Iran. *Water Environ. J.* **2012**, *26*, 445–454. [CrossRef]
14. Abbasi, T.; Abbasi, S.A. *Water Quality Indices*; Elsevier: Amsterdam, The Netherlands, 2012; pp. 353–356. [CrossRef]
15. Uddin, M.G.; Nash, S.; Olbert, A.I. A review of water quality index models and their use for assessing surface water quality. *Ecol. Indic.* **2021**, *122*, 107218. [CrossRef]
16. Wang, Q.; Li, S.; Jia, P.; Qi, C.; Ding, F. A review of surface water quality models. *Sci. World J.* **2013**, 231768. [CrossRef] [PubMed]
17. Loucks, D.P.; Van Beek, E. Water Quality Modeling and Prediction. In *Water Resource Systems Planning and Management*; Springer: Cham, Switzerland, 2017; pp. 417–446.
18. Obropta, C.C.; Niazi, M.; Kardos, J.S. Application of an environmental decision support system to a water quality trading program affected by surface water diversions. *Environ. Manag.* **2008**, *42*, 946–956. [CrossRef]
19. UNESCO. Water Resources Systems Planning and Management. 2005. Available online: https://ecommons.cornell.edu/bitstream/handle/1813/2804/12_chapter12.pdf?sequence=9&isAllowed=y (accessed on 5 July 2022).
20. Tsakiris, G.; Alexakis, D. Water quality models: An overview. *Eur. Water* **2012**, *37*, 33–46.
21. Ejigu, M.T. Overview of water quality modeling. *Cogent Eng.* **2021**, *8*, 1891711. [CrossRef]
22. Zheng, C.; Bennett, G.D. *Applied Contaminant Transport Modelling*; Van Nostrand Reinhold: New York, NY, USA, 1995.
23. Rode, M.; Suhr, U. Uncertainties in selected river water quality data. *Hydrol. Earth Syst. Sci.* **2007**, *11*, 863–874. [CrossRef]
24. Rode, M.; Arhonditsis, G.; Balin, D.; Kebede, T.; Krysanova, V.; Van Griensven, A.; Van Der Zee, S.E. New challenges in integrated water quality modelling. *Hydrol. Process.* **2010**, *24*, 3447–3461. [CrossRef]
25. Liu, Y.; Theller, L.O.; Pijanowski, B.C.; Engel, B.A. Optimal selection and placement of green infrastructure to reduce impacts of land use change and climate change on hydrology and water quality: An application to the Trail Creek Watershed, Indiana. *Sci. Total Environ.* **2016**, *553*, 149–163. [CrossRef]
26. Dujmović, J.; Allen, W.L., III. Soft computing logic decision making in strategic conservation planning for water quality protection. *Ecol. Inform.* **2021**, *61*, 101167. [CrossRef]
27. Ball, T. Management approaches to floodplain restoration and stakeholder engagement in the UK: A survey. *Ecohydrol. Hydrobiol.* **2008**, *8*, 273–280. [CrossRef]
28. Ravier, C.; Prost, L.; Jeuffroy, M.H.; Wezel, A.; Paravano, L.; Reau, R. Multi-criteria and multi-stakeholder assessment of cropping systems for a result-oriented water quality preservation action programme. *Land Use Policy* **2015**, *42*, 131–140. [CrossRef]
29. Shim, K.; Shin, S.; Kim, S.; Kim, Y.; Kim, G. Priority selection of water quality improvement through water quality data of tributaries of Nakdong River. *J. Korean Soc. Water Environ.* **2020**, *36*, 364–372. [CrossRef]
30. Angello, Z.A.; Behailu, B.M.; Tränckner, J. Selection of Optimum Pollution Load Reduction and Water Quality Improvement Approaches Using Scenario Based Water Quality Modeling in Little Akaki River, Ethiopia. *Water* **2021**, *13*, 584. [CrossRef]
31. Guo, T.; Gill, D.; Johengen, T.H.; Cardinale, B.L. What determines the public's support for water quality regulations to mitigate agricultural runoff ? *Environ. Sci. Policy* **2019**, *101*, 323–330. [CrossRef]
32. Kumar, A.; Subrahmanyam, G.; Mondal, R.; Cabral Pinto, M.M.S.; Shabnam, A.A.; Jigyasu, D.K.; Malyan, S.K.; Fagodiya, R.K.; Khan, S.A.; Yu, Z.G. Bio-remediation approaches for alleviation of cadmium contamination in natural resources. *Chemosphere* **2020**, *268*, 128855. [CrossRef]
33. Song, H.; Kumar, A.; Ding, Y.; Wang, J.; Zhang, Y. Removal of Cd^{2+} from wastewater by microorganism induced carbonate precipitation (MICP): An economic bioremediation approach. *Separ. Purif. Tech.* **2022**, *297*, 121540. [CrossRef]

34. Okumah, M.; Chapman, P.J.; Martin-Ortega, J.; Novo, P. Mitigating agricultural diffuse pollution: Uncovering the evidence base of the awareness–behaviour–water quality pathway. *Water* **2018**, *11*, 29. [CrossRef]
35. Bossio, D.A.; Fleck, J.A.; Scow, K.M.; Fujii, R. Alteration of soil microbial communities and water quality in restored wetlands. *Soil Biol. Biochem.* **2006**, *38*, 1223–1233. [CrossRef]

applied sciences

MDPI

Article

Ecosystem Services: A Systematic Literature Review and Future Dimension in Freshwater Ecosystems

Deeksha and Anoop Kumar Shukla *

Manipal School of Architecture and Planning, Manipal Academy of Higher Education, Manipal 576104, Karnataka, India
* Correspondence: anoopgeomatics@gmail.com

Abstract: Ecosystem services are part and parcel of human lives. It is of paramount importance to understand the interaction between these ecosystem services, as they are directly related to human life. In the modern era, quantification of ecosystem services (ES) is playing an important role in the proper understanding and efficient management of social–ecological systems. Even though a significant amount of literature is available to present on the topic, there is a need to build an adequate amount of knowledge repository. Hence, a systematic literature review method is used, in which research question and searching stages are defined. This review study is conducted on ecosystem services and remote-sensing-related keywords in the Scopus database. After a systematic analysis of the papers retrieved from the Elsevier, Scopus database, MDPI, and open source, a total of 140 primary articles were categorized according to their relationship with other ecosystem services, land use, land cover, and planning management. Major issue findings and important aspects have been analyzed and reported in each category. With this analysis and developments in the existing literature, we have potential areas for future research. Findings pointed out that regional or local-level ecosystem services-related work is immensely important, and a hotspot of current research aiming to understand the variability and spatiotemporal dynamics in terrestrial and aquatic ecosystems.

Keywords: ecosystem services; provisioning ecosystem services; regulating ecosystem services; cultural ecosystem services; supporting ecosystem services

Citation: Deeksha; Shukla, A.K. Ecosystem Services: A Systematic Literature Review and Future Dimension in Freshwater Ecosystems. *Appl. Sci.* **2022**, *12*, 8518. https://doi.org/10.3390/app12178518

Academic Editors: Amit Kuma, Santosh Subhash Palmate and Rituraj Shukla

Received: 11 June 2022
Accepted: 11 August 2022
Published: 25 August 2022

1. Introduction

The biophysical state of the ecosystem is affected by multiple elements and, simultaneously, by humans' ability to enjoy its services [1]. Furthermore, Anthropogenic and non-anthropogenic interventions can change the biochemical cycles and earth's energy equilibrium, in turn causing global warming, and future climate changes [2]. On the other hand, rapid urbanization degrades ecosystem services [1,3]. Gretchen, in his work, points out that the lifestyle of the people may be hampering the prosperity of ecological biodiversity at the expense of their descendants [4,5]. Talking about ecological biodiversity, the term ecosystem services is described by many authors as the process that helps to sustain human life, with the help of interaction between the natural ecosystem and the species [4,6]. Globally, human systems are supported by nature's contribution, i.e., ecosystem services, and [7,8] state that land use and land cover changes (LULC) induced by humans have increased over the last three decades, leading to changes from a natural setting to human-conquered land. Furthermore, according to Gómez-Baggethun & Barton [9] more than 50% of the world's inhabitants reside in urban areas and potentially receive benefits from the ecosystem services; the future projection of people living in urban areas is calculated to reach 66% by 2050. The main motive for this research is to find a way to conserve the existing ecosystem, with a background of arising global environmental issues.

Ecosystem Services

Ecosystem services (ES) are defined as the benefits people obtain from nature [1,4,6]. ES connects human well-being and natural systems to ecological and economic development to lay a platform between nature and society [10]. Land use and land cover changes increase the population rate and have a huge impact on ES, which is leading to its degradation [11–16]. Therefore, evaluation of ES has been the core subject of research in the academic section for years [17,18], and the recent past interventions also show the readiness of the study to inform policymakers in undertaking essential decisions in the policy making process along, with the integration of ecology, geography, and economy [19]. In the study [20], the authors put forward that the authenticity of the ecosystem cannot be based on human intervention alone; the same is considered authentic when the researcher considers both pristine and altered forms of the ecosystem, therefore understanding the change in fundamental characteristics of the ecosystem.

ES plays a vital part in constituting the well-being of an individual's life through security provision, meeting the basic needs for day-to-day life along with health and good social relationships with each other. Urban ecosystems are still a critical area of ES research, as half of the world's population dwells in urban areas. According to MEA, around 60% of global ES has been threatened or used inappropriately, and the same process is expected to continue essentially in the first half of the present century [1]. For this reason, recently, ES is significantly considered one of the vital aspects of land use planning and ecological environmental planning and management [19–32].

The interaction between the ESs can take place in two ways. The first is trade-offs, where an increase in the effect of one of the ES results in a decreased effect in other ES. The second is synergies, where the increase in the effect of one ES also leads to the increased effect of other ES [23,24]. When these relationships occur again across space and time, they are called ES bundles [25]. Understanding this relationship is rather critical, as it focuses on the relationship between ES by concentrating on inherent bundles rather than on discrete ES [26–28]. Studies by Bennett et al. say that the trade-offs and the synergies are caused by the interaction among various ecosystems, so the ecosystem services cannot be considered independent [29]. Braat & de Groot infer that the study of various ES is complex [30].

Ecosystems can be monitored at different levels; they can be studied at a global scale or regional scale, or local scale. Global-level studies are carried out worldwide, but researchers suggest studying the services at the local level, which gives us a better understanding of the situation, helping us to take up appropriate mitigation strategies at a regional level. This helps us achieve sustainable goals at the global level [28]. Although research studies by a wide range of scholars have shed light on the interaction of various ecosystem services in recent years, the amalgamation of our existing knowledge repository and gaps is still inadequate [31].

Various other frameworks emerged in the recent past for ES studies [33]. To account for the natural capital, Common International Classification of Ecosystem Services (CICES) integrates different criteria of various ESs. Closely capturing concepts that relate to nature's contribution to humans is the framework developed by Inter-governmental Science-Policy Platform on Biodiversity and Ecosystem Services (IPBES). To understand ES, Ref. [1] came up with a base framework for global ES study; therefore, ES can be classified into four categories: (i) provisioning ecosystem services, (ii) regulating ecosystem services, (iii) cultural ecosystem services, and (iv) supporting ecosystem services.

Provisioning ecosystem services (PES) are defined as the goods that can be directly extracted from nature and consumed, which have a certain market value. Examples of provisioning ecosystem services are water, food, wood, biofuels, etc. Stating the same, for the freshwater supply ecosystem service, it is necessary to have an ecosystem that is functioning in a good way [34]. Climatic factors such as precipitation, evaporation, and climate variability are the important components that control the water yield of the region [34]. Water yield has positive linkage between evapotranspiration and soil conservation [35], along with other components such as food production, timber, etc.

Regulating ecosystem services (RES) can be defined as the benefits that are drawn from the process of the ecosystem that modifies the condition that we are presently experiencing. Examples of the same will be climate changes, carbon storage, soil fertility, floods, etc. [29]. The study emphasizes the relationship in terms of trade-offs and/or synergies of regulating ecosystem with other ESs, due to which regulating ESs can be considered as one of the critical parameters for the assessment of ecological resilience [35]. Managing one ES parameter will improve synergies among other ES parameters, especially among carbon storage, low flow, biodiversity, etc. [36]. Carbon (C) storage is a key attribute in regulating the global service of climate regulation [9]. Practical implementation of C sequestration knowledge will take a back step in public policymaking due to the lack of effectiveness in translating scientific criteria [37]. Carbon sequestration acts as an important parameter in global climatic regulation [33]. Carbon is stored in four different layers in nature, i.e., aboveground biomass, belowground biomass, soil organic carbon, and dead matter storage. As carbon is stored (~70%) in the terrestrial ecosystem [38–40], its carbon dynamics potential could be affected in the future under rising carbon dioxide. Therefore, the carbon present in the soil considerably has a huge impact on the spatial and non-spatial data. Hence, long-and-medium term modeling taking into consideration of different LULC scenarios is the hotspot of current research, which helps policymakers with the mitigation strategies framework and decision-making process. This process can be re-scaled globally, regionally, and locally by co-relating different rationales to economic opportunities and regulatory policymaking.

Gómez-Baggethun et al. [41] introduced cultural ecosystem services (CES), but MEA studied and defined them as "the non-material or intangible benefits people obtain from the ecosystem either spiritually, through cognitive development, recreation, self-reflection or through experiencing aesthetically" [1]. In this ES, some of the services, such as recreations, have market value, whereas other services do not have the same. Functions fulfilling life information functionality are the different ways in which cultural ecosystems are included in the study [4,6,42]. Additionally, Sen & Guchhait simplify the definition by correlating humans' sociocultural practices with psychological development [43]. CES is also associated with the intangible benefits that people attain from nature due to the interaction [44,45]. Most of the studies on cultural ecosystem services deal with recreation services that are nature based and aesthetic [46], whereas not a great amount of study hqw been carried out on the spiritual value of landscape due to the limitation of modeling [47]. Supporting ecosystem services (SES) are the fundamental process of the ecosystem that supports life, such as photosynthesis, nutrient cycle, and evolution; this is a vital service that the ecosystem provides, which allows the rest of the ecosystem services to be delivered.

To achieve sustainable development of the city and conserve the ecosystem, it is necessary to understand each of the abovementioned ecosystem services and their interactions with the changing LULC [48]. This study can help provide future research perspectives and proper decision-making strategies. To offer the same, a literature review for each of the ecosystem services was carried out, providing a global perspective first and then elaborating on the studies conducted at the national level. It will provide an overview of the models and methods that are used for the quantitative study of ecosystem services in the limelight. The publications used various methods of quantitative assessments, such as spatial mapping, economic valuation, etc. Therefore, the main objective of our paper is to understand; (i) the global scenario of ES, (ii) where does ES study in India fall? (iii) what are the research gaps that could be studied, and a way forward in the same area.

2. Materials and Methods

2.1. Data Collection

The literature survey was carried out in November 2021, and data were collected from Science Direct, Scopus, MDPI, official reports, and Wiley. The main strategies of the basic literature review contained four different phases. First, to understand the total number of publications present, we used the keyword "Ecosystem services" to understand the pattern of the study on a yearly basis. Second, to understand works on different types of Ecosystem services, we used the keywords "Ecosystem services and Provisioning Ecosystem services". In the next search, using keywords "Ecosystem services and Regulating Ecosystem services", similarly, the search was carried out for cultural ES and supporting ES and their trend of publication for two decades. This was followed by the search to understand the types of models used to study ES, and also to figure out the types of models catering to ES. Additionally, a search was carried out to understand the trends of individual models serving various parameters of ES. Finally, the investigation was carried out to understand the ES publication in the Indian context using the search keyword "Ecosystem services and India".

After collection of the records, the initial step of analysis took place at various levels. The first step of the selection criteria was to select the papers which spoke about ES throughout the world. This was followed by the data range strategy of selection, wherein the collected data were segregated based on types of publication. In this step, most of the journal articles were selected over the books or conference papers, as the articles are periodical, and there was a high chance of understanding the current trend of publication. Books and conference proceedings were negated. The step was followed by the title and abstract search, wherein the non-related articles were negated after going through the article abstracts and title. This was followed by the criteria search, which considered the related variables of the study. As a result of this, 138 articles were extracted for this review. The assessment parameters for this review are based on the ecosystem services approach and include the date of publication, the context of the publications, the kind of data used/analyzed (qualitative or quantitative), as well as the spatial size of the study. Table 1 gives an insight into the criteria considered for the study.

Table 1. Criteria of Review with Feasible entries.

Criterion	Feasible Entry
Source	Science Direct database
Paper type	Research Paper
Data type	Quantitative data, Qualitative data
Spatial measurement scale	Local, regional, national, global scale
ES indicators	4 types of ES indicators

2.2. Data Analysis

The data analysis consisted the database of search records that was built from the data collection process. The study can be considered susceptible given the huge amount of data resent in the database. To avoid arbitrary comments, a systematic review was carried out on 140 selected papers that dug deeper to understand the knowledge base of the subject, which excluded book chapters, student theses, and reports. Regardless, the search output is considered authenticated and peer reviewed, as it was taken from the distinguished journal article database. The data were taken as a basis for future study (Figure 1).

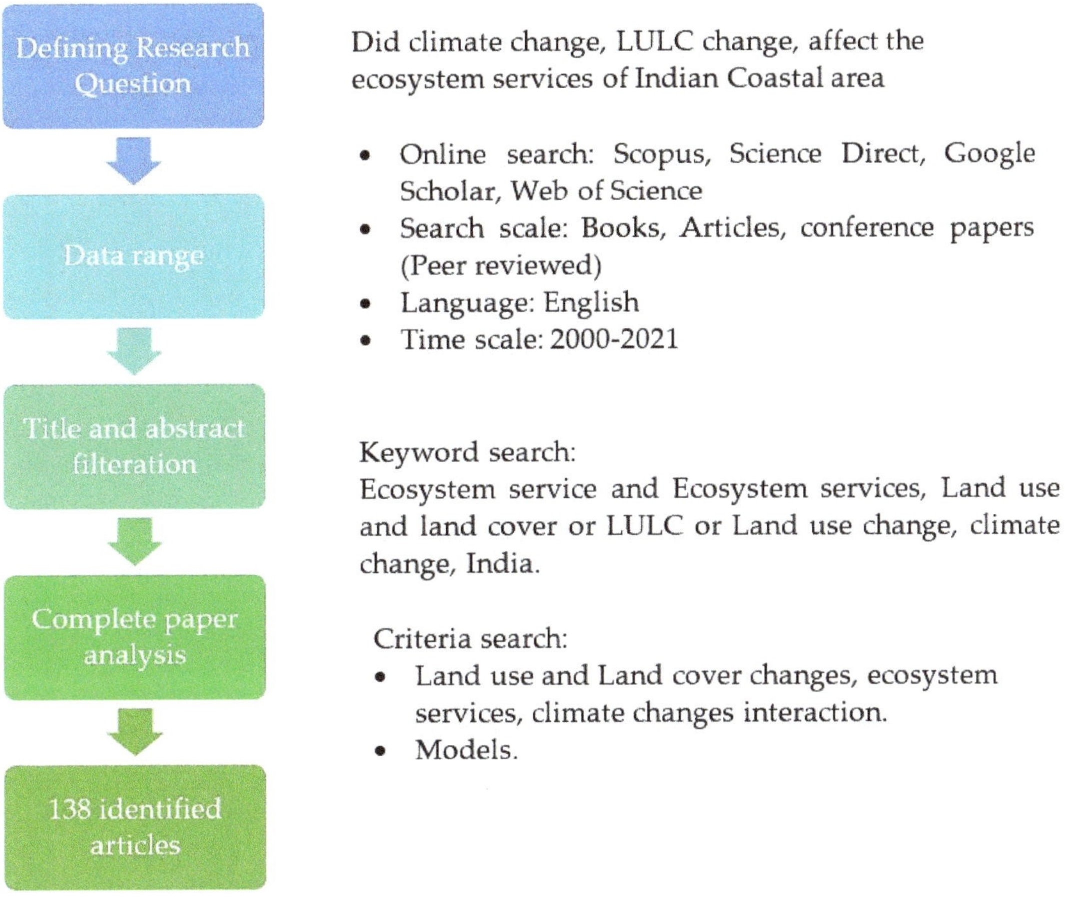

Figure 1. Showing the flow chart of the data analysis for ecosystem services.

3. Results

3.1. Mapping of Publication

Research publications until early 2000 were fewer in number (Figure 2). The reason was unfamiliarity with the subject; later, the work changed the whole lens on how the ecosystem was viewed [4,6]. From 2000 to 2005, we can find approximately two thousand papers on ecosystem services (Figure 2). Later, once the MEA [1] was published, we found a sudden rise in the graph, which denotes that universal attention was attained by ecosystem service research. Later on, it became one of the core research areas among academicians and scholars. From the years 2005 to 2010, we found publications that provided an insight into the trade-off and synergies among the ecosystem services. Following MEA, in 2010, TEEB [46] came up with a newer lens of added economic value to the ES. In 2011, CICES [49] gave common ground for all international works related to ecosystem services. Post-2015 marks a prolific change in the number of publications on ecosystem services, with the publication of the Sustainable Development Goals (SDG) 2030. These are aimed at making cities locally and globally sustainable due to the change in global climatic aspects.

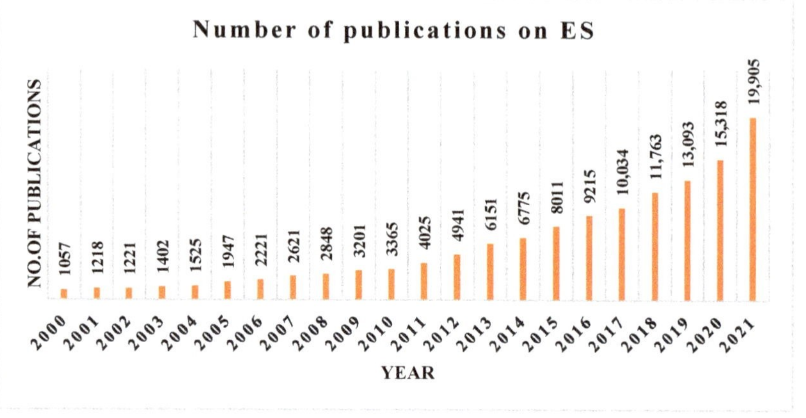

Figure 2. Number of publications on ecosystem services for two decades.

3.2. Chronological Publication on ES Papers

Ecosystem services papers published between the years 2000 and 2005 show that awareness of the subject was limited, wherein a critical understanding of the same was not present (Figure 3). Post MEA, the publications on the ecosystem services increased; today, we find 20,000 publications based on ESs (Figure 3). Sustainable development goals gave the necessary push required for the study and to make the cities more sustainable. MEA formed the ecosystem framework, along with four major categories.

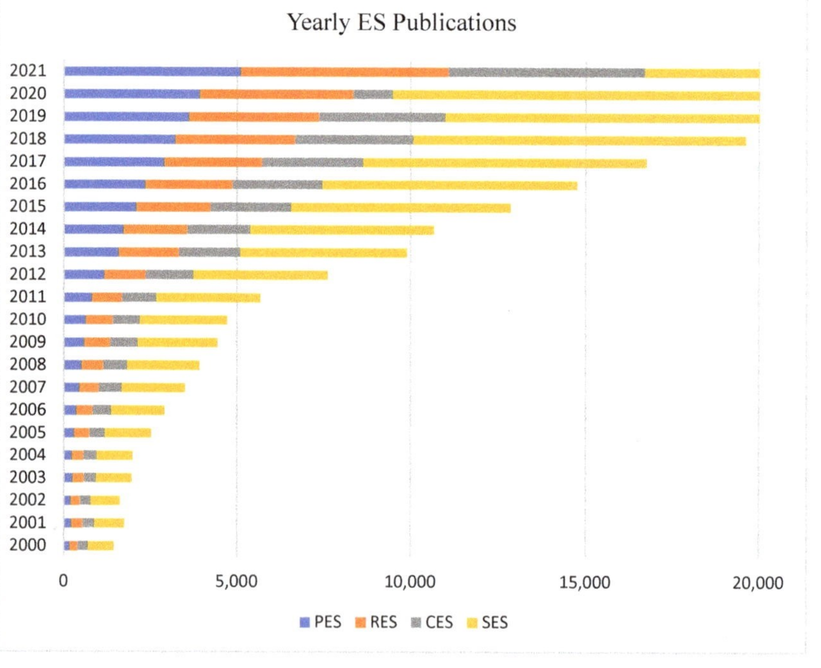

Figure 3. Number of publications on ecosystem services classification—chronology. Where PES indicates provisioning ecosystem services, RES is regulating ecosystem services, CES is cultural ecosystem services and SES is supporting ecosystem services.

The number of papers was classified into four ES categories (Figure 4). Sixteen percent of the papers discuss provisioning ecosystem services, basically focusing on agricultural products, freshwater bodies, food, etc. (Figure 3). The publication trend of PES is gradually increasing day by day and is more focused on the water and agriculture-related aspect, as it has a significant role to play in every human life today. Regulating ES has 17% of paper publications, mainly focusing on the vital aspects of the present-day scenario, i.e., climatic changes, carbon sequestration, floods, soil erosion, etc. The publication trend of RES falls in line with the provisioning ecosystem, as we find trade-offs and synergies among the ESs [29], so it is important to study critical aspects on the same basis (Figure 4).

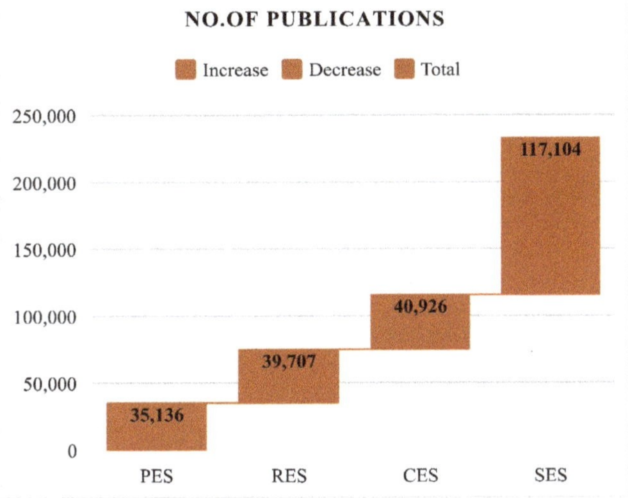

Figure 4. Number of publications on ecosystem services classification. Where PES denotes provisioning ecosystem services, RES is regulating ecosystem services, CES is cultural ecosystem services, and SES is supporting ecosystem services.

3.3. Models Used to Access Ecosystem Services

Modeling of the ES helps the researcher to quantify, spatially locate, and potentially evaluate the economic trends. Daily et al. [19] point out that this information plays a vital role in the decision making of urban planners, urban designers, and policymakers attempting to understand the effect of urban expansion on ES. In the present scenario, there is a proliferation of models and tools that helps us to map and access ES and vice versa [42,50,51].

Over time, numerous studies tried to simultaneously understand land use changes and their impact on ES, which helped designers and policymakers take appropriate steps to overcome the issue. To monitor LULC changes and ES changes, satellite images have been globally used as the most accurate tool [52,53]. Models are used to investigate the interactions (such as a trade-off, synergies, bundles/clusters, and flows) of ES, and deliberately put forward benefits that are enjoyed by humans for their well-being [54–56]. There is much importance given to enhancing ecosystem service management by objectively quantifying interactions among various ES [57].

Integrated Valuation of Environmental Services and Trade-offs (InVEST) is a globally accepted tool that was developed inside the Natural Capital Project [50,54–59]. The InVEST model can illustrate a spatially visualized map of the ESs. Comparing the InVEST model with other models, InVEST does not require any expertise; this model provides a nearly accurate assessment with limited demand of data input criteria, and is relevant in understanding the areas dealing with ecological processes [50,51,60]. InVEST model is a useful tool for assessing small-scale and local studies which give relevant and credible results for LULC and ES [58]. The InVEST toolbox is used to determine nearly 14 ES for supply changes using user-defined base setups like land use land cover and climatic changes [61,62].

The Soil and Water Assessment Tool (SWAT) is considered universally to simulate hydrological processes [63]. Further, the model has the flexibility in a spatial discretization that evaluates the space, locally, regionally, and globally. Like this, a decent number of models are used to assess ES changes; some of them are ARIES, LUCI, CA- Markov, SLEUTH, CLUES, etc.

The study of ecosystem services is performed quantitatively using mapping and modeling techniques. Researchers also have used a combination of models to assess ES, such as a combination of model mapping ES (such as InVEST, SWAT, ARIES) and model mapping urban expansion. With the help of the statistical model, the mapping was carried out. Urban expansion models such as LUSD–urban (Land Use Scenario Dynamics–urban) [64] help in a multi-scale simulation of urban expansion, LUSD–urban along with Cellular Automata (CA) and system dynamics models signifies micro-scale evolutionary factors and macro-scale resource constraints. This model has undergone certain iterations in recent years, with improved accuracy and an average kappa index [65]. The other models are SLEUTH (slope, land use, exclusion, urban extent, transportation, and hill shade) [66], CLUE-S (the Conversion of Land Use and its Effects at Small regional extent) [67]. Statistical models such as correlation analysis [13,68], regression analysis [28], and root mean square deviation were used [69]. This combination of models is efficient at forming the correlation among a few variables but is not considered to be functionally viable. The most celebrated models are InVEST, ARIES, and SWAT.]. Figure 5 gives a brief idea of various models used by researchers.

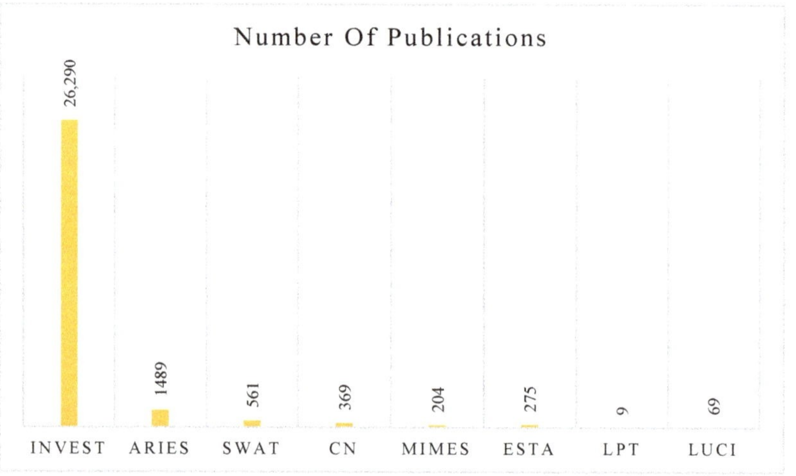

Figure 5. Number of publications on assessment models of ecosystem services.

The most commonly used base data are LULC, soil data, terrain data, and hydrological data. This gives a whole picture of different criteria such as habitats, soil types, vegetation class, and biomes. According to Metzger et al., the above data are used as ecosystem services indicators [70]. Adding on the same data can be used for valuation and spatial estimation of ecosystem services [71]; the other types of data used for ES assessment are census data; climatic data such as precipitation data, which is used for water yield assessment; and a digital elevation model (DEM); this is used for hydrology assessment [72].

The models used to access ESs spatiotemporally are InVEST [60], SWAT [61], ARIES (Artificial Intelligence for Ecosystem Services) [73], LUCI, etc. According to the publication trend from 2000 till 2021, we find the InVEST model is being used extensively due to its input data criteria; it uses open source data that are freely available, with a mapping/modeling scale of 30 m × 30 m. This model helps us to access multiple ecosystem services, (water quality, soil erosion, carbon sequestration, biodiversity conservation, nutrients, agricultural produce, etc.) [72].

3.4. World ES Publication Status

Ecosystem service study post [1] has been proliferated due to rapid urbanization, causing temporal changes such as climatic variation, global warming, etc. We find work on ecosystem services is mainly carried out in the developed countries. It is of prime importance to study the changing trajectory of spatial settings in developing countries, as the population increase demands changes in the land use and land cover dynamics. The interrelation between the ESs and human activities play a critical role in global climatic conditions. Publication trends in the world ES scenario is showcased in Figures 6 and 7.

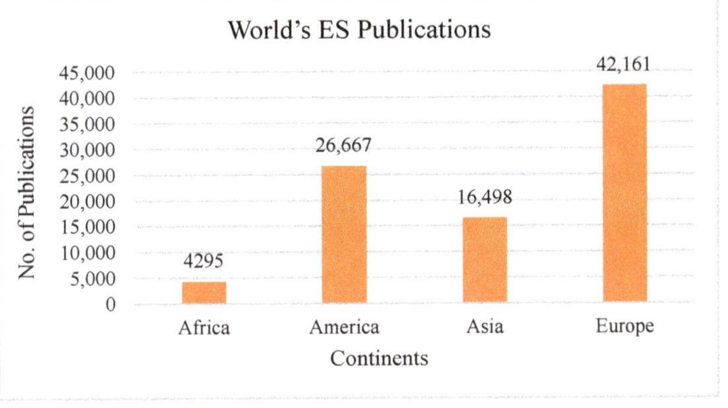

Figure 6. Trend of world's ecosystem service publications.

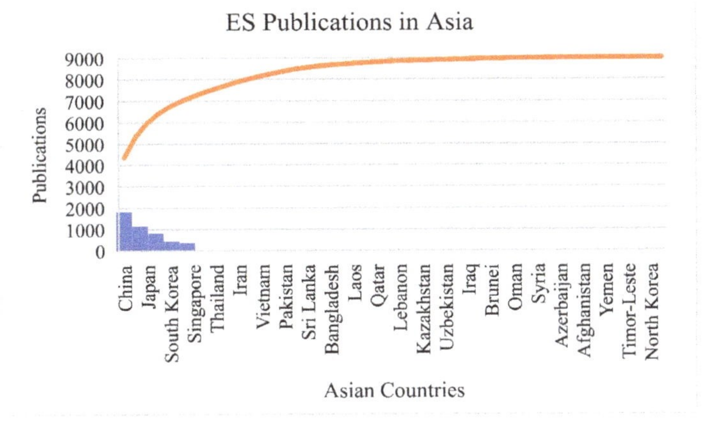

Figure 7. Trend of Asia's ecosystem service publications.

In the present scenario, we find ES studied globally in three ways: (1) estimating the physical quantity of services provided [74]; this is primary work that is carried out, to understand the influence of LULC change on ES, as well as its impact on the climatic aspect; (2) the economic value [75], which is needed to understand the income that is generated due to the interaction of ES, and also helps in estimating the economy lost due to the deterioration of ES, and (3) the basic benefit transfer method [6].

3.5. ES Publication Status of India

The publication trend of ecosystem services in the Indian context gives us a brief idea of the present knowledge gap. After the search carried out in the Scopus database, we found hardly 200 papers published on ESs (Figure 8). India is a very diverse country regarding its spatial, temporal, and cultural aspects. It will be of prime importance for the study of ES to bring about awareness of ESs' influence on recent global temporal changes. India is a developing country; hence, it has experienced a lot of spatiotemporal changes in recent decades.

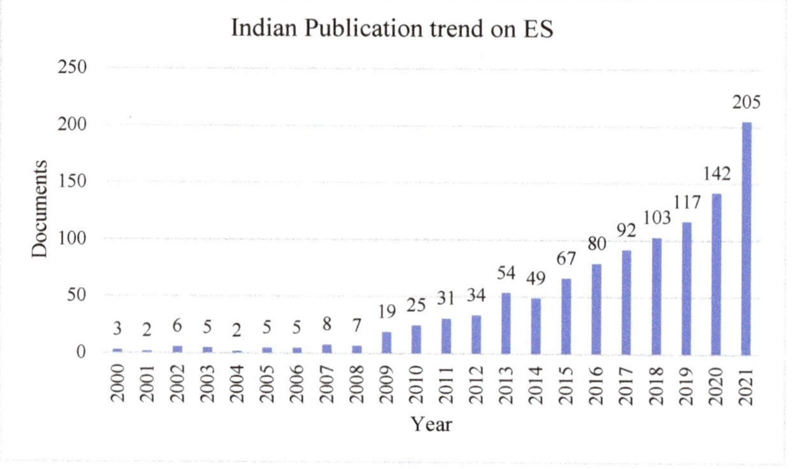

Figure 8. Indian Ecosystem service Publication trends.

India is peninsular; it is surrounded by the sea on three sides. With this being said, mangrove plays an important role in protecting the coastal region. Table 2 gives a brief idea on the studies conducted on ES in India. Giri et al. studied the status of mangrove forests in Southeast Asia [76]. Prasad et al. [77] studied the rate of degradation of seagrass impacting regulating ES due to human activities, whereas Edward et al. gave insight into methods of restoring seagrass [78]. The study conducted by [79] to understand the spatiotemporal dynamics in the mid-sized town of Telangana using statistical methods showed unsustainable growth trends among LULC variables, making study of the patterns vital. Studies were conducted to understand the degradation and rate of sedimentation of wetlands in the Western Himalayan region of Himachal Pradesh, showing a large-scale unregulated development causing the damage to ES [80]. Furthermore, Sannigrahi et al. [81] measured 17 ESs; Sannigrahi et al. [82] showed that climatic factors, biophysical factors, and environmental stress significantly affect the ESs in the Sundarbans region. The seasonal variation was captured using GHG (Green House Gas) on carbon pools in the degraded Sundarbans region [83]. Talukdar et al. [84] demonstrated the relationship between LULC and changes in ES; later on, Das et al. [85] shed light on decreasing ecosystem health in the lower Gangetic Plain region. Stakeholder participation plays a vital role in conserving ES. Sinclair et al. showed the willingness of the stakeholders to maintain the same [86]. ESs by the world hotspot region of Western Ghats played a vital role, elaborating on the impact of LULC on the ecological hotspot region and ES of Western Ghats [87,88]. Water richness and wetland habitable suitability criteria are important for understanding the habitat suitability of a populated region [89]. Further, there are stresses related to the dynamics of soil carbon in alternative cropping techniques [90,91]. Shah et al. [92] came up with the framework to understand the ecosystem services with a comprehensive view of common resources used by policymakers to attain sustainability.

Table 2. Studies conducted on ESs in India.

Author (s) and Study Year	Region	ES Categories Studied (P, R, S, C)	Number of ES Assessed	Grain	Methodology Used to Access ES
Giri et al. [76]	South Asia, Pakistan, Bangladesh, and India (Goa and Sundarbans)	S	1	Goa–23.5 m Sundarbans–30 m	GIS and ERDAS (unsupervised, ISODATA)
Srinivasarao et al. [91]	Semi-arid tropics of India	P, R	2		Survey
Shah & Garg [92]			Cascading Framework		
Sahani & Rahavaswamy [79]	Khammam, Telangana	P, R, C	3	Municipality	GIS with LULC base and Shannon's entropy
Ramachandra et al. [88]	Western Ghats, Karnataka	S	1	160,000 km^2 Mountain	GIS with Markov Analysis
Shukla et al. [14]	Upper Ganga basin	P	1	22,292 km^2 Wetland	GIS with Pearson's correlation and multiple linear regression
Prasad et al. [77]	Palk Bay	S	1	330 km^2 Wetland	Sampling design
Malik & Rai [80]	Himachal Pradesh	P, R	2	12,562 km^2 Wetland	GIS and ERDAS Imagine, Socioecological interviews
Sannigrahi et al. [81]	Sundarbans	P, R, S, C	17	4264 km^2	GIS using LULC analysis and ESV
Sannigrahi et al. [82]	Sundarbans	P, R, S	5	4264 km^2	NPP with InVEST and CA-Markov
Shukla et al. [13]	Upper Ganga basin	P	1	22,292 km^2 Wetland	Soil and Water Assessment Tool (SWAT) with MANOVA
Padhy et al. [83]	Sundarbans	R	2	4264 km^2	Sampling design
Talukdar et al. [84]	Lower Gangetic plain	P, R, S, C	17		GIS using LULC analysis and ESV
Ramachandra & Bharath [87]	Western Ghats, Karnataka	P, R, S	3	160,000 km^2 Mountain	GIS with Markov Analysis
Debanshi & Pal [89]	Ganges-Brahmaputra delta	P, S	3	6358.21 km^2 Wetland	Artificial Neural Network (ANN) and Support Vector Machine (SVM)
Babu et al. [90]	Mizoram	R	2	22.0 Mha	Sampling design
Das et al. [85]	Kolkata	P, R, S, C	17	Metropolitan area	GIS using LULC analysis and EH
Sinclair et al. [86]	Ashtamudi lake Ramsar, Kerala	P, S	3	56 km^2 Wetland	Questionnaire and sampling, econometric model
Sen & Guchhait [43]	Bardhaman	C	5	Municipality	field survey with semi-structured questionnaire
Pathak et al. [93]	Upper Ganga basin	P	1	22,292 km^2 Wetland	mean Lumped Zhang model with InVEST

Where P denotes provisioning ES, R is regulating ES, S is supporting ES, and C is cultural ES.

4. Discussion

4.1. Contribution of ES and Global Issues

According to the study, Millenium Ecosystem Assessment [1] inferred that 15 of 24 ecosystem services had degraded globally. Anthropogenic activities are the main reason for 60% of the deterioration of provisioning ecosystem services [1,94–96]. On a global scale, we projected the impact of LULC on ecosystem services and concluded that changes in LULC can deteriorate the ES [97].

Bennett et al. [29] infer that trade-off and/or synergies do take place between regulation ecosystem services and other ESs, considering this the main determinant to access the ecological changes [35]. In the global scenario of the past five decades, due to an increase in the population, demand, and usage of water, intensive agricultural produce, industrialization, and economic growth, Ref. [98] pinpoints that the use of water has tripled. Studies conducted by [14,99–101] prove that there is a significant role played by LULC on the water yield. A trade-off relationship between ES was found by Zhang et al. [102] between provision ES and soil conservation. The study conducted by Yi et al. [103] found that there is a significant connection between carbon storage and soil sediment retention in an urban watershed and river basins. However, it was also proved that the synergies and trade-offs occur in different scenarios [104]. Hence, a study analyzing various ESs is vital to urban management, planning, and policy decision making [28,105].

Literature proves that there is a direct relationship between the carbon and soil-based ESs [106–108]; Rodríguez et al. [109] proved the existence of a positive correlation between aboveground carbon storage with water regulation and supply. However, knowledge related to the potential of the coastal belt to regulate climate change and emission levels due to different anthropogenic activities is not available. Nevertheless, it is critical to access the influence of carbon sequestration on climate changes by different LULC on susceptible areas that are sensitive to changing processes. Many research scholars have studied the influence of LULC on carbon stock and climatic changes [110]. Following this study, we find [111] established the relationship between precipitation variation and LULC and its influence on ES

The study of soil erosion has caught the attention of researchers recently, and studies have been conducted globally on various scales [112]. The study conducted by Vaezi et al. [113] showed the result of ecosystem services hampered due to soil erosion, desertification, etc. Additionally, due to the presence of spatial heterogeneity, Ref. [114] explains the importance of soil-related study at various scale dynamics. This helps us to understand the effect of soil and its trade-off and/or synergies at various scales, as demonstrated by [115].

The major issue found globally today, as summarized by [116–119], shows that ES have been critically impacted due to the intense interaction between the ecosystem and humans at a regional scale, and this has to be looked after with immediate effect. This is also important according to the study [120], which found that the relationship between ES is spatially heterogeneous. Some researchers have studied the LULC change on a smaller scale can transform into synergy in spatiotemporal distribution at a larger scale [121–123]. Sun et al. [104] stated that studies at the future level at a regional scale should be designed to improve various scenarios in a detailed way to cater to the local situation and policy planning. Researchers such [82,124,125] found that there was a significant influence of climatic factors on the ESs variations. Therefore, Refs. [126,127] suggest that the effective planning management strategy is to incorporate ES bundles and hotspots in the decision-making process.

4.2. Way forward to ES Research in India

Based on the literature review conducted, there is a need to account for ecosystem services on different LULC in Indian scenarios, varying in urban settings and geolocation of the urban areas. On a national scale, we have the work of LULC changes giving an insight into the change dynamics in land use. There is empirical evidence that shows there are evident changes in the structure of ES due to urban expansion, which will lead to the degradation of the same [65]. However, little is known about the intensity of ES losses at a regional scale in the Indian context, due to the influence of LULC. It is also important to understand the influence of climatic factors on ES at a regional scale, providing better service to society.

The reason to select the ESs is that first, ESs such as carbon sequestration, water yield, and soil retention are the common focus areas for research study, as they represent the ESs subset [128–130]. Second, the quantitative methods of spatially analyzing the driving forces of these ESs can be reinforced by the efficient availability of large-scale data [130,131]. Vallet et al. [132] state that the study of ES interaction is important when it comes to questioning the usefulness of various criteria to come up with appropriate decision-making needs and expectations. From the literature survey, it is seen that provision, regulation, and support of ESs are threatened [27].

Even though there is a prolific amount of research on ES, there must be in-depth knowledge about the relationship and tradeoff among various ES [27], which should also be explored in terms of emerging climatic changes [27]. Additionally, Refs. [36,84] suggests that research should be conducted on every site-specific scenario leading to informed design management strategies, which in turn elevates ES benefits. The authors of [132] investigated the relationship between ES and urbanization, and concluded that for LULC, topography has a greater influence on ESs than urbanization.

5. Conclusions

Understanding the changes in ESs and their relationship with the help of spatially explicit methods could be helpful for the study to be conducted. In the present review, research dynamics of ES in the global scenario are given, and are then narrowed down to the national scale of the Indian scenario between 2000 and 2021. This analysis is based on 138 articles gathered from the databases of Science Direct, etc., with the help of bibliometric statistics such as keywords, countries, and outcomes. Additionally, ecosystem types, geographical location of the studies conducted, and assessment and valuation methods are in the limelight. The number of publications in the Indian context is gradually increasing. We find a steady increase in the publication trend post-2015. We find publications focusing on the study of two or more ES categories. Crop production and water yield have focused on provisioning ecosystem services; carbon sequestration, soil conservation for regulating ecosystem services; biodiversity conservation, along with the nutrient cycle, for supporting ecosystem services; psychological behavior, and quality of life in cultural ecosystems. We find these studies have been completed, since these are considered major determinants ruling these ESs along with the help of readily available research methodological framework. On the other hand, due to lack of methods and data, it is difficult to map the remaining ESs. In the Indian context, water, carbon, and soil play a major role in improving the socioeconomical aspect of ES.

India is a peninsular country with a wide variety of physical landscapes, including croplands, woods, grasslands, deserts, rivers, lakes, deltas, shelves, oceans, mountains, plateaus, basins, and islands. With its existing rapid economic growth and massive urbanization, India has become increasingly vulnerable to both natural disasters, such as droughts and floods, and human-caused ecological disasters, such as deforestation, salinization, erosion, and water, air, and soil pollution. The ES transdisciplinary paradigm serves as a useful framework for analyzing diversified natural assets and addressing environmental issues through integrated ecosystem management. However, the dataset built in this review work is not comprehensive; it can serve as a foundation for future studies, with the hope of creating a complete ES research database at the national level.

Author Contributions: Conceptualization, D. and A.K.S.; methodology, D. and A.K.S.; validation, D. and A.K.S.; formal analysis, D. and A.K.S.; writing—original draft preparation, D. and A.K.S.; writing—review and editing, A.K.S.; supervision, A.K.S. All authors have read and agreed to the published version of the manuscript.

Funding: This research received no external funding.

Institutional Review Board Statement: Not applicable.

Informed Consent Statement: Not applicable.

Data Availability Statement: The data that support the findings of this study are available from the corresponding author, upon reasonable request.

Acknowledgments: The authors are thankful to the Scopus database, Science Direct, Web of Science, and Google Scholar for providing the ecosystem services-related publications. The authors are also thankful to the anonymous reviewers for their constructive suggestions as to how to improve the quality of the present work.

Conflicts of Interest: This manuscript has not been published or presented elsewhere in part or entirety and is not under consideration by another journal. There are no conflicts of interest to declare.

References

1. Millennium Ecosystem Assessment (MEA). *Ecosystems and Human Well-Being: Synthesis*; Island Press: Washington, DC, USA, 2005; ISBN 1597260401.
2. Song, X.-P.; Hansen, M.C.; Stehman, S.V.; Potapov, P.V.; Tyukavina, A.; Vermote, E.F.; Townshend, J.R. Global Land Change from 1982 to 2016. *Nature* **2018**, *560*, 639–643. [CrossRef] [PubMed]
3. Barbier, E.B.; Hacker, S.D.; Kennedy, C.; Koch, E.W.; Stier, A.C.; Silliman, B.R. The Value of Estuarine and Coastal Ecosystem Services. *Ecol. Monogr.* **2011**, *81*, 169–193. [CrossRef]
4. Daily, G.C. Daily_1997_Natures-Services-Chapter-1. In *Societal Dependence on Natural Ecosystems*; Daily, G.C., Ed.; Island Press: Washington, DC, USA, 1997.
5. Guo, Z.; Zhang, L.; Li, Y. Increased Dependence of Humans on Ecosystem Services and Biodiversity. *PLoS ONE* **2010**, *5*, 13113. [CrossRef]
6. Costanza, R.; d'Arge, R.; de Groot, R.; Farber, S.; Grasso, M.; Hannon, B.; Limburg, K.; Naeem, S.; O'Neill, R.V.; Paruelo, J.; et al. The Value of the World's Ecosystem Services and Natural Capital. *Nature* **1997**, *387*, 253–260. [CrossRef]
7. Pascual, U.; Balvanera, P.; Díaz, S.; Pataki, G.; Roth, E.; Stenseke, M.; Watson, R.T.; Başak Dessane, E.; Islar, M.; Kelemen, E.; et al. Valuing Nature's Contributions to People: The IPBES Approach. *Curr. Opin. Environ. Sustain.* **2017**, *26*, 7–16. [CrossRef]
8. Joshi, N.; Baumann, M.; Ehammer, A.; Fensholt, R.; Grogan, K.; Hostert, P.; Jepsen, M.R.; Kuemmerle, T.; Meyfroidt, P.; Mitchard, E.T.A.; et al. A Review of the Application of Optical and Radar Remote Sensing Data Fusion to Land Use Mapping and Monitoring. *Remote Sens.* **2016**, *8*, 70. [CrossRef]
9. Gómez-Baggethun, E.; Barton, D.N. Classifying and Valuing Ecosystem Services for Urban Planning. *Ecol. Econ.* **2013**, *86*, 235–245. [CrossRef]
10. Xue, L.; Zhu, B.; Wu, Y.; Wei, G.; Liao, S.; Yang, C.; Wang, J.; Zhang, H.; Ren, L.; Han, Q. Dynamic Projection of Ecological Risk in the Manas River Basin Based on Terrain Gradients. *Sci. Total Environ.* **2019**, *653*, 283–293. [CrossRef]
11. Wang, L.J.; Ma, S.; Zhao, Y.G.; Zhang, J.C. Ecological Restoration Projects Did Not Increase the Value of All Ecosystem Services in Northeast China. *For. Ecol. Manag.* **2021**, *495*, 119340. [CrossRef]
12. Kumar, P.; Thakur, P.K.; Bansod, B.K.S.; Debnath, S.K. Groundwater: A Regional Resource and a Regional Governance. *Environ. Dev. Sustain.* **2018**, *20*, 1133–1151. [CrossRef]
13. Shukla, A.K.; Ojha, C.S.P.; Garg, R.D.; Shukla, S.; Pal, L. Influence of Spatial Urbanization on Hydrological Components of the Upper Ganga River Basin, India. *J. Hazard. Toxic Radioact. Waste* **2020**, *24*, 04020028. [CrossRef]
14. Shukla, A.K.; Ojha, C.S.P.; Mijic, A.; Buytaert, W.; Pathak, S.; Garg, R.D.; Shukla, S. Population Growth, Land Use and Land Cover Transformations, and Water Quality Nexus in the Upper Ganga River Basin. *Hydrol. Earth Syst. Sci.* **2018**, *22*, 4745–4770. [CrossRef]
15. Peters, M.K.; Hemp, A.; Appelhans, T.; Becker, J.N.; Behler, C.; Classen, A.; Detsch, F.; Ensslin, A.; Ferger, S.W.; Frederiksen, S.B.; et al. Climate–Land-Use Interactions Shape Tropical Mountain Biodiversity and Ecosystem Functions. *Nature* **2019**, *568*, 88–92. [CrossRef] [PubMed]
16. Wu, C.; Chen, B.; Huang, X.; Dennis Wei, Y.H. Effect of Land-Use Change and Optimization on the Ecosystem Service Values of Jiangsu Province, China. *Ecol. Indic.* **2020**, *117*, 106507. [CrossRef]
17. Fisher, B.; Turner, R.K.; Morling, P. Defining and Classifying Ecosystem Services for Decision Making. *Ecol. Econ.* **2009**, *68*, 643–653. [CrossRef]
18. Kragt, M.E.; Robertson, M.J. Quantifying Ecosystem Services Trade-Offs from Agricultural Practices. *Ecol. Econ.* **2014**, *102*, 147–157. [CrossRef]
19. Daily, G.C.; Polasky, S.; Goldstein, J.; Kareiva, P.M.; Mooney, H.A.; Pejchar, L.; Ricketts, T.H.; Salzman, J.; Shallenberger, R. Ecosystem Services in Decision Making: Time to Deliver—Enhanced Reader. *Front. Ecol. Environ.* **2009**, *7*, 21–28. [CrossRef]
20. Xia, M.; Jia, K.; Wang, X.; Bai, X.; Li, C.; Zhao, W.; Hu, X.; Cherubini, F. A Framework for Regional Ecosystem Authenticity Evaluation—A Case Study on the Qinghai-Tibet Plateau of China. *Glob. Ecol. Conserv.* **2021**, *31*, e01849. [CrossRef]
21. Costanza, R.; de Groot, R.; Braat, L.; Kubiszewski, I.; Fioramonti, L.; Sutton, P.; Farber, S.; Grasso, M. Twenty Years of Ecosystem Services: How Far Have We Come and How Far Do We Still Need to Go? *Ecosyst. Serv.* **2017**, *28*, 1–16. [CrossRef]
22. Sattler, C.; Loft, L.; Mann, C.; Meyer, C. Methods in Ecosystem Services Governance Analysis: An Introduction. *Ecosyst. Serv.* **2018**, *34*, 155–168. [CrossRef]

23. Han, H.-Q.; Liu, Y.; Gao, H.-J.; Zhang, Y.-J.; Wang, Z.; Chen, X.-Q. Tradeoffs and Synergies between Ecosystem Services: A Comparison of the Karst and Non-Karst Area. *J. Mt. Sci.* **2020**, *17*, 1221–1234. [CrossRef]
24. Dade, M.C.; Mitchell, M.G.E.; McAlpine, C.A.; Rhodes, J.R. Assessing Ecosystem Service Trade-Offs and Synergies: The Need for a More Mechanistic Approach. *Ambio* **2019**, *48*, 1116–1128. [CrossRef] [PubMed]
25. Cord, A.F.; Bartkowski, B.; Beckmann, M.; Dittrich, A.; Hermans-Neumann, K.; Kaim, A.; Lienhoop, N.; Locher-Krause, K.; Priess, J.; Schröter-Schlaack, C.; et al. Towards Systematic Analyses of Ecosystem Service Trade-Offs and Synergies: Main Concepts, Methods and the Road Ahead. *Ecosyst. Serv.* **2017**, *28*, 264–272. [CrossRef]
26. Turner, K.G.; Odgaard, M.V.; Bøcher, P.K.; Dalgaard, T.; Svenning, J.C. Bundling Ecosystem Services in Denmark: Trade-Offs and Synergies in a Cultural Landscape. *Landsc. Urban Plan.* **2014**, *125*, 89–104. [CrossRef]
27. Jiang, C.; Zhang, H.; Tang, Z.; Labzovskii, L. Evaluating the Coupling Effects of Climate Variability and Vegetation Restoration on Ecosystems of the Loess Plateau, China. *Land Use Policy* **2017**, *69*, 134–148. [CrossRef]
28. Jia, X.; Fu, B.; Feng, X.; Hou, G.; Liu, Y.; Wang, X. The Tradeoff and Synergy between Ecosystem Services in the Grain-for-Green Areas in Northern Shaanxi, China. *Ecol. Indic.* **2014**, *43*, 103–113. [CrossRef]
29. Bennett, E.M.; Peterson, G.D.; Gordon, L.J. Understanding Relationships among Multiple Ecosystem Services. *Ecol. Lett.* **2009**, *12*, 1394–1404. [CrossRef]
30. Braat, L.C.; de Groot, R. The Ecosystem Services Agenda:Bridging the Worlds of Natural Science and Economics, Conservation and Development, and Public and Private Policy. *Ecosyst. Serv.* **2012**, *1*, 4–15. [CrossRef]
31. Lee, H.; Lautenbach, S. A Quantitative Review of Relationships between Ecosystem Services. *Ecol. Indic.* **2016**, *66*, 340–351. [CrossRef]
32. Li, J.; Zhang, C.; Zhu, S. Relative Contributions of Climate and Land-Use Change to Ecosystem Services in Arid Inland Basins. *J. Clean. Prod.* **2021**, *298*, 126844. [CrossRef]
33. Zhang, L.; Cheng, L.; Chiew, F.; Fu, B. Understanding the Impacts of Climate and Landuse Change on Water Yield. *Curr. Opin. Environ. Sustain.* **2018**, *33*, 167–174. [CrossRef]
34. Abera, W.; Tamene, L.; Kassawmar, T.; Mulatu, K.; Kassa, H.; Verchot, L.; Quintero, M. Impacts of Land Use and Land Cover Dynamics on Ecosystem Services in the Yayo Coffee Forest Biosphere Reserve, Southwestern Ethiopia. *Ecosyst. Serv.* **2021**, *50*, 101338. [CrossRef]
35. Bennett, E.M.; Cumming, G.S.; Peterson, G.D. A Systems Model Approach to Determining Resilience Surrogates for Case Studies. *Ecosystems* **2005**, *8*, 945–957. [CrossRef]
36. Carter Berry, Z.; Jones, K.W.; Gomez Aguilar, L.R.; Congalton, R.G.; Holwerda, F.; Kolka, R.; Looker, N.; Lopez Ramirez, S.M.; Manson, R.; Mayer, A.; et al. Evaluating Ecosystem Service Trade-Offs along a Land-Use Intensification Gradient in Central Veracruz, Mexico. *Ecosyst. Serv.* **2020**, *45*, 101181. [CrossRef]
37. Garrastazú, M.C.; Mendonça, S.D.; Horokoski, T.T.; Cardoso, D.J.; Rosot, M.A.D.; Nimmo, E.R.; Lacerda, A.E.B. Carbon Sequestration and Riparian Zones: Assessing the Impacts of Changing Regulatory Practices in Southern Brazil. *Land Use Policy* **2015**, *42*, 329–339. [CrossRef]
38. Delibas, M.; Tezer, A.; Kuzniecow Bacchin, T. Towards Embedding Soil Ecosystem Services in Spatial Planning. *Cities* **2021**, *113*, 103150. [CrossRef]
39. Kumar, A.; Kumar, M.; Pandey, R.; Yu, Z.G.; Yu, Z.G.; Cabral-Pintod, M. Forest soil nutrient stocks along with an altitudinal range of Uttarakhand Himalayas: An aid to Nature Based Climate Solutions. *CATENA* **2021**, *207*, 105667. [CrossRef]
40. Rai, P.; Vineeta; Shukla, G.; Manohar, A.K.; Bhat, J.A.; Kumar, A.; Kumar, M. Carbon Storage of Single Tree and Mixed Tree Dominant Species Stands in a Reserve Forest—Case Study of the Eastern Sub-Himalayan Region of India. *Land* **2021**, *10*, 435. [CrossRef]
41. Kumar, A.; Kumar, M. Estimation of Biomass and Soil Carbon Stock in the Hydroelectric Catchment of India and its Implementation to Climate Change. *J. Sustain. For.* **2020**, *36*. [CrossRef]
42. de Groot, R.; Brander, L.; van der Ploeg, S.; Costanza, R.; Bernard, F.; Braat, L.; Christie, M.; Crossman, N.; Ghermandi, A.; Hein, L.; et al. Global Estimates of the Value of Ecosystems and Their Services in Monetary Units. *Ecosyst. Serv.* **2012**, *1*, 50–61. [CrossRef]
43. Sen, S.; Guchhait, S.K. Urban Green Space in India: Perception of Cultural Ecosystem Services and Psychology of Situatedness and Connectedness. *Ecol. Indic.* **2021**, *123*, 107338. [CrossRef]
44. Ko, H.; Son, Y. Perceptions of Cultural Ecosystem Services in Urban Green Spaces: A Case Study in Gwacheon, Republic of Korea. *Ecol. Indic.* **2018**, *91*, 299–306. [CrossRef]
45. Chan, K.M.A.; Satterfield, T.; Goldstein, J. Rethinking Ecosystem Services to Better Address and Navigate Cultural Values. *Ecol. Econ.* **2012**, *74*, 8–18. [CrossRef]
46. TEEB. The Economics of Ecosystems and Biodiversity. In *The Economics of Ecosystems and Biodiversity Ecological and Economic Foundations*; Kumar, P., Ed.; Earthscan: London, UK; Washington, DC, USA, 2010.
47. Nelson, E.; Bhagabati, N.; Ennaanay, D.; Lonsdorf, E.; Pennington, D.; Sharma, M. Modeling Terrestrial Ecosystem Services. In *Encyclopedia of Biodiversity*, 2nd ed.; Elsevier Inc.: Amsterdam, The Netherlands, 2013; pp. 347–361, ISBN 9780123847195.
48. Hasan, S.S.; Zhen, L.; Miah, M.G.; Ahamed, T.; Samie, A. Impact of Land Use Change on Ecosystem Services: A Review. *Environ. Dev.* **2020**, *34*, 100527. [CrossRef]

49. Haines-Young, R.; Potschin, M. *Common International Classification of Ecosystem Services (CICES)*; European Environment Agency: Copenhagen, Denmark, 2011.
50. Signorello, G.; Marzo, A.; Prato, C.; Sturiale, G.; de Salvo, M. Assessing the Hidden Impacts of Hypothetical Eruption Events at Mount Etna. *Environ. Sustain. Indic.* **2020**, *8*, 100056. [CrossRef]
51. Bagstad, K.J.; Semmens, D.J.; Winthrop, R. Comparing Approaches to Spatially Explicit Ecosystem Service Modeling: A Case Study from the San Pedro River, Arizona. *Ecosyst. Serv.* **2013**, *5*, 40–50. [CrossRef]
52. Verma, P.; Raghubanshi, A.S. Rural Development and Land Use Land Cover Change in a Rapidly Developing Agrarian South Asian Landscape. *Remote Sens. Appl. Soc. Environ.* **2019**, *14*, 138–147. [CrossRef]
53. Kam, J.; Jayanthi, S.C.; Raghavswamy, V. Spatio-temporal analysis of land use in urban mumbai-using multi-sensor satellite data and gis techniques. *J. Indian Soc. Remote Sens.* **2006**, *34*, 385–396.
54. Posner, S.; Verutes, G.; Koh, I.; Denu, D.; Ricketts, T. Global Use of Ecosystem Service Models. *Ecosyst. Serv.* **2016**, *17*, 131–141. [CrossRef]
55. Malinga, R.; Gordon, L.J.; Jewitt, G.; Lindborg, R. Mapping Ecosystem Services across Scales and Continents—A Review. *Ecosyst. Serv.* **2015**, *13*, 57–63. [CrossRef]
56. Dang, A.N.; Jackson, B.M.; Benavidez, R.; Tomscha, S.A. Review of Ecosystem Service Assessments: Pathways for Policy Integration in Southeast Asia. *Ecosyst. Serv.* **2021**, *49*, 101266. [CrossRef]
57. Mouchet, M.A.; Lamarque, P.; Martín-López, B.; Crouzat, E.; Gos, P.; Byczek, C.; Lavorel, S. An Interdisciplinary Methodological Guide for Quantifying Associations between Ecosystem Services. *Glob. Environ. Change* **2014**, *28*, 298–308. [CrossRef]
58. Bogdan, S.-M.; Pătru-Stupariu, I.; Zaharia, L. The Assessment of Regulatory Ecosystem Services: The Case of the Sediment Retention Service in a Mountain Landscape in the Southern Romanian Carpathians. *Procedia Environ. Sci.* **2016**, *32*, 12–27. [CrossRef]
59. Tallis, H.; Polasky, S. Mapping and Valuing Ecosystem Services as an Approach for Conservation and Natural-Resource Management. *Ann. NY Acad. Sci.* **2009**, *1162*, 265–283. [CrossRef] [PubMed]
60. Bangash, R.F.; Passuello, A.; Sanchez-Canales, M.; Terrado, M.; López, A.; Elorza, F.J.; Ziv, G.; Acuña, V.; Schuhmacher, M. Ecosystem Services in Mediterranean River Basin: Climate Change Impact on Water Provisioning and Erosion Control. *Sci. Total Environ.* **2013**, *458–460*, 246–255. [CrossRef] [PubMed]
61. Fang, Z.; Bai, Y.; Jiang, B.; Alatalo, J.M.; Liu, G.; Wang, H. Quantifying Variations in Ecosystem Services in Altitude-Associated Vegetation Types in a Tropical Region of China. *Sci. Total Environ.* **2020**, *726*, 138565. [CrossRef]
62. Sharp, R.; Tallis, H.T.; Ricketts, T.; Guerry, A.D.; Wood, S.A.; Chaplin-Kramer, R.; Nelson, E.; Ennaanay, D.; Wolny, S.; Olwero, N.; et al. *InVEST User's Guide InVEST User's Guide Integrated Valuation of Ecosystem Services and Tradeoffs*; The Natural Capital Project: Stanford, CA, USA, 2014. Available online: https://scholar.google.com/citations?view_op=view_citation&hl=en&user=yPREfmsAAAAJ&citation_for_view=yPREfmsAAAAJ:3fE2CSJIrl8C (accessed on 10 June 2022).
63. Notter, B.; Hurni, H.; Wiesmann, U.; Abbaspour, K.C. Modelling Water Provision as an Ecosystem Service in a Large East African River Basin. *Hydrol. Earth Syst. Sci.* **2012**, *16*, 69–86. [CrossRef]
64. He, C.; Okada, N.; Zhang, Q.; Shi, P.; Zhang, J. Modeling Urban Expansion Scenarios by Coupling Cellular Automata Model and System Dynamic Model in Beijing, China. *Appl. Geogr.* **2006**, *26*, 323–345. [CrossRef]
65. Xie, W.; Huang, Q.; He, C.; Zhao, X. Projecting the Impacts of Urban Expansion on Simultaneous Losses of Ecosystem Services: A Case Study in Beijing, China. *Ecol. Indic.* **2018**, *84*, 183–193. [CrossRef]
66. Clarke, K.C.; Hoppen, S.; Gaydos, L. A Self-Modifying Cellular Automaton Model of Historical Urbanization in the San Francisco Bay Area. *Environ. Plan. B Plan. Des.* **1997**, *24*, 247–261. [CrossRef]
67. Verburg, P.H.; Soepboer, W.; Veldkamp, A.; Limpiada, R.; Espaldon, V.; Mastura, S.S.A. Modeling the Spatial Dynamics of Regional Land Use: The CLUE-S Model. *Environ. Manag.* **2002**, *30*, 391–405. [CrossRef] [PubMed]
68. Raudsepp-Hearne, C.; Peterson, G.D.; Bennett, E.M. Ecosystem Service Bundles for Analyzing Tradeoffs in Diverse Landscapes. *Proc. Natl. Acad. Sci. USA* **2010**, *107*, 5242–5247. [CrossRef] [PubMed]
69. Lu, N.; Fu, B.; Jin, T.; Chang, R. Trade-off Analyses of Multiple Ecosystem Services by Plantations along a Precipitation Gradient across Loess Plateau Landscapes. *Landsc. Ecol.* **2014**, *29*, 1697–1708. [CrossRef]
70. Metzger, M.J.; Rounsevell, M.; Acosta-Michlik, L.; Leemans, R.; Schröter, D. The vulnerability of ecosystem services to land use change. *Agric. Ecosyst. Environ.* **2006**, *114*, 69–85. [CrossRef]
71. Sutton, P.C.; Costanza, R. Global estimates of market and non-market values derived from nighttime satellite imagery, land cover, and ecosystem service valuation. *Ecol. Econ.* **2002**, *41*, 509–527. [CrossRef]
72. Nelson, E.; Mendoza, G.; Regetz, J.; Polasky, S.; Tallis, H.; Cameron, D.; Chan, K.M.; Daily, G.C.; Goldstein, J.; Kareiva, P.M.; et al. Modeling Multiple Ecosystem Services, Biodiversity Conservation, Commodity Production, and Tradeoffs at Landscape Scales. *Front. Ecol. Environ.* **2009**, *7*, 4–11. [CrossRef]
73. Villa, F.; Bagstad, K.J.; Voigt, B.; Johnson, G.W.; Portela, R.; Honzák, M.; Batker, D. A Methodology for Adaptable and Robust Ecosystem Services Assessment. *PLoS ONE* **2014**, *9*, e91001. [CrossRef] [PubMed]
74. Jiang, C.; Wang, F.; Zhang, H.; Dong, X. Quantifying Changes in Multiple Ecosystem Services during 2000–2012 on the Loess Plateau, China, as a Result of Climate Variability and Ecological Restoration. *Ecol. Eng.* **2016**, *97*, 258–271. [CrossRef]
75. Han, X.; Yu, J.; Shi, L.; Zhao, X.; Wang, J. Spatiotemporal Evolution of Ecosystem Service Values in an Area Dominated by Vegetation Restoration: Quantification and Mechanisms. *Ecol. Indic.* **2021**, *131*, 108191. [CrossRef]

76. Giri, C.; Long, J.; Abbas, S.; Murali, R.M.; Qamer, F.M.; Pengra, B.; Thau, D. Distribution and Dynamics of Mangrove Forests of South Asia. *J. Environ. Manag.* **2015**, *148*, 101–111. [CrossRef] [PubMed]
77. Prasad, M.H.K.; Ganguly, D.; Paneerselvam, A.; Ramesh, R.; Purvaja, R. Seagrass Litter Decomposition: An Additional Nutrient Source to Shallow Coastal Waters. *Environ. Monit. Assess.* **2019**, *191*, 5. [CrossRef] [PubMed]
78. Edward, J.K.P.; Raj, K.D.; Mathews, G.; Kumar, P.D.; Arasamuthu, A.; D' Souza, N.; Bilgi, D.S. Seagrass Restoration in Gulf of Mannar, Tamil Nadu, Southeast India: A Viable Management Tool. *Environ. Monit. Assess.* **2019**, *191*, 430. [CrossRef] [PubMed]
79. Sahani, S.; Raghavaswamy, V. Analyzing Urban Landscape with City Biodiversity Index for Sustainable Urban Growth. *Environ. Monit. Assess.* **2018**, *190*, 471. [CrossRef]
80. Malik, M.; Rai, S.C. Drivers of Land Use/Cover Change and Its Impact on Pong Dam Wetland. *Environ. Monit. Assess.* **2019**, *191*, 203. [CrossRef] [PubMed]
81. Sannigrahi, S.; Chakraborti, S.; Joshi, P.K.; Keesstra, S.; Sen, S.; Paul, S.K.; Kreuter, U.; Sutton, P.C.; Jha, S.; Dang, K.B. Ecosystem Service Value Assessment of a Natural Reserve Region for Strengthening Protection and Conservation. *J. Environ. Manag.* **2019**, *244*, 208–227. [CrossRef]
82. Sannigrahi, S.; Zhang, Q.; Pilla, F.; Joshi, P.K.; Basu, B.; Keesstra, S.; Roy, P.S.; Wang, Y.; Sutton, P.C.; Chakraborti, S.; et al. Responses of Ecosystem Services to Natural and Anthropogenic Forcings: A Spatial Regression Based Assessment in the World's Largest Mangrove Ecosystem. *Sci. Total Environ.* **2020**, *715*, 137004. [CrossRef] [PubMed]
83. Padhy, S.R.; Bhattacharyya, P.; Dash, P.K.; Reddy, C.S.; Chakraborty, A.; Pathak, H. Seasonal Fluctuation in Three Mode of Greenhouse Gases Emission in Relation to Soil Labile Carbon Pools in Degraded Mangrove, Sundarban, India. *Sci. Total Environ.* **2020**, *705*, 135909. [CrossRef] [PubMed]
84. Talukdar, S.; Singha, P.; Shahfahad; Mahato, S.; Praveen, B.; Rahman, A. Dynamics of Ecosystem Services (ESs) in Response to Land Use Land Cover (LU/LC) Changes in the Lower Gangetic Plain of India. *Ecol. Indic.* **2020**, *112*, 106121. [CrossRef]
85. Das, M.; Das, A.; Pereira, P.; Mandal, A. Exploring the Spatio-Temporal Dynamics of Ecosystem Health: A Study on a Rapidly Urbanizing Metropolitan Area of Lower Gangetic Plain, India. *Ecol. Indic.* **2021**, *125*, 107584. [CrossRef]
86. Sinclair, M.; Vishnu Sagar, M.K.; Knudsen, C.; Sabu, J.; Ghermandi, A. Economic Appraisal of Ecosystem Services and Restoration Scenarios in a Tropical Coastal Ramsar Wetland in India. *Ecosyst. Serv.* **2021**, *47*, 101236. [CrossRef]
87. Ramachandra, T.V.; Bharath, S. Carbon Sequestration Potential of the Forest Ecosystems in the Western Ghats, a Global Biodiversity Hotspot. *Nat. Resour. Res.* **2020**, *29*, 2753–2771. [CrossRef]
88. Ramachandra, T.V.; Bharath, S.; Gupta, N. Modelling Landscape Dynamics with LST in Protected Areas of Western Ghats, Karnataka. *J. Environ. Manag.* **2018**, *206*, 1253–1262. [CrossRef]
89. Debanshi, S.; Pal, S. Modelling Water Richness and Habitat Suitability of the Wetlands and Measuring Their Spatial Linkages in Mature Ganges Delta of India. *J. Environ. Manag.* **2020**, *271*, 110956. [CrossRef]
90. Babu, S.; Singh, R.; Avasthe, R.K.; Yadav, G.S.; Mohapatra, K.P.; Selvan, T.; Das, A.; Singh, V.K.; Valente, D.; Petrosillo, I. Soil Carbon Dynamics in Indian Himalayan Intensified Organic Rice-Based Cropping Sequences. *Ecol. Indic.* **2020**, *114*, 106292. [CrossRef]
91. Srinivasarao, C.; Lal, R.; Kundu, S.; Babu, M.B.B.P.; Venkateswarlu, B.; Singh, A.K. Soil Carbon Sequestration in Rainfed Production Systems in the Semiarid Tropics of India. *Sci. Total Environ.* **2014**, *487*, 587–603. [CrossRef]
92. Shah, A.; Garg, A. Urban Commons Service Generation, Delivery, and Management: A Conceptual Framework. *Ecol. Econ.* **2017**, *135*, 280–287. [CrossRef]
93. Pathak, S.; Ojha, C.S.P.; Garg, R.D. Applicability of the InVEST Model for Estimating Water Yield in Upper Ganga Basin. Society of Earth Scientists Series. In *The Ganga River Basin: A Hydrometeorological Approach*; Chauhan, M.S., Ojha, C.S.P., Eds.; Springer: Cham, Switzerland, 2021; pp. 219–231.
94. Yang, Y.; Zheng, H.; Kong, L.; Huang, B.; Xu, W.; Ouyang, Z. Mapping Ecosystem Services Bundles to Detect High- and Low-Value Ecosystem Services Areas for Land Use Management. *J. Clean. Prod.* **2019**, *225*, 11–17. [CrossRef]
95. Jew, E.; Burdekin, O.; Dougill, A.J.; Sallu, S.M. Rapid Land Use Change Threatens Provisioning Ecosystem Services in Miombo Woodlands. *Nat. Resour. Forum* **2019**, *43*, 56–70. [CrossRef]
96. Jopke, C.; Kreyling, J.; Maes, J.; Koellner, T. Interactions among Ecosystem Services across Europe: Bagplots and Cumulative Correlation Coefficients Reveal Synergies, Trade-Offs, and Regional Patterns. *Ecol. Indic.* **2015**, *49*, 46–52. [CrossRef]
97. Nelson, E.; Sander, H.; Hawthorne, P.; Conte, M.; Ennaanay, D.; Wolny, S.; Manson, S.; Polasky, S. Projecting Global Land-Use Change and Its Effect on Ecosystem Service Provision and Biodiversity with Simple Models. *PLoS ONE* **2010**, *5*, e14327. [CrossRef] [PubMed]
98. Bennear, L.S.; Olmstead, S.M. The Impacts of the "Right to Know": Information Disclosure and the Violation of Drinking Water Standards. *J. Environ. Econ. Manag.* **2008**, *56*, 117–130. [CrossRef]
99. Paudyal, K.; Baral, H.; Bhandari, S.P.; Bhandari, A.; Keenan, R.J. Spatial Assessment of the Impact of Land Use and Land Cover Change on Supply of Ecosystem Services in Phewa Watershed, Nepal. *Ecosyst. Serv.* **2019**, *36*, 100895. [CrossRef]
100. Farley, K.A.; Jobbágy, E.G.; Jackson, R.B. Effects of Afforestation on Water Yield: A Global Synthesis with Implications for Policy. *Glob. Change Biol.* **2005**, *11*, 1565–1576. [CrossRef]
101. Rimal, B.; Sharma, R.; Kunwar, R.; Keshtkar, H.; Stork, N.E.; Rijal, S.; Rahman, S.A.; Baral, H. Effects of Land Use and Land Cover Change on Ecosystem Services in the Koshi River Basin, Eastern Nepal. *Ecosyst. Serv.* **2019**, *38*, 100963. [CrossRef]

102. Zhang, Y.; Lu, X.; Liu, B.; Wu, D.; Fu, G.; Zhao, Y.; Sun, P. Spatial Relationships between Ecosystem Services and Socioecological Drivers across a Large-Scale Region: A Case Study in the Yellow River Basin. *Sci. Total Environ.* **2021**, *766*, 142480. [CrossRef]
103. Yi, H.; Güneralp, B.; Kreuter, U.P.; Güneralp, İ.; Filippi, A.M. Spatial and Temporal Changes in Biodiversity and Ecosystem Services in the San Antonio River Basin, Texas, from 1984 to 2010. *Sci. Total Environ.* **2018**, *619–620*, 1259–1271. [CrossRef]
104. Sun, X.; Li, F.; Sun, X.; Li, F. Spatiotemporal Assessment and Trade-Offs of Multiple Ecosystem Services Based on Land Use Changes in Zengcheng, China. *Sci. Total Environ.* **2017**, *609*, 1569–1581. [CrossRef]
105. Feng, Q.; Zhao, W.; Fu, B.; Ding, J.; Wang, S. Ecosystem Service Trade-Offs and Their Influencing Factors: A Case Study in the Loess Plateau of China. *Sci. Total Environ.* **2017**, *607–608*, 1250–1263. [CrossRef]
106. Izquierdo, A.E.; Clark, M.L. Spatial Analysis of Conservation Priorities Based on Ecosystem Services in the Atlantic Forest Region of Misiones, Argentina. *Forests* **2012**, *3*, 764–786. [CrossRef]
107. Sahoo, U.K.; Tripathi, O.P.; Nath, A.J.; Deb, S.; Das, D.J.; Gupta, A.; Devi, N.B.; Charturvedi, S.S.; Singh, S.L.; Kumar, A.; et al. Quantifying tree diversity, carbon stocks and sequestration potential for diverse land-uses in northeast India. *Front. Environ. Sci.* **2021**. [CrossRef]
108. Kumar, M.; Kumar, A.; Thakur, T.K.; Pandey, R.; Shaboo, U.K. Soil Organic Carbon Estimation along an Altitudinal Gradient of Chir-Pine forests of Garhwal Himalaya, India: A field inventory to remote sensing approach. *Land Degrad. Dev.* **2022**. [CrossRef]
109. Rodríguez, N.; Armenteras, D.; Retana, J. National Ecosystems Services Priorities for Planning Carbon and Water Resource Management in Colombia. *Land Use Policy* **2015**, *42*, 609–618. [CrossRef]
110. Tan, Z.; Liu, S.; Tieszen, L.L.; Tachie-Obeng, E. Simulated Dynamics of Carbon Stocks Driven by Changes in Land Use, Management and Climate in a Tropical Moist Ecosystem of Ghana. *Agric. Ecosyst. Environ.* **2009**, *130*, 171–176. [CrossRef]
111. Chiang, L.C.; Lin, Y.P.; Huang, T.; Schmeller, D.S.; Verburg, P.H.; Liu, Y.L.; Ding, T.S. Simulation of Ecosystem Service Responses to Multiple Disturbances from an Earthquake and Several Typhoons. *Landsc. Urban Plan.* **2014**, *122*, 41–55. [CrossRef]
112. Anache, J.A.A.; Flanagan, D.C.; Srivastava, A.; Wendland, E.C. Land Use and Climate Change Impacts on Runoff and Soil Erosion at the Hillslope Scale in the Brazilian Cerrado. *Sci. Total Environ.* **2018**, *622–623*, 140–151. [CrossRef] [PubMed]
113. Vaezi, A.R.; Ahmadi, M.; Cerdà, A. Contribution of Raindrop Impact to the Change of Soil Physical Properties and Water Erosion under Semi-Arid Rainfalls. *Sci. Total Environ.* **2017**, *583*, 382–392. [CrossRef]
114. Wei, W.; Chen, L.; Yang, L.; Fu, B.; Sun, R. Spatial Scale Effects of Water Erosion Dynamics: Complexities, Variabilities, and Uncertainties. *Chin. Geogr. Sci.* **2012**, *22*, 127–143. [CrossRef]
115. Wen, X.; Zhen, L. Soil Erosion Control Practices in the Chinese Loess Plateau: A Systematic Review. *Environ. Dev.* **2020**, *34*, 100493. [CrossRef]
116. Li, J.; Zhou, Z.X. Natural and Human Impacts on Ecosystem Services in Guanzhong—Tianshui Economic Region of China. *Environ. Sci. Pollut. Res.* **2016**, *23*, 6803–6815. [CrossRef]
117. Liao, C.; Yue, Y.; Wang, K.; Fensholt, R.; Tong, X.; Brandt, M. Ecological Restoration Enhances Ecosystem Health in the Karst Regions of Southwest China. *Ecol. Indic.* **2018**, *90*, 416–425. [CrossRef]
118. Peng, J.; Tian, L.; Liu, Y.; Zhao, M.; Hu, Y.; Wu, J. Ecosystem Services Response to Urbanization in Metropolitan Areas: Thresholds Identification. *Sci. Total Environ.* **2017**, *607–608*, 706–714. [CrossRef] [PubMed]
119. Han, R.; Feng, C.C.; Xu, N.; Guo, L. Spatial Heterogeneous Relationship between Ecosystem Services and Human Disturbances: A Case Study in Chuandong, China. *Sci. Total Environ.* **2020**, *721*, 137818. [CrossRef]
120. Zhang, Z.; Liu, Y.; Wang, Y.; Liu, Y.; Zhang, Y.; Zhang, Y. What Factors Affect the Synergy and Tradeoff between Ecosystem Services, and How, from a Geospatial Perspective? *J. Clean. Prod.* **2020**, *257*, 120454. [CrossRef]
121. Raudsepp-Hearne, C.; Peterson, G.D. Scale and Ecosystem Services: How Do Observation, Management, and Analysis Shift with Scale—Lessons from Québec. *Ecol. Soc.* **2016**, *21*, 16. [CrossRef]
122. de Groot, R.S.; Alkemade, R.; Braat, L.; Hein, L.; Willemen, L. Challenges in Integrating the Concept of Ecosystem Services and Values in Landscape Planning, Management and Decision Making. *Ecol. Complex.* **2010**, *7*, 260–272. [CrossRef]
123. Bai, Y.; Chen, Y.; Alatalo, J.M.; Yang, Z.; Jiang, B. Scale Effects on the Relationships between Land Characteristics and Ecosystem Services- a Case Study in Taihu Lake Basin, China. *Sci. Total Environ.* **2020**, *716*, 137083. [CrossRef] [PubMed]
124. Clerici, N.; Cote-Navarro, F.; Escobedo, F.J.; Rubiano, K.; Villegas, J.C. Spatio-Temporal and Cumulative Effects of Land Use-Land Cover and Climate Change on Two Ecosystem Services in the Colombian Andes. *Sci. Total Environ.* **2019**, *685*, 1181–1192. [CrossRef] [PubMed]
125. Ma, S.; Wang, L.J.; Jiang, J.; Chu, L.; Zhang, J.C. Threshold Effect of Ecosystem Services in Response to Climate Change and Vegetation Coverage Change in the Qinghai-Tibet Plateau Ecological Shelter. *J. Clean. Prod.* **2021**, *318*, 128592. [CrossRef]
126. Yohannes, H.; Soromessa, T.; Argaw, M.; Warkineh, B. Spatio-Temporal Changes in Ecosystem Service Bundles and Hotspots in Beressa Watershed of the Ethiopian Highlands: Implications for Landscape Management. *Environ. Chall.* **2021**, *5*, 100324. [CrossRef]
127. Li, Y.; Zhang, L.; Qiu, J.; Yan, J.; Wan, L.; Wang, P.; Hu, N.; Cheng, W.; Fu, B. Spatially Explicit Quantification of the Interactions among Ecosystem Services. *Landsc. Ecol.* **2017**, *32*, 1181–1199. [CrossRef]
128. Jiang, C.; Li, D.; Wang, D.; Zhang, L. Quantification and Assessment of Changes in Ecosystem Service in the Three-River Headwaters Region, China as a Result of Climate Variability and Land Cover Change. *Ecol. Indic.* **2016**, *66*, 199–211. [CrossRef]
129. Khan, M.; Sharma, A.; Goyal, M.K. Assessment of Future Water Provisioning and Sediment Load under Climate and LULC Change Scenarios in a Peninsular River Basin, India. *Hydrol. Sci. J.* **2019**, *64*, 405–419. [CrossRef]

130. Jianying, X.; Jixing, C.; Yanxu, L. Partitioned Responses of Ecosystem Services and Their Tradeoffs to Human Activities in the Belt and Road Region. *J. Clean. Prod.* **2020**, *276*, 123205. [CrossRef]
131. Naidoo, R.; Balmford, A.; Costanza, R.; Fisher, B.; Green, R.E.; Lehner, B.; Malcolm, T.R.; Ricketts, T.H. Global Mapping of Ecosystem Services and Conservation Priorities. *Proc. Natl. Acad. Sci. USA* **2008**, *105*, 9495–9500. [CrossRef]
132. Zhang, Y.; Liu, Y.; Zhang, Y.; Liu, Y.; Zhang, G.; Chen, Y. On the Spatial Relationship between Ecosystem Services and Urbanization: A Case Study in Wuhan, China. *Sci. Total Environ.* **2018**, *637–638*, 780–790. [CrossRef]

applied sciences

MDPI

Article

Remote Sensing of the Water Quality Parameters for a Shallow Dam Reservoir

Andrzej Bielski * and Cezary Toś

Faculty of Environmental Engineering and Energy, Cracow University of Technology, Warszawska 24, 31-155 Cracow, Poland; cezary.tos@pk.edu.pl

* Correspondence: abielski@riad.usk.pk.edu.pl

Abstract: This study examines the chlorophyll a content and turbidity in the shallow dam reservoir of Lake Dobczyce. The analysis of satellite images for thirteen wavelength ranges enabled the selection of wavelengths applicable for a remote determination of chlorophyll a and turbidity. The selection was completed as the test of the significance of the coefficients in the equation, which calculates the values of the parameters on the basis of reflectance. The reflectance of the reservoir surface differs from the reflectance of individual water components, and the overlapping of spectral curves makes it difficult to isolate the significant reflectance. In the case of Lake Dobczyce, the significant reflectance was for wavelengths 665, 705, 740, and 842 nm (chlorophyll a) and for wavelengths 705, 740, and 783 nm (turbidity). In the model, the natural logarithm of chlorophyll a or turbidity was a linear combination of the natural log reflectance and the squares of those logarithms. A lake surface reflectance also includes the bottom reflectance. The reflectance obtained from the Sentinel-2 satellite was corrected with a bottom reflectance determined using the Lambert–Beer equation. The reflectance of a given surface may vary with the position of both the satellite and the sun, atmospheric pollution, and other factors. Correction of reflectance from satellite measurements was performed, as reflectance changes for the reference surface; the reference reflectance was assumed as the first reflectance of the reference surface observed during the study. The models helped to develop the maps of turbidity and chlorophyll a content in the lake.

Keywords: Sentinel-2; chlorophyll; turbidity; lake; concentration modeling of contaminants

Citation: Bielski, A.; Toś, C. Remote Sensing of the Water Quality Parameters for a Shallow Dam Reservoir. *Appl. Sci.* **2022**, *12*, 6734. https://doi.org/10.3390/app12136734

Academic Editors: Amit Kumar, Santosh Subhash Palmate and Rituraj Shukla

Received: 25 May 2022
Accepted: 1 July 2022
Published: 2 July 2022

1. Introduction

The chlorophyll content in surface waters is related to nutrients, such as nitrogen and phosphorus, and serves as one of many indicators of eutrophication. High amounts of these elements in water contribute to the excessive growth of algae, resulting in poor water quality. This topic is especially urgent in surface water intakes used for municipal or industrial consumers. The chlorophyll content in water can be determined by the traditional laboratory method, based on acetone extraction and absorbance measurement [1,2]. However, to track down changes in the chlorophyll content in rivers or lakes, and to develop concentration maps, a large number of analyses would have to be done; it would be a time-consuming and ineffective approach.

Satellite images of water surfaces allow for faster and more cost-effective estimation of chlorophyll content in water [3]. Another advantage of remote sensing (teledetection) is a spatial analysis of chlorophyll concentrations from in situ data are collected at particular points [4]. There are, however, a number of problems regarding the remote sensing of chlorophyll faces in surface waters. Water regime [5], as well as the reservoir's depth [6], are associated with the chemical composition and content of biological elements in water. Radiometric and atmospheric corrections also play an important role in the case of satellite data. The correction models, such as ATCOR [7], Second Simulation of a Satellite Signal in the Solar Spectrum (6SV), Acolite, or Sen2cor are available and used in these studies. All these concerns have prompted the use of empirical [8,9] or semi-analytical approaches in the development of teledetection methods for monitoring chlorophyll content in inland waters. These methods are based on the physics of the interactions of radiation with water and its compounds [10,11]. In recent years, neural networks and machine learning methods have been employed in the research on chlorophyll detection [8,12].

Empirical methods, though well suited to local conditions [13], require quite a lot of data. On the other hand, semi-analytical methods are more universal, but provide less accurate results. Deep learning methods are still in the early stage of development, and they require a large amount of heterogeneous teaching data; their complexity delays their implementation in small subjects, such as water reservoirs.

The research focused on the automation of the estimation of turbidity (in nephelometric units, NTU) and chlorophyll content in water taken from shallow dam reservoirs used for drinking purposes. In such reservoirs, water is usually classified as case 2, where optical properties are a function of at least three water components, i.e., phytoplankton, suspended sediments, and colored dissolved organic matter [14]. Such objects, with their firm positions, usually exhibit multiple time observation series of basic water quality indicators (turbidity, chlorophyll). The described remote water quality research favors the empirical method, which, however, must include an optical interaction between the main pollution components, i.e., mineral suspension and phytoplankton, and the shallow tank. Therefore, the authors offer a combination of statistical models that include these elements. The models have been verified using Sentinel 2 data in Lake Dobczyce; the lake serves as the drinking water reservoir for Krakow, Poland.

2. Data, Methods, and Techniques

2.1. Remote Sensing Methods for a Chlorophyll Content

In surface waters, algae and bacteria contain many pigments that can be analyzed using spectral methods. Listed according to color, these are, e.g., chlorophylls a, b, c, c1, c2, d, e, f, and g—green; carotene—orange; xanthophyll—yellow; phycoerythrin—red; phycocyanin—blue; and fucoxanthin—brown. The dominant pigments in photosynthetic organs are chlorophyll a (blue-green) and chlorophyll b (yellow-green).

Chlorophyll a is the most frequently used indicator of surface water quality. The surface spectral reflectance curves for water with different chlorophyll a concentrations are presented in Figure 1. Blue and far-red light ranges are strongly absorbed by chlorophyll, while the reflectance peaks are recorded at the wavelengths of approximately 566 and 688 nm (Figure 1).

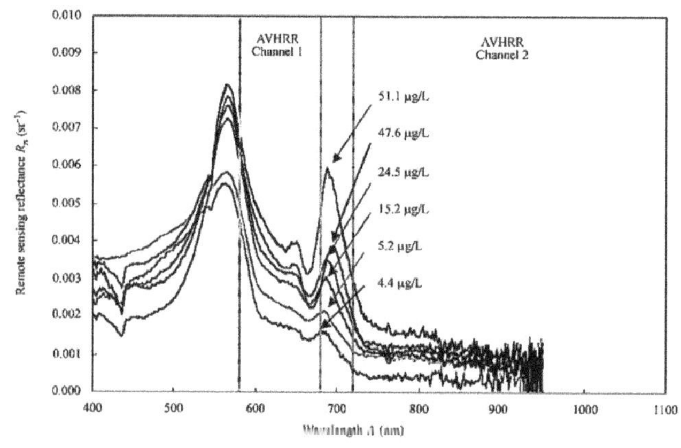

Figure 1. Surface remote sensing reflectance spectrum for the waters with different concentration of chlorophyll a, collected in the study area (119°52′–119°54′ E, 26°16′–26°19′ N) at 10:00–15:40, 2 June 2003. Reprinted with permission from [15].

Mineral and organic suspensions may pose a serious difficulty in studying the concentration of chlorophyll a in surface waters. Both suspensions result in the reflectance peaks at approximately 550, 712.5, and 805 nm (Figure 2), so a shift in the peaks only takes place relative to the peaks for chlorophyll a. The suspensions may also enhance the reflectance in relation to that of chlorophyll a and therefore, models describing the relationship between chlorophyll a and the reflectance should consider several wavelengths.

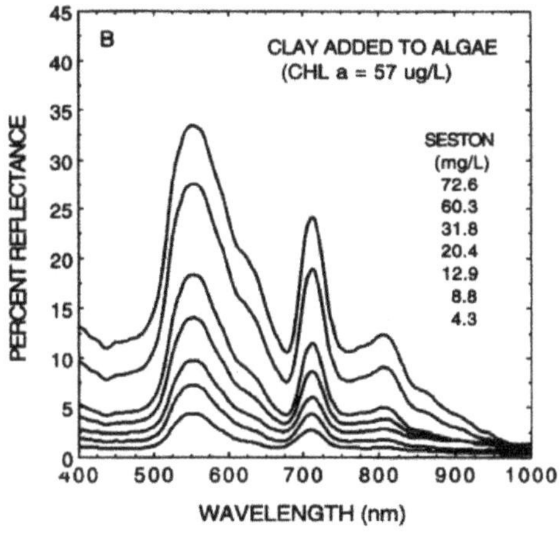

Figure 2. Relative contributions of chlorophyll and suspended sediment to a reflectance spectra of the surface water, based on in situ laboratory measurements made 1 m above the water surface by the authors of [16,17]. Reprinted with permission from [16,17].

The presence of clay or dusty particles in water results in a specific spectrogram (Figures 3 and 4) [18]. High reflectance in the range of 580–690 nm and reflectance near 810 nm will strongly distort the chlorophyll a spectrogram. The first range is characteristic for a specific type of suspension. In the case of lake bottom sediments that are composed of clay or dusty suspensions, they will have a similar spectrogram. The reflectance pattern around 810 nm is similar for clay and dusty suspensions and therefore, the reflectance around 810 nm can be used to estimate the suspension concentration.

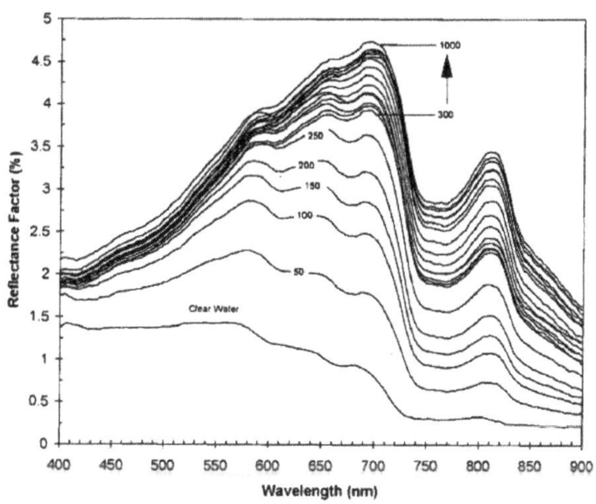

Figure 3. Reflectance for water with a clay suspension [g/m^3]. Reprinted with permission from [18].

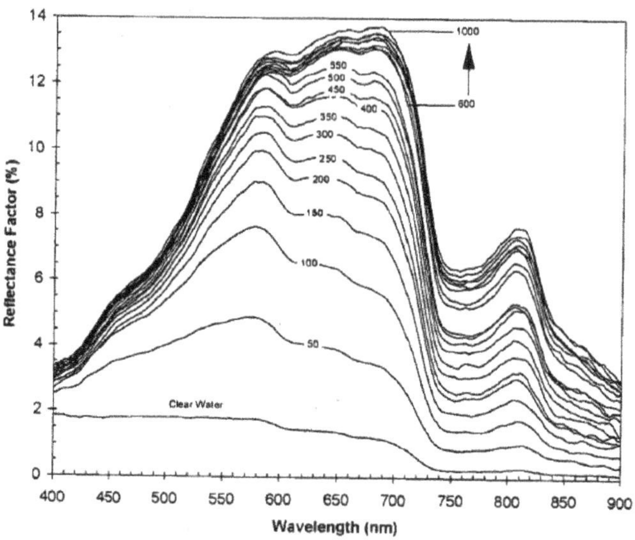

Figure 4. Reflectance for water with a dusty suspension [g/m^3]. Reprinted with permission from [18].

Mathematical models using the reflectance R to determine the concentration of $C_{chl\text{-}a}$ chlorophyll a in water are developed for a given type of surface water. These are the most abundant types of empirical models. Models such as GlobColour or Modis Aqua can be used globally or adapted to local reservoir conditions. They can take the form of polynomial dependence, as shown in the research of [19,20], dependence, as a form of the products or quotients of expressions [21,22], or a logarithmic form [23,24]. The estimation of model parameters can be carried out in different ways—using multiple linear recreation, support vector machine regression (SVR), or genetic algorithms. Examples of models used to determine the chlorophyll a content in the water are presented in Table 1.

Table 1. Typical mathematical formulas to calculate chlorophyll a concentrations.

Formula	Source
$C_{chl\text{-}a}=(23.09 \pm 0.98) + (117.42 \pm 2.49)\cdot(R^{-1}{}_{660\text{–}670} - R^{-1}{}_{700\text{–}730})\cdot R_{740\text{–}760}$ $C_{chl\text{-}a} = -(16.2 \pm 1.8) + (136.3 \pm 3.2)\cdot(R^{-1}{}_{662\text{–}672} \cdot R_{743\text{–}753})$	(Gitelson et al., 2006) [21]
$C_{chl\text{-}a} = (0.74\cdot R_{681} + [(681\ \text{nm} - 665\ \text{nm})/(681\ \text{nm} - 620\ \text{nm})]\cdot(R_{620} - R_{681}) - R_{665}$	(Shen et al., 2010) [22]
$\log_{10}(C_{chl\text{-}a}) = (0.32978 + 2.6465X + 1.9988X^2 + 0.5708X^3 + 3.033X^4)$ $X = \log_{10}(\text{Max}(R_{443nm}, R_{486nm})/R_{551nm})$	(Son et al., 2020) [23]
$C_{chl\text{-}a}$ [mg/L] $= 1.67 + 299\ R_{438} - 33.1\ R_{675} - 7217\ R_{438}R_{675} - 14022\ R^2{}_{438} - 973\ R^2{}_{675} + 373702\ R^2{}_{438}R_{675} + 112440\ R_{438}R^2{}_{675} - 3317051\ R^2{}_{438}R^2{}_{675}$ $C_{chl\text{-}a}$ [mg/L] $= 3.45 + 66.2\ R_{438} - 100\ R_{550} - 3.9\ R_{675} + 5349\ R_{438}R_{550} - 16643\ R_{550}R_{675} - 12682\ R_{438}R_{675} + 3077\ R^2{}_{438} + 5209\ R^2{}_{550} + 15992\ R^2{}_{675}$	(Johan et al., 2018) [20]
$\log_{10}(C_{chl\text{-}a}) = (0.366 - 3.067X + 1.930X^2 + 6.049X^3 - 1.532\ X^4)$ $X = \log_{10}(\text{Max}(R_{443nm}/R_{555nm}, R_{490nm}/R_{555nm}, R_{510nm}/R_{555nm}))$	GlobColour (Johnson et al., 2013) [25] (Diouf et al., 2018) [24]
$\log_{10}(C_{chl\text{-}a}) = (0.6994 - 2.0384X + 0.4656RX^2 + 0.4337X^3)$ $X = \log_{10}(\text{Max}(Rrs_{(443/555)}, Rrs_{(490=555)})$ $C_{chl\text{-}a} = [\text{mg/m}^3]$	Modis-Aqua (Johnson et al., 2013) [24]

The models are characterized by a different precision of results. Johnson et al., 2013 [25], testing global models GlobColour or Modis Aqua, obtained low R^2 determination coefficients of 0.25–0.51; on the other hand, for local conditions, this indicator may reach up to the value 0.96, with an error for RMSE chlorophyll a content of 0.07 mg chl/m^3, determined on the basis of a few data covering 5 days of composite images for the year 2014 [24].

Empirical models that are focused on determining the chlorophyll a concentration take into account the variability of the aquatic environment in terms of the content of other substances in a very limited way, which may affect the accuracy of the results obtained.

A possible solution to this problem in the case of inland waters may be the construction of seasonal models (used for imaging registered at specific seasons of the year) [26]. Another more widely used technique is the use of bioptic methods based on the modeling of light-water interaction, such as Hydrolight [27], or the newer 2SeaColor [28], adapted to water with high turbidity.

In the case of the second model, the determination indicator was equal 0.71, and model error RMSE was 6.23 mg chl/m^3 [28]. Such models may include bottom reflectance [29,30], which is extremely important for shallow tanks, as pointed out by Hicks et al. 2013 [31].

In the works (Li et al., 2017, 2018) [32,33], a semi-analytical method for colored dissolved organic matter (CDOM) was presented. This method includes bottom effect and is accurate (RMSE = 0.17 mg chl/m^3 and R^2 = 0.87); it also uses multispectral data.

The extremely promising result for turbid water research was the model developed by Gitelson et al. [21]. Finding three ranges of wavelengths guaranteeing the least error to estimate the chlorophyll a content enabled an accuracy of 7.8 mg chl a/m^3 in the range of 1.2 to 236 mg chl a/m^3, medium relative error 18.3%, with a high determination factor $R^2 \approx 0.96$. The version of the model with two wave length ranges provided much worse results. Despite the high accuracy of in situ research for low concentrations (below 10 mg chl a/m^3), there are relative errors reaching almost 100%.

Remote bathymetrical tests of shallow water, conducted by Lee et al. [34,35], yield the possibility of linking the effect of bottom reflectance and chlorophyll concentration. The developed model includes many different reflectances, including bottom reflectance and the absorption of radiation for various water components, such as phytoplankton pigments. However, the model is focused on the remote determination of depth. The number of model parameters is very large, making it troublesome to simultaneously determine the data for several wavelengths. Therefore, the absorption of radiation by phytoplankton is an input parameter for this model, not an output enabling the determination of chlorophyll concentration. In the case of oligotrophic sea waters, the computing models are simpler. Computational difficulties appear when many water components with spectrally different properties affect the calculation of the content of a given water component. This usually requires the use of the reflectances of many water components for many wave lengths. Particularly large complications appear in the case of shallow and turbid water.

The model of chlorophyll a content estimation in a reservoir used for drinking water intake was presented in this work. This application model is characterized by high accuracy, taking into account the bottom effect and interactions between chlorophyll and mineral or organic suspension.

2.2. Area of Study

The research was carried out in the area of Lake Dobczyce, located in the Myślenice poviat, in the Lesser Poland voivodeship of Poland. It is a dam reservoir, constructed in 1986 by damming the Raba river with a 30 m high and 617 m long dam. The reservoir area is approximately 10.7 km^2, and its total capacity is 127 million m^3. The reservoir serves as a source of drinking water. The lake water quality is determined by the quality of the Raba river, which varies due to anthropogenic activities and precipitation. These factors contribute to significant fluctuations in the concentrations of nutrients, mineral and organic suspensions, and other substances. The depth of the lake is just over twenty-six meters (Figure 5).

Figure 5. Depth map of Lake Dobczyce.

In the case of Dobczyce Lake, it is difficult to determine its range, especially in the western shallow region. This problem was solved by using information on the area elevation of the water surface and lidar altitude data (ALS), which, due to its high density, was subjected to resampling from the 5 × 5 m network using the MIN method, suggested by Śliwiński et al., 2022 [36], as the most suitable method for this type of task.

2.3. Data for a Remote Sensing Research

Images from Sentinel 2 (L1C product), registered from 13 April 2016 to 31 December 2021, were used for the teledetection quality modeling of water in Lake Dobczyce. The choice of specific images was determined by the dates of water quality tests in situ. A significant limitation in the number of images resulted from the rejection of images in which clouds or ice covered the lake.

The content of chlorophyll a and suspended solids (as turbidity) in the water was measured, in the period from 2016 to 2021, in the laboratory of the Waterworks of the City of Krakow, located at the water intake. The study covered the period from 2016 to 2021. Chlorophyll a was analyzed in acetone extracts using the monochromatic spectrophotometric method, with correction for pheopigment a [2,37]; absorbance was measured on a Hach DR 4000 U spectrophotometer, while turbidity was measured on a TL 2360 spectrophotometer, according to the standard methods [38].

2.4. Spectral Reflectance Curves

To compare the data recorded by Sentinel-2 satellites at different times, they were normalized to the conditions above the reference surface. The reference area was established as a fragment of the lakeside dam crest; it shows a low reflectance, and its properties are stable in time. The air composition over the reference surface was assumed to be the same as that over the lake surface. To correct lake surface reflectance obtained from satellite data, the special corrective parameter was introduced. It is the ratio of the reflectance registered on 13 April 2016 (13 April 2016, base date) in the reference field and the actual reflectance at a given day in the field. The normalized reflectance for the lake surface was the product of the corrective parameter and the reflectance of a given point on the lake surface at a given date. Such normalization is considered as an atmospheric correction. It made it possible to accommodate changes in air quality over the reference surface and the lake surface.

Spectral curves for the fragment of the dam crest are presented in Figure 6.

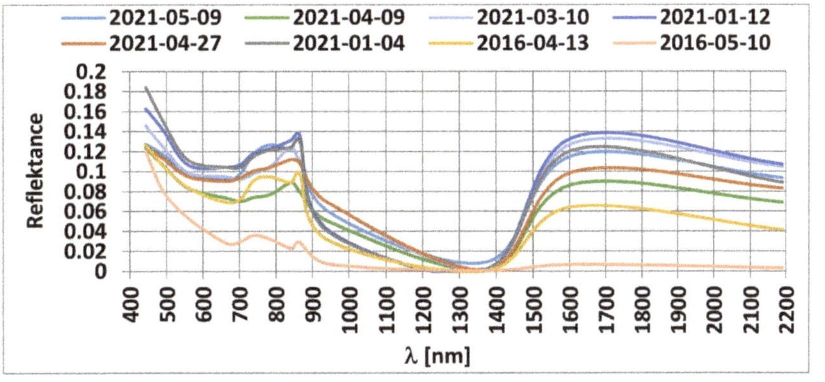

Figure 6. Some spectral curves for the fragment of the dam crest.

The base reflection curve for the fragment of the dam crest (reference area) is dated 13 April 2016 (Figure 6). Other spectral curves have a similar shape, but pass through points with different reflectance values, even though the properties of the dam crest surface have not changed; this was due to changes in the atmosphere composition over time.

The specific course of the spectral curves for Lake Dobczyce (Figure 7) results from the combination (superposition) of partial spectral curves for the water surface, water column, and bottom of the reservoir. Backscattering in a blue band also results in high reflectance (Figure 7). It should be noted that even at a high concentration of, e.g., montmorillonite suspensions (500 g/m^3), the spectral curve does not show strong local extremes for wavelengths over 440 nm (Figure 8). Such a concentration corresponds to a turbidity of 500 NTU, rarely found in surface waters (mostly after heavy rains). Starting from approximately 440 nm, the spectral curve drops down (Figure 8), as shown in the curves in Figure 7. At concentrations exceeding 500 g/m^3, the maximum, corresponding to 440 nm, shifts towards longer wavelengths. The spectral curves may take completely different shapes than the ones shown in Figures 7 and 8 for other types of the mineral suspensions.

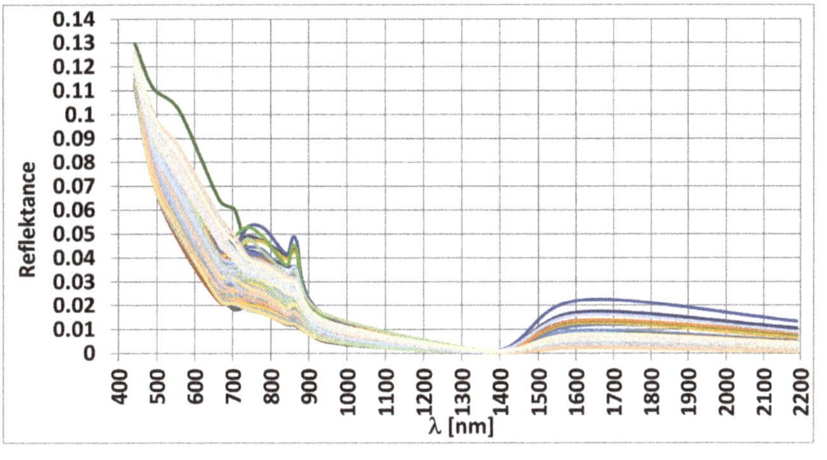

Figure 7. Reflectance at different measuring points of Lake Dobczyce in the years 2016–2021, after a normalization of the original reflectance from the satellite.

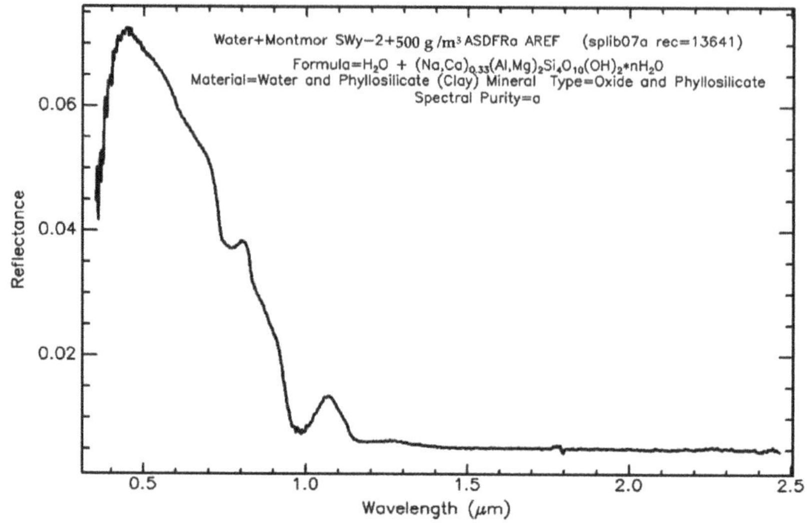

Figure 8. Spectral curve for the montmorillonite suspensions in water, USGS Spectral Library Version 7, https://crustal.usgs.gov/speclab/QueryAll07a.php?quick_filter=water (accessed on 1 January 2022). Reprinted with permission from [39].

The spectral curves for the chlorophyll solution can also differ significantly. At low concentrations (2.97 mg chl/m^3) (Figure 9), there are no strong local extremes on the spectral curve above 400 nm. Characteristic local extremes appear at higher concentrations (7.609 mg chl/m^3) (Figure 10) and are similar to the curves in Figure 2. Therefore, it may be concluded that concentration of water components (chlorophyll, type of suspension) will have a decisive influence on the shape of the spectral curves.

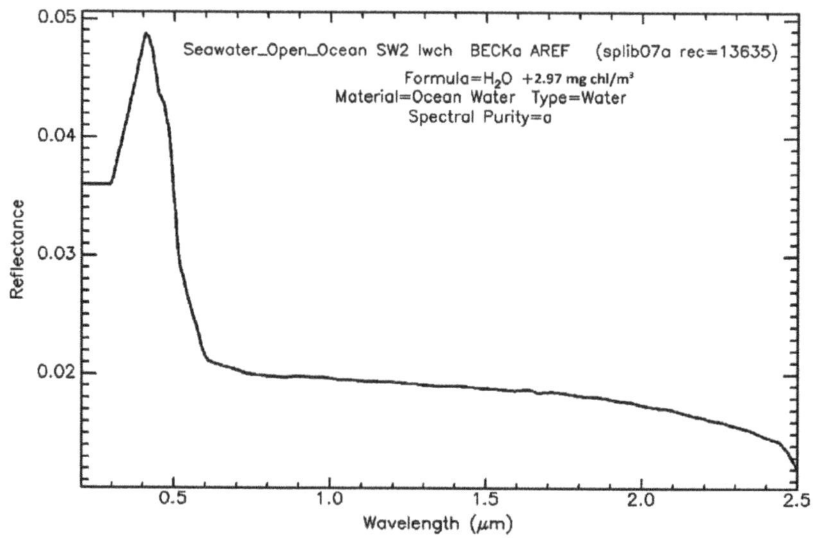

Figure 9. Spectral curve of chlorophyll in water (2.97 mg chl/m^3), USGS Spectral Library Version 7, https://crustal.usgs.gov/speclab/QueryAll07a.php?quick_filter=water (accessed on 1 January 2022). Reprinted with permission from [39].

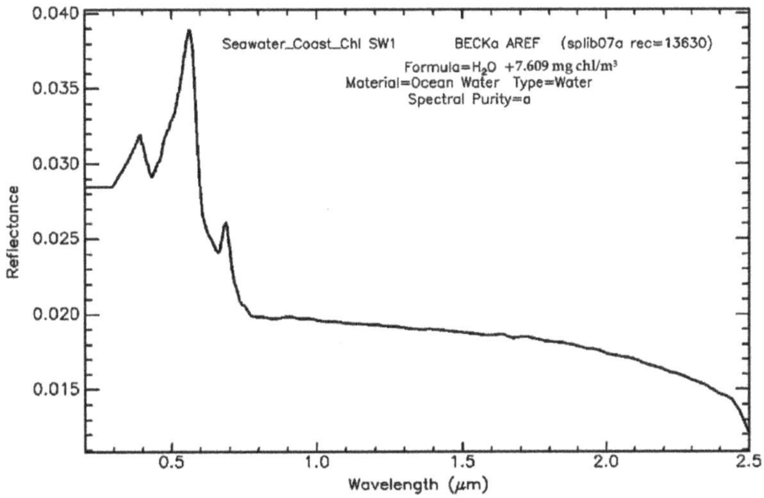

Figure 10. Spectral curve of chlorophyll in water (7.609 mg chl/m^3), USGS Spectral Library Version 7, https://crustal.usgs.gov/speclab/QueryAll07a.php?quick_filter=water (accessed on 1 January 2022). Reprinted with permission from [39].

2.5. Model

To compute concentrations of chlorophyll a (C_{chl}) from reflectance R obtained from satellite images for different wavelengths, a mathematical relationship has to be developed that meets the rules for logical concentration values for extremely high or low reflectance values. Empirical models, which are a linear combination of partial formulas, are risky to use because they cannot determine model parameters for all possible R values. The risk arises from a finite number of observational data, which usually does not include information on extreme values of R. Therefore, it may happen that C_{chl} calculated from the R values obtained from the new photos will be illogical, or even negative. Generally, extrapolation of such a model beyond the R values used to determine the model parameters yields highly ambiguous results; that is why models originating from the theory of dimensional analysis are considered to be safer. These types of models are usually the product of function modules at the appropriate power. They may also generate poor results outside of the R values used in the model estimation, but at least they guarantee that the results are positive.

The initial model was defined as:

$$C_{chl} = \alpha_0 R_{443}^{\alpha_1} R_{490}^{\alpha_2} R_{560}^{\alpha_3} R_{665}^{\alpha_4} R_{705}^{\alpha_5} R_{740}^{\alpha_6} R_{783}^{\alpha_7} R_{842}^{\alpha_8} R_{865}^{\alpha_9} R_{945}^{\alpha_{10}} R_{1375}^{\alpha_{11}} R_{1610}^{\alpha_{12}} R_{2190}^{\alpha_{13}} \quad (1)$$

where:

α_0 to α_{13}—model coefficients;

$R_{\ldots}$—radiation reflectance for wavelengths: 443, 490, 560, 665, 705, 740, 783, 842, 865, 945, 1375, 1610, and 2190 nm, determined from the satellite data; and

C_{chl}—concentration of chlorophyll a [mg/m^3].

The logarithmic Equation (1) leads to a linear relationship for the logarithms of reflectance R. The spectral curves of chlorophyll a (Figure 1) and the suspensions (Figures 2–4) show that the reflectance for wavelengths over 865 nm and shorter than 490 nm will not provide significant information on the concentration of chlorophyll a. Therefore, the logarithmic form of Equation (1), after simplification, takes the form:

$$\begin{aligned} \ln C_{chl} = a_0 + a_1 \ln R_{490} + a_2 \ln R_{560} + a_3 \ln R_{665} + a_4 \ln R_{705} + \\ a_5 \ln R_{740} + a_6 \ln R_{783} + a_7 \ln R_{842} + a_8 \ln R_{865} \end{aligned} \quad (2)$$

where:

a_0 to a_8—model coefficients,

$R_{\ldots}$.—radiation reflectance for wavelengths: 490, 560, 665, 705, 740, 783, 842, and 865 nm, determined from the satellite data.

To make the model more general, additional components, like the squares of the reflectance log, were introduced:

$$\begin{aligned}\mathbf{ln}C_{chl} = {} & a_0 + a_1\mathbf{ln}R_{490} + a_2\mathbf{ln}R_{650} + a_3\mathbf{ln}R_{665} + a_4\mathbf{ln}R_{705} + \\ & a_5\mathbf{ln}R_{740} + a_6\mathbf{ln}R_{783} + a_7\mathbf{ln}R_{842} + a_8\mathbf{ln}R_{865} + a_9(\mathbf{ln}R_{490})^2 \\ & a_{10}(\mathbf{ln}R_{560})^2 + a_{11}(\mathbf{ln}R_{665})^2 + a_{12}(\mathbf{ln}R_{705})^2 + a_{13}(\mathbf{ln}R_{740})^2 + \\ & a_{14}(\mathbf{ln}R_{783})^2 + a_{15}(\mathbf{ln}R_{842})^2 + a_{16}(\mathbf{ln}R_{865})^2\end{aligned} \quad (3)$$

The regression analysis of Equation (3) enabled the selection of such coefficients (from a_1 to a_{16}) that showed a probability lower than the significance coefficient 0.05 in the Student's t-distribution (significance test of the equation coefficients). It was found that the most important coefficients are related to the wavelengths: 665, 705, 740, and 842 nm. These are the lengths approximately corresponding to the local minimum and maximum on the spectral curve of chlorophyll a (Figures 1 and 2) and the local minimum and maximum on the spectral curve of the suspensions (Figures 2–4). In the case of the Sentinel satellite, there were no 760 and 810 nm wavelength channels corresponding to the local minimum and maximum on the spectral curve of the suspensions (Figures 2–4); hence, the 740 and 842 nm channels turned out to be statistically significant. Eventually, the model looked as follows:

$$\begin{aligned}\mathbf{ln}C_{chl} = {} & a_0 + a_3\mathbf{ln}R_{665} + a_4\mathbf{ln}R_{705} + a_5\mathbf{ln}R_{740} + a_7\mathbf{ln}R_{842} + \\ & a_{11}(\mathbf{ln}R_{665})^2 + a_{12}(\mathbf{ln}R_{705})^2 + a_{13}(\mathbf{ln}R_{740})^2 + a_{15}(\mathbf{ln}R_{842})^2\end{aligned} \quad (4)$$

To eliminate the influence of the lake bottom reflectance, some reflectance corrections in Equation (4) were required. The effect of light reflection by the lake bottom in four light wavelengths is shown in Figure 11. However, just the reflectance corrections for 665 nm and 705 nm are sufficient for a good model accuracy. Of course, the reflectance for 740 nm and 842 nm may also be corrected, but their impact is negligible for the quality of the model (4). Generally, the shorter the wave, the stronger the bottom reflectance. Such observation is also confirmed by the spectral curves (Figure 7), which show that the shorter the wave, the higher the reflectance.

Following the Lambert–Beer law, absorbance ($-\mathbf{ln}(I/I_0)$) of the medium for UV, Vis, and IR radiation is proportional to an optical path length l and a concentration of substance C. This means that a light intensity I in a water column decreases exponentially:

$$I = I_0 \cdot \mathbf{exp}(-k \cdot C \cdot l) \quad (5)$$

I_0—incident light intensity [W/m^2];

k—absorption coefficient [m^3/(g·m)];

C—concentration of a radiation absorber [g/m^3];

l—optical path length [m].

Assuming that the lake bottom reflectance R_b is approximately proportional to the intensity of the radiation reaching the depth of l = h, the reflectance over a water surface, recorded by a photosensitive sensor, would be described by the relationship:

$$R_b = \frac{\alpha_b \cdot I_0 \cdot \mathbf{exp}(-k \cdot C \cdot 2h)}{I_0} = \alpha_b \cdot \mathbf{exp}(-k \cdot C \cdot 2h) \quad (6)$$

α_b—coefficient of the radiation reflectance through the bottom;

C—concentration of all substances (e.g., chlorophyll a, mineral suspensions, organic suspensions, water) responsible for absorption of radiation [g/m^3], i.e., water turbidity [NTU];

h—depth (actual length of an optical path, when shooting close to the zenith, is approximately equal to the depth h) [m].

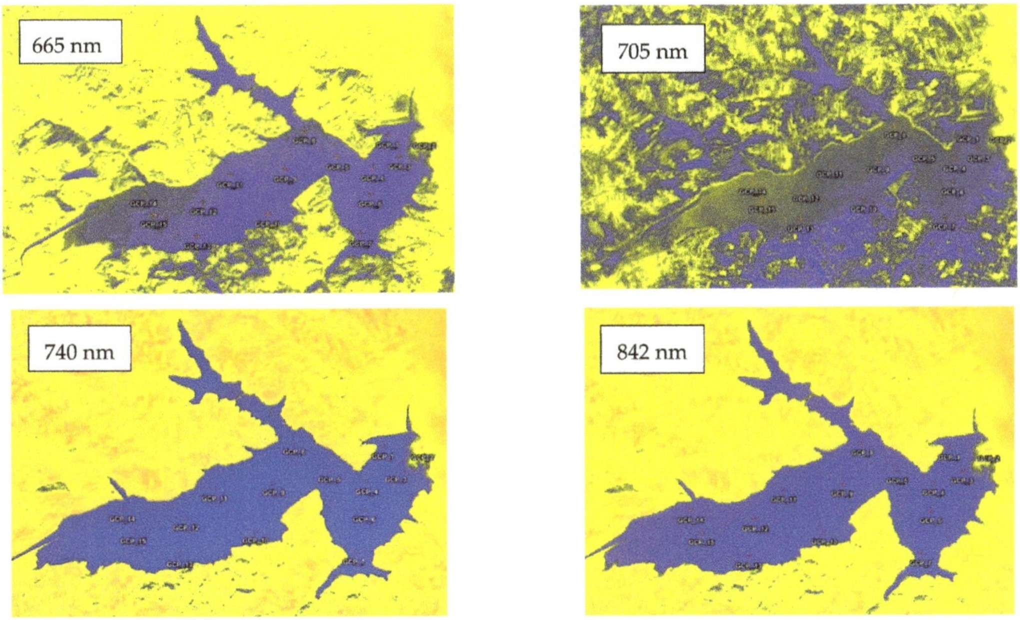

Figure 11. Images of Lake Dobczyce at four wavelengths (color representation: blue, yellow, violet, with the same color saturation in each photo).

The number 2 in Formula (6) means that radiation passes twice through the water layer.

The reflectance R, recorded by the satellite, is approximately equal to the sum of the four principal reflectances:

$$R = \frac{\alpha_s I_0 + I_{chl} + I_m + \alpha_b (I_0 - \alpha_s I_0)\mathbf{exp}(-k \cdot C \cdot 2h)}{I_0} = \\ \alpha_s + \frac{I_{chl}}{I_0} + \frac{I_m}{I_0} + \alpha_b (1 - \alpha_s)\mathbf{exp}(-k \cdot C \cdot 2h) = \\ \alpha_s + R_{chl} + R_m + \alpha_b (1 - \alpha_s)\mathbf{exp}(-k \cdot C \cdot 2h) \quad (7)$$

α_s—coefficient of the radiation reflectance for the water surface (water surface reflectance);

R_{chl}—reflectance related to chlorophyll a in water, as the total effect of radiation;

I_{chl}—reflected at different water depths;

R_m—reflectance related to the substances, other than chlorophyll a, in water, as the total effect of radiation I_m reflected at different water depths;

$\alpha_b (1 - \alpha_s)\mathbf{exp}(-k \cdot C \cdot 2h)$—reflectance of the lake bottom R_b.

Intensity of radiation I_{chl}, I_m reflected from water at different depths is defined as the depth integration from the derivative of the radiation intensity at different depths, with respect to the depth defined by (5); intensity I_0 must be reduced by the intensity of the radiation reflected from the water surface $\alpha_s I_0$. Therefore:

$$I_{chl} = -\int_0^h \frac{d(\alpha_{chl} C_{chl} (I_0 - \alpha_s I_0)\mathbf{exp}(-k \cdot C \cdot 2l))}{dl} dl = \\ \alpha_{chl} C_{chl} (I_0 - \alpha_s I_0)(1 - \mathbf{exp}(-k \cdot C \cdot 2h)) \quad (8)$$

$$I_m = -\int_0^h \frac{d(\alpha_m C_m (I_0 - \alpha_s I_0)\mathbf{exp}(-k \cdot C \cdot 2l))}{dl} dl = \\ \alpha_m C_m (I_0 - \alpha_s I_0)(1 - \mathbf{exp}(-k \cdot C \cdot 2h)) \quad (9)$$

C_{chl}—concentration of chlorophyll a [mg/m^3];

α_{chl}—coefficient of the radiation reflectance for chlorophyll a [m^3/mg];

C_m—concentration of non-chlorophyll a substances in water responsible for light reflection [g/m^3];

α_m—coefficient of the radiation reflectance for substances other than chlorophyll a [m^3/g].

If h and/or C are sufficiently high, then $\mathbf{exp}(-k{\cdot}C{\cdot}2h) << 1$ and:

$$I_{chl} \approx \alpha_{chl}C_{chl}(I_0 - \alpha_s I_0) \Rightarrow R_{chl} \approx \alpha_{chl}C_{chl}(1 - \alpha_s) \tag{10}$$

$$I_m \approx \alpha_m C_m(I_0 - \alpha_s I_0) \Rightarrow R_m \approx \alpha_m C_m(1 - \alpha_s) \tag{11}$$

In model (4), the $R \ldots$ reflectance for different wavelengths is the computational reflectance R_{calc}, which are formally sums $R_{chl,\,\ldots} + R_{m,\,\ldots}$ for different wavelengths. Therefore, each reflectance in model (4) should be corrected and replaced with the computational reflectance for different wavelengths:

$$\begin{aligned} R_{\ldots} \leftarrow R_{calc,\,\ldots} = R_{chl,\,\ldots} + R_{m,\,\ldots} = \\ R_{\ldots} - \alpha_{s,\,\ldots} - \alpha_{b,\,\ldots}(1 - \alpha_{s,\,\ldots})\mathbf{exp}(-k_{\ldots}{\cdot}C{\cdot}2h) \end{aligned} \tag{12}$$

$\square_{\ldots}$ —the index dots refer to different wavelengths;

$R_{\ldots}$ —radiation reflectance for different wavelengths determined from satellite;

$R_{calc,\,\ldots}$—computational reflectance for different wavelengths.

The coefficient α_s (or reflectance) for different wavelengths is small if compared to other reflectance, and therefore should be neglected, so:

$$\begin{aligned} R_{\ldots} \leftarrow R_{calc,\,\ldots} = R_{chl,\,\ldots} + R_{m,\,\ldots} = \\ R_{\ldots} - \alpha_{b,\,\ldots}\mathbf{exp}(-k_{\ldots}{\cdot}C{\cdot}2h) \end{aligned} \tag{13}$$

A reflectance correction (13) in Equation (4) gives the equation describing the concentration of chlorophyll a as:

$$\begin{aligned} C_{chl} = \mathbf{exp}[&a_0 + a_3\mathbf{ln}(R_{665} - \alpha_{b,720mu665}\mathbf{exp}(-k_{665}{\cdot}C{\cdot}2h)) + \\ &a_4\mathbf{ln}(R_{705} - \alpha_{b,720mu705}\mathbf{exp}(-k_{705}{\cdot}C{\cdot}2h)) + \\ &a_5\mathbf{ln}R_{740} + a_7\mathbf{ln}R_{842} + \\ &a_{11}(\mathbf{ln}(R_{665} - \alpha_{b,720mu665}\mathbf{exp}(-k_{665}{\cdot}C{\cdot}2h)))^2 + \\ &a_{12}(\mathbf{ln}(R_{705} - \alpha_{b,720mu705}\mathbf{exp}(-k_{705}{\cdot}C{\cdot}2h)))^2 + \\ &a_{13}(\mathbf{ln}R_{740})^2 + a_{15}(\mathbf{ln}R_{842})^2] \end{aligned} \tag{14}$$

C_{chl}—chlorophyll a concentration [mg/m^3];

$\mathbf{exp}(a_0)$—model's coefficient [mg/m^3].

Assume the following parameter values for Equations (7)–(9) relating to one wavelength: $\alpha_s = 0.002$, $\alpha_m = 0.001$, $\alpha_b = 0.3$, $\alpha_{chl} = 0.001$, $k = 0.05$ $(NTU{\cdot}m)^{-1}$, $C_m = C \approx 5$ NTU (in practice, water turbidity is assumed as C) and $C_{chl} = \{1, 5, 10, 20, 50\}$ mg/m^3. From the Equations (7)–(9), the reflectance R can be determined, as a function of the depth h (Figure 12). Knowing R and the parameters of Equations (7)–(9), the concentration of chlorophyll a C_{chl} can be again determined. These would be horizontal lines of constant values $C_{chl} = \{1, 5, 10, 20, 50\}$ mg/m^3 for all depths h. In order to check the quality of the model (14), we write it for a single wavelength:

$$\begin{aligned} C_{chl} = \mathbf{exp}[&a_0 + a_3\mathbf{ln}(R - \alpha_b^*\mathbf{exp}(-k^*{\cdot}C{\cdot}2h)) + \\ &a_{11}(\mathbf{ln}(R - \alpha_b^*\mathbf{exp}(-k^*{\cdot}C{\cdot}2h)))^2] \end{aligned} \tag{15}$$

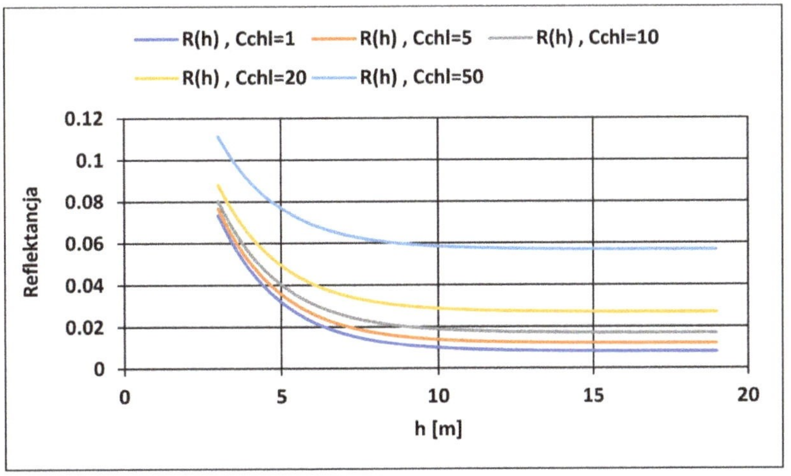

Figure 12. Changes in reflectance R as a function of depth h for the sensor above the water surface at different concentrations of chlorophyll a C_{chl} = {1, 5, 10, 20, 50} mg/m^3.

The parameters of model (15) were determined on the basis of changes in reflectance R as a function of h, determined from Equations (7)–(9). The following parameters were obtained: $a_0 = -0.994654$, $a_3 = -3.91129$, $\alpha_b^* = 0.293404$, $k^* = 0.04999945$ (NTU·m)$^{-1}$, $a_{11} = -0.767181$. Parameters: α_b^* and k^* are almost equal to $\alpha_b = 0.3$, $k = 0.05$ (NTU·m)$^{-1}$ used in Equations (7)–(9). Formally, α_b^* should be equal to $\alpha_s + \alpha_b(1 - \alpha_s) = 0.002 + 0.3 \cdot (1 - 0.002) = 0.3014$, as in Equation (12). The compliance of the parameters α_b^* and k^* with the parameters α_b and k confirms that a correction (13) for the bottom reflectance in the models (14) and (15) was needed. Model (15) is less accurate for shallow depths and higher concentrations of chlorophyll a C_{chl} (Figure 13), along with model (14).

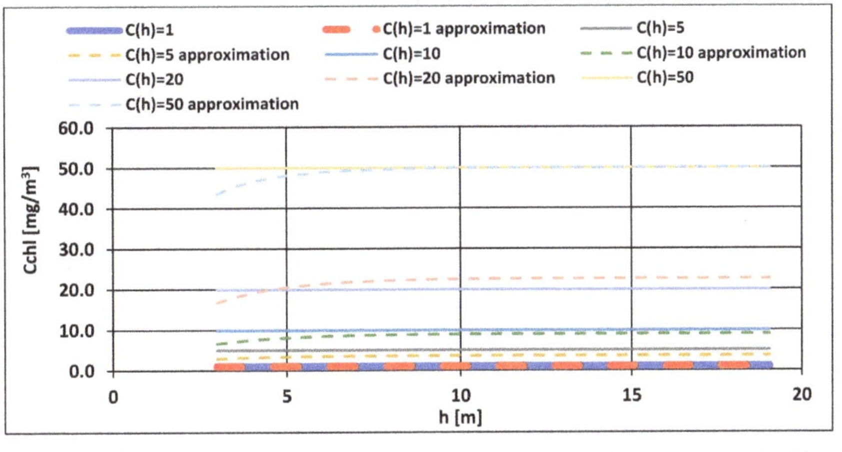

Figure 13. Chlorophyll a C_{chl} concentrations as a function of depth h for Equations (7)–(9) and model (15) (approximation).

It would be difficult to use a combination of Equations (7)–(9) to calculate chlorophyll a concentrations on the basis of the reflectance recorded by a satellite; such reflectance is a combination of both a chlorophyll a reflectance and the reflectance of other substances present in the water. Such calculations were performed for the wavelengths of 665 nm and 705 nm. However, accurate parameters of models (7)–(9) for one wavelength and known turbidity could not be found. Therefore, model (14) has been proposed as the one that that includes the reflectance for different wavelengths.

3. Results

3.1. Model Parameters for Chlorophyll a

Based on the measurements of the concentration of chlorophyll a C_{chl}, turbidity C, lake depth h, and the reflectance $R \ldots$ for wavelengths: 665, 705, 740, and 842 nm, the model parameters were determined (14). Due to the large number of parameters:

$$a_0\ ,\ a_3\ ,\ \alpha_{b,\,665}\ ,\ k_{665}\ ,\ a_4\ ,\ \alpha_{b,\,705}\ ,\ k_{705}\ ,\ a_5\ ,\ a_7\ ,\ a_{11}\ ,\ a_{12}\ ,\ a_{13}\ ,\ a_{15}$$

the two-stage procedure was introduced. At first, a preliminary estimate of the parameters: $a_0\ ,\ a_3\ ,\ a_4\ ,\ a_5\ ,\ a_7\ ,\ a_{11}\ ,\ a_{12}\ ,\ a_{13}\ ,\ a_{15}$ was produced for the logarithm of r (4) using the least square method, i.e., minimizing the sum of squares of deviations between the logarithm of the measured chlorophyll concentration C_{chl} and the logarithm of the concentration calculated from model (4). Then, other model parameters were found, assuming the parameter values previously determined for model (4) for the calculations. All parameters were determined by minimizing the sum of squared deviations between the measured C_{chl} concentration and the one calculated from model (14); the correlation coefficient was 0.944. The values of the model parameters (14) are summarized in Table 2, col. 2. The model fit is shown in Figure 14. If turbidity was calculated from model (17), the parameters of model (14) would have been provided, as shown in Table 2, col. 3. The correlation coefficient was 0.925. The fit of model (14), while using model (17), is shown in Figure 15. In both cases, the model's fit to the measurements was good. If there is no detailed data on turbidity and the parameter shows little variability, the average value of turbidity can be used in the calculations.

Table 2. Model parameters (14).

Parameter	Value at Known Turbidity	Value at Turbidity Calculated from Model (17)	Units
1	2	3	4
a_0	−32.25907512	−75.68466110	ln(mg/m^3)
a_3	−43.55040094	−72.72855795	–
$\alpha_{b,\,665}$	0.015293030	0.006693771	–
k_{665}	0.007665104	0.011140396	(NTU·m)$^{-1}$
a_4	42.36776151	62.29811979	–
$\alpha_{b,\,705}$	0.010635655	0.008960131	–
k_{705}	0.003498578	0.000530822	(NTU·m)$^{-1}$
a_5	−23.86732103	−47.45269805	–
a_7	4.731529537	9.465054057	–
a_{11}	−5.149196218	−9.056892001	–
a_{12}	4.533167105	6.641324249	–
a_{13}	−2.345421643	−5.388690882	–
a_{15}	0.193746731	0.802524849	–

3.2. Maps of Chlorophyll a Concentrations for Lake Dobczyce

The map of chlorophyll a concentrations (Figure 16) was developed from satellite images using models (14) and (17). Long retention times observed in Lake Dobczyce (average 157 days), contribute to the algae growth and high chlorophyll a concentrations; higher concentrations are noted in the middle of the lake. In this region, towards the dam, low flow velocities are also observed (Figure 17); therefore, concentrations slightly decrease due to sedimentation of the suspended solids and algae. There are stagnant zones at the northern and southern banks of the lake, where flow velocities are very low (Figure 17). Such conditions promote algae growth, and the observed concentration of chlorophyll a is high. Moreover, in the lake branches, where a water exchange is low and the retention times exceed the average one, high concentrations of chlorophyll a are observed (Figure 16).

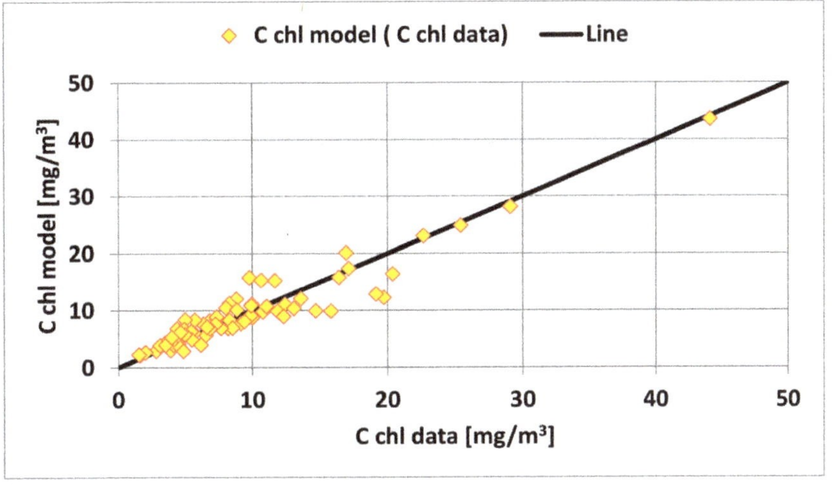

Figure 14. Model fit (C_{chl} model) (14) to the measured concentrations of chlorophyl-a (C_{chl} data) at the known turbidity values.

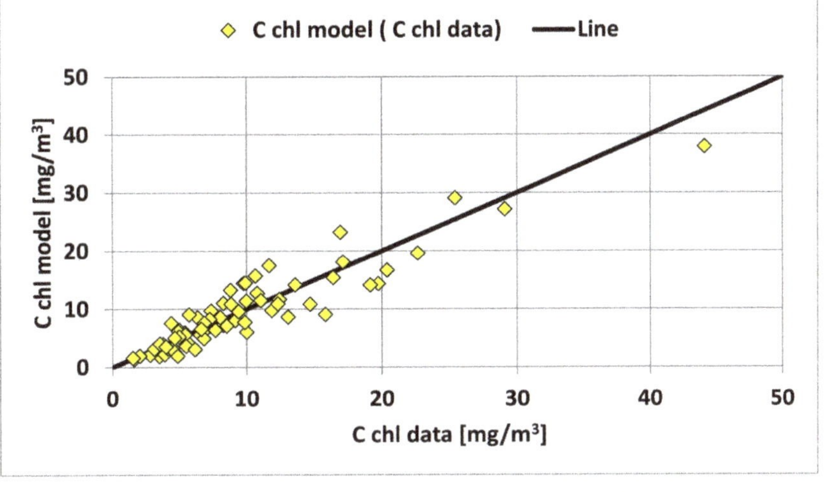

Figure 15. Model fit (C_{chl} model) (14) to the measured concentrations of chlorophyl-a (C_{chl} data) for turbidity model (17).

In October, water temperatures in the lake are low (around 13 °C), and the algae growth slows down. At that time, higher concentrations of algae could be found only in the northern branch of the lake (stagnant zone) and at the southern shores, where low flow velocities (Figure 17) are responsible for stagnant zones (Figure 18). The chlorophyll a model showed a good fit to the measurement data (Figure 18, table).

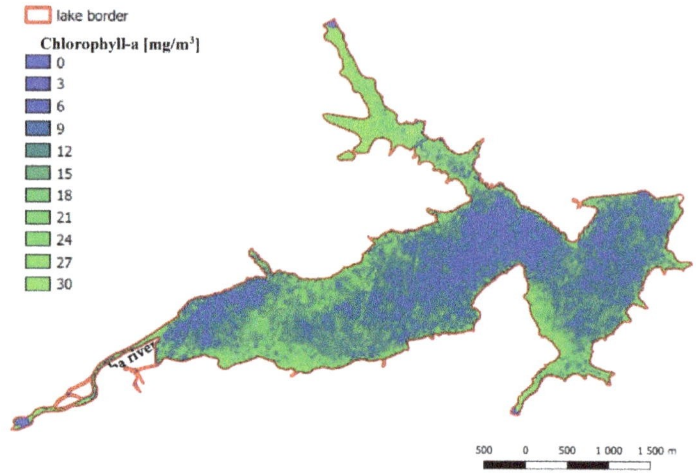

Figure 16. Chlorophyll concentration in Lake Dobczyce 9 May2021 (imagery by Sentinel 2).

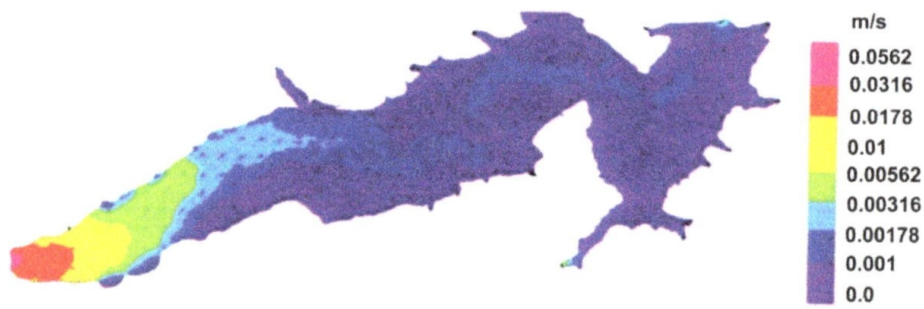

Figure 17. Two-dimensional field of the average vertical velocity in the main part of Lake Dobczyce (without the northern bay), with the total flow of 10 m^3/s and no-wind conditions (model RMA2). Reprinted with permission from [40].

3.3. *Models Parameters for Turbidity*

The turbidity C model was developed similarly to the chlorophyll a model. At first, it was determined which factors from the range of a_0–a_{16} are significant for turbidity in Equation (3). Then, the significance test was used for the a_1–a_{16} coefficients at a significance level of 0.05 using the Student's t-distribution. It was found that the most important coefficients are related to the wavelengths: 705, 740, and 783 nm; these lengths approximately correspond to the local minimum and maximum on the spectral curve of suspensions (Figures 2–4). The initial form of the equation was as follows:

$$\begin{aligned}\mathbf{ln}C = a_{C,0} + a_{C,4}\mathbf{ln}R_{705} + a_{C,5}\mathbf{ln}R_{740} + a_{C,6}\mathbf{ln}R_{783} + \\ a_{C,12}(\mathbf{ln}R_{705})^2 + a_{C,13}(\mathbf{ln}R_{740})^2 + a_{C,14}(\mathbf{ln}R_{783})^2\end{aligned} \tag{16}$$

C—turbidity [NTU];
$a_{\ldots}$—model coefficients;
$R_{\ldots}$—radiation reflectance for wavelengths: 705, 740, and 783 nm.
Taking into account the reflectance of the lake bottom leads to a relationship:

$$\begin{aligned}C = \mathbf{exp}\ \ a_{C,0} + a_{C,4}\mathbf{ln}(R_{705} - \alpha_{C,b,\,705}\mathbf{exp}(-k_{C,705}\cdot C\cdot 2h)) + \\ a_{C,5}\mathbf{ln}R_{740} + a_{C,6}\mathbf{ln}R_{783} + \\ a_{C,12}\left(\mathbf{ln}\left(R_{705} - \alpha_{C,b,\,705}\mathbf{exp}\left(-k_{C,705}\cdot C\cdot 2h\right)\right)\right)^2 + \\ a_{C,13}(\mathbf{ln}R_{740})^2 + a_{C,14}(\mathbf{ln}R_{783})^2\end{aligned} \tag{17}$$

$\mathbf{exp}(a_{C,0})$—model's coefficient [NTU].

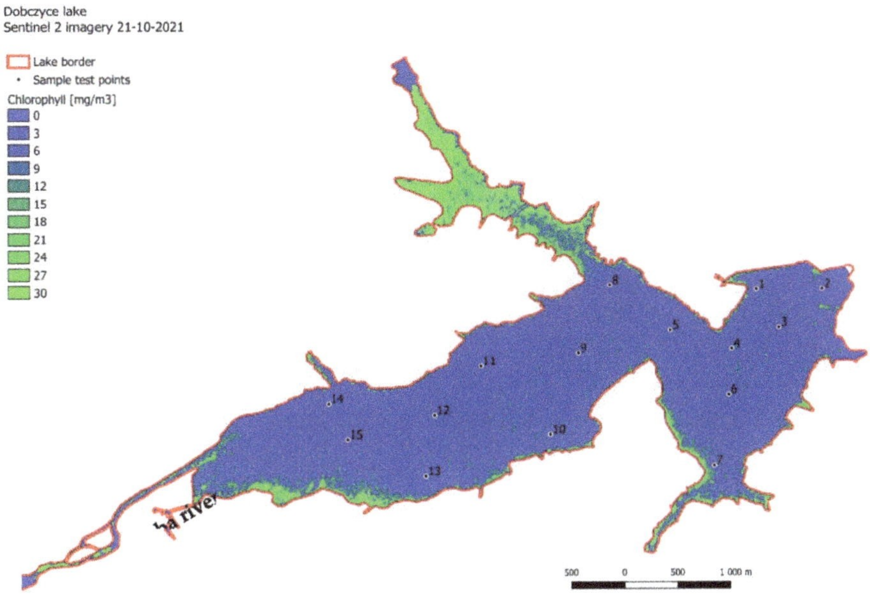

Point	Sample chl.-a [mg/m³]	Model chl.-a [mg/m³]
1	4.63	5.02
3	3.96	3.54
4	4.96	5.24
5	6.61	6.59
6	5.48	3.66
7	3.51	4.04
8	6.15	3.10
9	4.85	1.96
10	5.42	5.63
11	4.90	6.19
12	4.60	2.68
13	4.11	3.25
14	4.48	3.36
15	3.87	2.24

Figure 18. Chlorophyll concentrations in Lake Dobczyce, 21 October 2021 (imagery by Sentinel 2).

The effect of light reflection by the lake bottom is shown in Figure 11. It appears that it is sufficient to consider the reflectance correction for 705 nm to obtain a satisfactory accuracy of model (17). Of course, it is also possible to correct the reflectance for 740 nm and 783 nm, but the overall improvement of the model quality would be negligible.

Equation (17) is implicit due to turbidity C. The C value can be obtained by successive approximations, i.e., inserting in $\mathbf{exp}\left(-k_{C,705}\cdot C\cdot 2h\right)$ values C from the previous approximation; after several attempts (e.g., 4), the value C becomes reasonably accurate. Another method involves searching for the function's zero, i.e., the difference of the right side of Equation (17) and C. Here, the regula falsi method of searching for the function's zero can be used.

Based on the measurement data of: turbidity C, lake depth h, and the reflectance R... for wavelengths 705, 740, and 783 nm, the parameters of model (17) were determined. Due to the large number of parameters:

$$a_{C,0}\ ,\ a_{C,4}\ ,\ \alpha_{C,b,705}\ ,\ k_{C,705}\ ,\ a_{C,5}\ ,\ a_{C,6}\ ,\ a_{C,12}\ ,\ a_{C,13}\ ,\ a_{C,14}$$

the two-stage procedure was employed. First, the values of some parameters were initially estimated: $a_{C,0}$, $a_{C,4}$, $a_{C,5}$, $a_{C,6}$, $a_{C,12}$, $a_{C,13}$, $a_{C,14}$, and then all the others were estimated, while correcting the values of the pre-estimated parameters. The parameters were determined using the least squares method (the best fit) for turbidity C and turbidity from model (17). The model parameters are summarized in Table 3. The correlation coefficient was 0.939, and the model fit is shown in Figure 19. A good fit of the model to the measurement data was obtained.

3.4. Maps of Turbidity for Lake Dobczyce

The water turbidity map for the lake (Figure 20) was developed from the satellite images and model (17). Long retention times in Lake Dobczyce (average 157 days) favor the sedimentation of suspended particles. Therefore, the water turbidity in the middle of the lake, towards the dam, where there are low velocities (Figure 17), is lower than turbidity close to the place where the Raba river enters the lake (Figure 20). In the branches of the lake, where water exchange is low and retention times are longer, low turbidity may be attributed to good sedimentation conditions (Figure 20); turbidity is also low on the south-eastern shores of the lake, where the flow velocities are low.

Table 3. Values of the model (17) parameters.

Parameter	Value	Unit
$a_{C,0}$	−28.87376959	ln(NTU)
$a_{C,4}$	13.12134743	–
$\alpha_{C,b,705}$	0.014648999	–
$k_{C,705}$	0.0000434391	$(\text{NTU}\cdot\text{m})^{-1}$
$a_{C,5}$	−59.94110417	–
$a_{C,6}$	26.75417218	–
$a_{C,12}$	1.323875637	–
$a_{C,13}$	−8.392820945	–
$a_{C,14}$	3.936558098	–

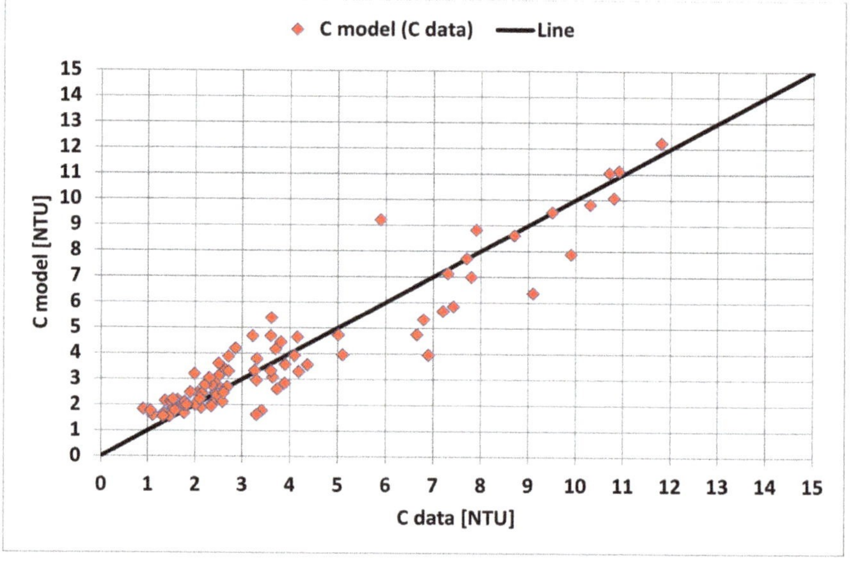

Figure 19. Model (17) fit (C model) to the measured turbidity (C data).

In October (Figure 21) and in May (Figure 20), turbidity decreased along the lake towards the dam. Moreover, in the northern branch, sedimentation contributed to a lower turbidity (Figure 21). The turbidity model showed a good fit with the measurements (Figure 21, table); however, at point 2 (Figure 21), the turbidity calculations were poor due to clouds obscuring the view.

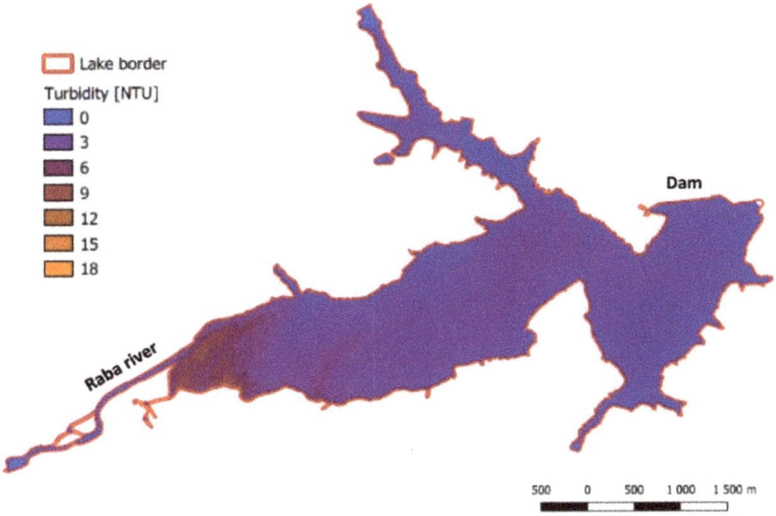

Figure 20. Water turbidity of Lake Dobczyce 9 May 2021 (imagery by Sentinel 2).

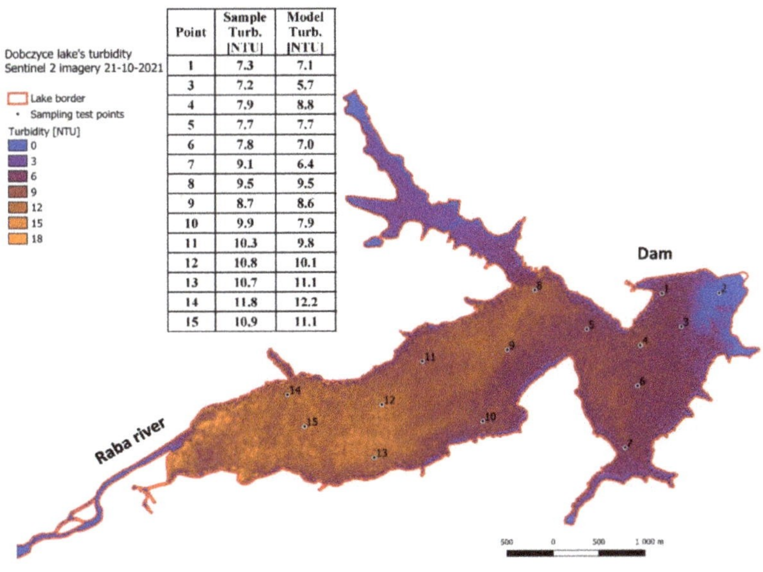

Point	Sample Turb. [NTU]	Model Turb. [NTU]
1	7.3	7.1
3	7.2	5.7
4	7.9	8.8
5	7.7	7.7
6	7.8	7.0
7	9.1	6.4
8	9.5	9.5
9	8.7	8.6
10	9.9	7.9
11	10.3	9.8
12	10.8	10.1
13	10.7	11.1
14	11.8	12.2
15	10.9	11.1

Figure 21. Water turbidity of Lake Dobczyce, 21 October 2021 (imagery by Sentinel 2).

4. Discussion

The authors developed the model to calculate the concentration of chlorophyll a and turbidity in the water. In the case of Lake Dobczyce, the chlorophyll a model takes into account reflectance corresponding to the middle wavelengths 665, 705, 740, and 842 nm, while the bottom effect is related to the wavelengths 665 and 705 nm. The model for turbidity considers the reflectance corresponding to the middle wavelengths 705, 740, and 783 nm, while the bottom effect is related to the wavelength 705 nm. To eliminate a minor reflectance related to other wavelengths from the pseudo-linear model (3), the Student's *t*-test was been used for both chlorophyll a and turbidity. It cannot be predicted in advance whether the models for different lakes will always use the reflectance for the same wavelengths. If so, the model coefficients may probably be different due to different water characteristic and other properties of the bottom sediments. In the case of Lake Dobczyce, all of the discriminants that are quotients of reflectance differences or reflectance quotients (Table 1) were not used in the models because they did not improve their quality.

The average relative error of the model for chlorophyll a is about 0.216, while the average error is about 2.01 mg Chl a/m^3. Graphs for errors are shown in Figures 22 and 23.

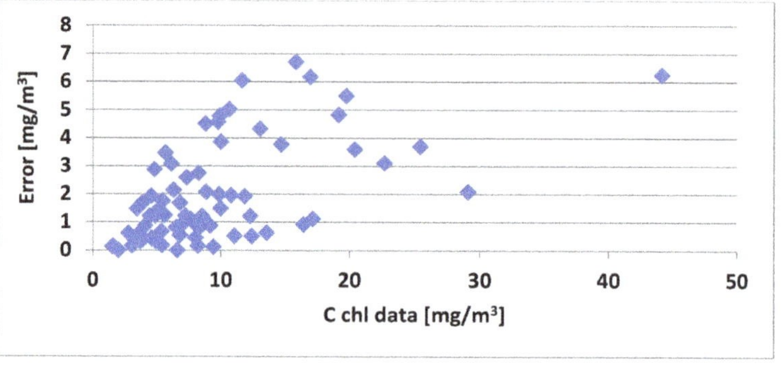

Figure 22. Model errors for chlorophyll a versus measurement data.

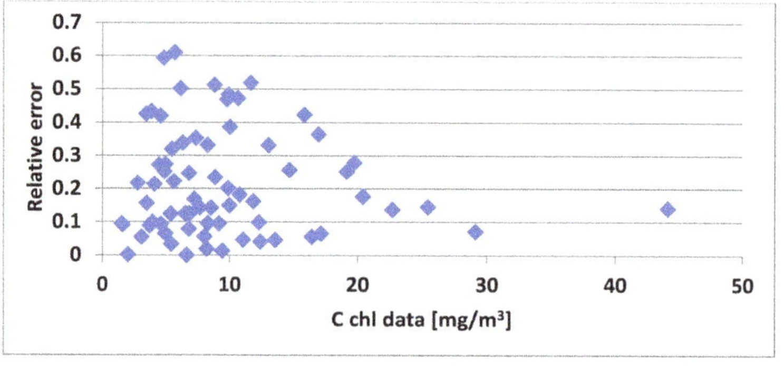

Figure 23. Model relative errors for chlorophyll a versus measurement data.

The greatest errors of the model relate to concentrations of about 15 mg Chl a/m^3 (Figure 22) and the highest relative errors concentrations of about 6 mg Chl a/m^3 (Figure 23). For low and high chlorophyll a concentrations, the relative errors are the smallest. Sometimes, the existence of greater errors is caused by the unrepresentativeness of point measurement relative to the area represented by one raster. In addition, a greater number of data with average values of chlorophyll a concentration increases the likelihood of a greater error.

The reflectance corrections for 665 nm were smaller than for 705 nm (Figure 24). This may be due to the fact that the wavelength of emissions for chlorophyll a is 663 nm, after it was stimulated using a 430 nm radiation. Backscattering with a length of 665 nm, recorded by the satellite, therefore had to be more characteristic of chlorophyll a than for other substances contained in water. Backscattering with a length of 705 nm is characteristic of minerals, in particular montmorillonite (Figure 8), and this likely explains why the reflectance corrections were greater. Reflectances for other wavelengths, occurring in the model for chlorophyll a, can be characteristic of other substances contained in water and atmosphere ingredients.

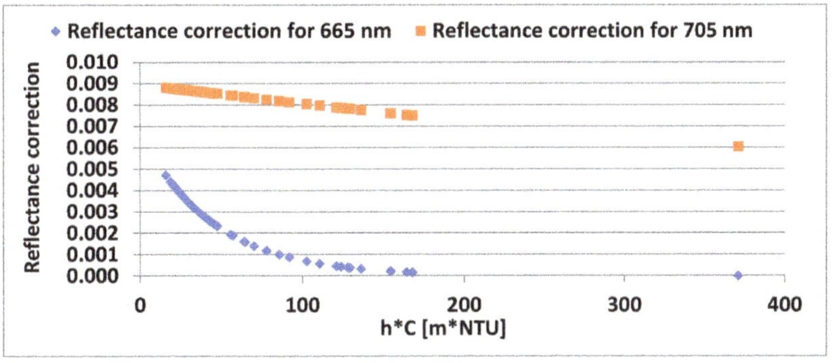

Figure 24. The reflection corrections graph versus the product of the depth and turbidity for the chlorophyll a model.

The average relative error of the model for turbidity is about 0.184, while the average error is about 0.629 NTU. Graphs of errors are shown in Figures 25 and 26.

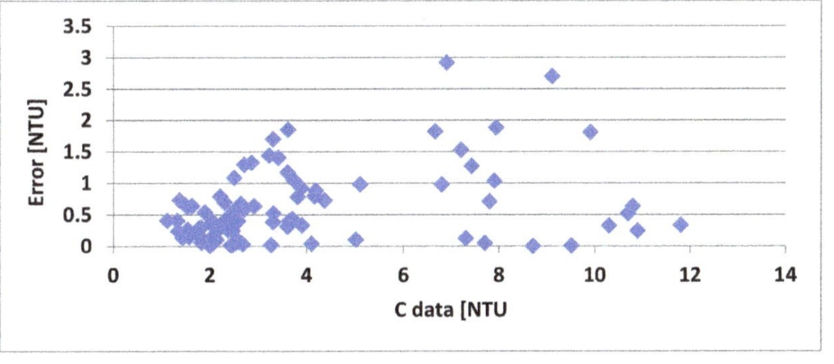

Figure 25. Model errors for turbidity versus measurement data.

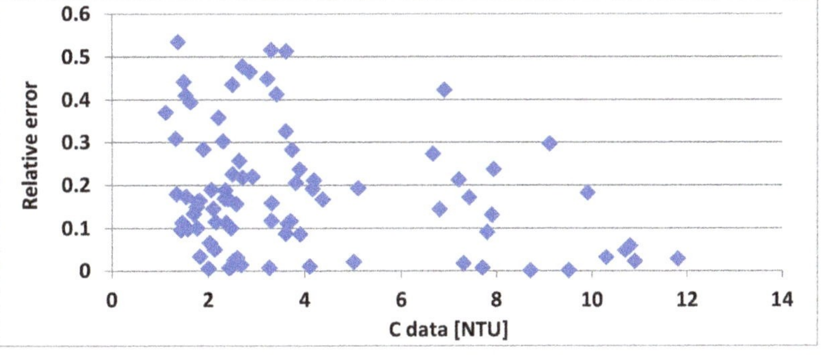

Figure 26. Model relative errors for turbidity versus measurement data.

The greatest errors for the turbidity model concern the turbidity of about 7 NTU (Figure 25) and the greatest relative errors of turbidity of about 2–4 NTU (Figure 26). Sometimes, the existence of greater errors is caused by the unrepresentativeness of point measurement relative to the area represented by one raster. In addition, a greater number of data with average turbidity increases the likelihood of a greater error.

The reflectance corrections for 705 nm for the turbidity model were approximately constant and amounted to around 0.0146.

The remote determination of water quality parameters requires corrections of the reflectance measured by the satellite, which may pose some problems. In urbanized areas or areas close to industrial agglomerations, standard corrections do not give good results. The changes in composition of the atmosphere (content of moisture, dust, and other pollutants) may require different corrections. Therefore, it becomes necessary to determine a reference surface that enables a reflectance correction. This can be any surface with spectral properties constant in time. In this case, a fragment of the dam crest served as the reference surface. The base reflectance was the first reference surface reflectance in a series of measurements. The changes in the reflectance of the reference surface in relation to the base surface made it possible to correct the reflectance of the lake surface.

5. Conclusions

- The large number of different models used to calculate chlorophyll a concentrations and turbidity in water means that there is no one universal model for use with different bodies of water.
- Models developed as the product of the reflectance powers (after logarithm) undergo statistical analysis to eliminate irrelevant components. This procedure simplifies the model. The final model may take into account the reflectances of many wavelengths, which are then eliminated by the statistical test at the significance level, e.g., 0.05 (Student's t-distribution).
- The specific physic-chemical and biological composition of the water in a given reservoir may result in significant differences between the spectral curves for individual water components.
- Shallow reservoirs require corrections for the bottom reflectance. It is impossible to predict in advance at what depth and for which wavelengths the bottom reflectance is already negligible. The analysis of satellite images helps to determine whether the bottom reflectance at a given wavelength is insignificant (poor bottom visibility); this simplifies the model.
- It is difficult or even impossible to use a combination of physical equations (type (7)–(9)) to calculate the concentration of chlorophyll a based on the reflectance recorded by the satellite; such reflectance is a mix of the chlorophyll a reflectance and the reflectance of many other substances present in the water. The authors could not find parameters of the model types (7)–(9) for just one wavelength and the known turbidity and maintain a satisfactory accuracy. Therefore, they proposed model (14) that took into account the reflectance for different wavelengths.
- Reflectances corresponding to the middle wave range 665, 705, 740, and 842 nm have been used in the model of chlorophyll concentration, while the effect of lake bottom interaction is associated with wavelengths 665 and 705 nm.
- Reflectances corresponding to the middle wave range 705, 740, and 783 nm have been used in the model of turbidity, while the effect of lake bottom interaction is associated with wavelength 705 nm.
- The described models use the reflectances normalized to the conditions prevailing at a specific moment over the reference surface.

Author Contributions: Conceptualization, A.B.; Formal analysis, A.B.; Software, C.T.; Validation, A.B.; Visualization, C.T. All authors have read and agreed to the published version of the manuscript.

Funding: This research received no external funding.

Informed Consent Statement: Informed consent was obtained from all subjects involved in the study.

Acknowledgments: The authors thank the Waterworks of the City of Krakow for providing assistance in the physic-chemical analyses of water samples and for sharing the data regarding the water quality of Lake Dobczyce.

Conflicts of Interest: The authors declare no conflict of interest.

References

1. Clesceri, L.S.; Greenberg, A.E.; Eaton, A.D. *Standard Methods for the Examination of Water and Wastewater*, 20th ed.; APHA American Public Health Association: Washington, DC, USA, 1998.
2. Aminot, A.; Rey, F. *Standard Procedure for the Determination of Chlorophyll a by Spectroscopic Methods*; International Council for the Exploration of the Sea: Copenhagen, Denmark, 2000; ISSN 0903-2606. Available online: https://www.researchgate.net/publication/242385183_Standard_procedure_for_the_determination_of_chlorophyll_a_by_spectroscopic_methods (accessed on 29 June 2022).
3. Miller, H.M.; Sexton, N.R.; Koontz, L.; Loomis, J.; Koontz, S.R.; Hermans, C. USGS Open-File Report 2011-1031: The Users, Uses, and Value of Landsat and Other Moderate-Resolution Satellite Imagery in the United States—Executive Report. Available online: https://pubs.usgs.gov/of/2011/1031/ (accessed on 29 June 2022).
4. Allan, M.G.; Hamilton, D.P.; Hicks, B.J.; Brabyn, L. Landsat Remote Sensing of Chlorophyll a Concentrations in Central North Island Lakes of New Zealand. *Int. J. Remote Sens.* **2011**, *32*, 2037–2055. [CrossRef]
5. Ventura, D.L.T.; Martinez, J.-M.; de Attayde, J.L.; Martins, E.S.P.R.; Brandini, N.; Moreira, L.S. Long-Term Series of Chlorophyll-a Concentration in Brazilian Semiarid Lakes from Modis Imagery. *Water* **2022**, *14*, 400. [CrossRef]
6. Albert, A.; Gege, P. Inversion of Irradiance and Remote Sensing Reflectance in Shallow Water between 400 and 800 Nm for Calculations of Water and Bottom Properties. *Appl. Opt.* **2006**, *45*, 2331–2343. [CrossRef] [PubMed]
7. Richter, R.; Schläpfer, D. *Atmospheric/Topographic Correction for Satellite Imagery*; DLR Report DLR-IB 565–601: Wessling, Germany, 2005.
8. Song, K. Water Quality Monitoring Using Landsat Themate Mapper Data with Empirical Algorithms in Chagan Lake, China. *J. Appl. Remote Sens.* **2011**, *5*, 053506. [CrossRef]
9. Lim, J.; Choi, M. Assessment of Water Quality Based on Landsat 8 Operational Land Imager Associated with Human Activities in Korea. *Environ. Monit. Assess.* **2015**, *187*, 384. [CrossRef]
10. Le, C.; Li, Y.; Zha, Y.; Sun, D.; Huang, C.; Lu, H. A Four-Band Semi-Analytical Model for Estimating Chlorophyll a in Highly Turbid Lakes: The Case of Taihu Lake, China. *Remote Sens. Environ.* **2009**, *113*, 1175–1182. [CrossRef]
11. Kudela, R.M.; Palacios, S.L.; Austerberry, D.C.; Accorsi, E.K.; Guild, L.S.; Torres-Perez, J. Application of Hyperspectral Remote Sensing to Cyanobacterial Blooms in Inland Waters. *Remote Sens. Environ.* **2015**, *167*, 196–205. [CrossRef]
12. Peterson, K.T.; Sagan, V.; Sloan, J.J. Deep Learning-Based Water Quality Estimation and Anomaly Detection Using Landsat-8/Sentinel-2 Virtual Constellation and Cloud Computing. *GISci. Remote Sens.* **2020**, *57*, 510–525. [CrossRef]
13. Sagan, V.; Peterson, K.T.; Maimaitijiang, M.; Sidike, P.; Sloan, J.; Greeling, B.A.; Maalouf, S.; Adams, C. Monitoring Inland Water Quality Using Remote Sensing: Potential and Limitations of Spectral Indices, Bio-Optical Simulations, Machine Learning, and Cloud Computing. *Earth-Sci. Rev.* **2020**, *205*, 103187. [CrossRef]
14. IOCCG. *Remote Sensing of Inherent Optical Properties: Fundamentals, Tests of Algorithms, and Applications*; International Ocean Colour Coordinating Group (IOCCG): Dartmouth, NS, Canada, 2006; ISSN 1098-6030. Available online: https://www.ioccg.org/reports/report5.pdf (accessed on 1 January 2022).
15. Li, Y.; Shang, S.; Zhang, C.; Ma, X.; Huang, L.; Wu, J.; Zeng, Y. Remote Sensing of Algal Blooms Using a Turbidity-Free Function for near-Infrared and Red Signals. *Chin. Sci. Bull.* **2006**, *51*, 464–471. [CrossRef]
16. Schalles, J.F.; Schiebe, F.R.; Starks, P.J.; Troeger, W.W. Estimation of Algal and Suspended Sediment Loads (singly and Combined) Using Hyperspectral Sensors and Integrated Mesocosm Experiments. In Proceedings of the Fourth International Conference on Remote Sensing for Marine and Coastal Environments, Orlando, FL, USA, 17–19 March 1997; pp. 111–120.
17. Ritchie, J.C.; Zimba, P.V.; Everitt, J.H. Remote Sensing Techniques to Assess Water Quality. *Photogramm. Eng. Remote Sens.* **2003**, *69*, 695–704. [CrossRef]
18. Lodhi, M.A.; Rundquist, D.C.; Han, L.; Kuzila, M.S. The Potential for Remote Sensing of Loess Soils Suspended in Surface Waters. *J. Am. Water Resour. Assoc.* **1997**, *33*, 111–117. [CrossRef]
19. Alparslan, E.; Aydöner, C.; Tufekci, V.; Tüfekci, H. Water Quality Assessment at Ömerli Dam Using Remote Sensing Techniques. *Environ. Monit. Assess.* **2007**, *135*, 391. [CrossRef] [PubMed]
20. Johan, F.B.; Jafri, M.Z.B.; San, L.H.; Omar, W.M.W.; Ho, T.C. Chlorophyll *a* Concentration of Fresh Water Phytoplankton Analysed by Algorithmic Based Spectroscopy. *J. Phys. Conf. Ser.* **2018**, *1083*, 012015. [CrossRef]
21. Gitelson, A.A.; Dall'Olmo, G.; Moses, W.; Rundquist, D.C.; Barrow, T.; Fisher, T.R.; Gurlin, D.; Holz, J. A Simple Semi-Analytical Model for Remote Estimation of Chlorophyll-a in Turbid Waters: Validation. *Remote Sens. Environ.* **2008**, *112*, 3582–3593. [CrossRef]
22. Shen, F.; Zhou, Y.-X.; Li, D.-J.; Zhu, W.-J.; Salama, M.S. Medium Resolution Imaging Spectrometer (MERIS) Estimation of Chlorophyll-*a* Concentration in the Turbid Sediment-Laden Waters of the Changjiang (Yangtze) Estuary. *Int. J. Remote Sens.* **2010**, *31*, 4635–4650. [CrossRef]
23. Son, S.; Wang, M. Water Quality Properties Derived from VIIRS Measurements in the Great Lakes. *Remote Sens.* **2020**, *12*, 1605. [CrossRef]
24. Diouf, D.; Seck, D. Modeling the Chlorophyll-a from Sea Surface Reflectance in West Africa by Deep Learning Methods: A Comparison of Multiple Algorithms. *Int. J. Artif. Intell. Appl.* **2019**, *10*, 33–40. [CrossRef]

25. Johnson, R.; Strutton, P.G.; Wright, S.W.; McMinn, A.; Meiners, K.M. Three Improved Satellite Chlorophyll Algorithms for the Southern Ocean. *J. Geophys. Res. Ocean.* **2013**, *118*, 3694–3703. [CrossRef]
26. Hansen, C.H.; Williams, G.P.; Adjei, Z.; Barlow, A.; Nelson, E.J.; Miller, A.W. Reservoir Water Quality Monitoring Using Remote Sensing with Seasonal Models: Case Study of Five Central-Utah Reservoirs. *Lake Reserv. Manag.* **2015**, *31*, 225–240. [CrossRef]
27. Mobley, C.D. Hydrolight 3.1 Users' Guide. 1996. Available online: https://apps.dtic.mil/sti/pdfs/ADA356954.pdf (accessed on 1 January 2022).
28. Salama, M.S.; Verhoef, W. Two-Stream Remote Sensing Model for Water Quality Mapping: 2SeaColor. *Remote Sens. Environ.* **2015**, *157*, 111–122. [CrossRef]
29. Cannizzaro, J.P.; Carder, K.L. Estimating Chlorophyll a Concentrations from Remote-Sensing Reflectance in Optically Shallow Waters. *Remote Sens. Environ.* **2006**, *101*, 13–24. [CrossRef]
30. Werdell, P.J.; Jeremy Werdell, P.; Roesler, C.S. Remote Assessment of Benthic Substrate Composition in Shallow Waters Using Multispectral Reflectance. *Limnol. Oceanogr.* **2003**, *48*, 557–567. [CrossRef]
31. Hicks, B.J.; Stichbury, G.A.; Brabyn, L.K.; Allan, M.G.; Ashraf, S. Hindcasting Water Clarity from Landsat Satellite Images of Unmonitored Shallow Lakes in the Waikato Region, New Zealand. *Environ. Monit. Assess.* **2013**, *185*, 7245–7261. [CrossRef] [PubMed]
32. Li, J.; Yu, Q.; Tian, Y.Q.; Becker, B.L. Remote Sensing Estimation of Colored Dissolved Organic Matter (CDOM) in Optically Shallow Waters. *ISPRS J. Photogramm. Remote Sens.* **2017**, *128*, 98–110. [CrossRef]
33. Li, J.; Yu, Q.; Tian, Y.Q.; Becker, B.L.; Siqueira, P.; Torbick, N. Spatio-Temporal Variations of CDOM in Shallow Inland Waters from a Semi-Analytical Inversion of Landsat-8. *Remote Sens. Environ.* **2018**, *218*, 189–200. [CrossRef]
34. Lee, Z.; Carder, K.L.; Mobley, C.D.; Steward, R.G.; Patch, J.S. Hyperspectral Remote Sensing for Shallow Waters. 2. Deriving Bottom Depths and Water Properties by Optimization. *Appl. Opt.* **1999**, *38*, 3831–3843. [CrossRef]
35. Lee, Z.; Carder, K.L.; Mobley, C.D.; Steward, R.G.; Patch, J.S. Hyperspectral Remote Sensing for Shallow Waters. I. A Semianalytical Model. *Appl. Opt.* **1998**, *37*, 6329–6338. [CrossRef]
36. Śliwiński, D.; Konieczna, A.; Roman, K. Geostatistical Resampling of LiDAR-Derived DEM in Wide Resolution Range for Modelling in SWAT: A Case Study of Zgłowiączka River (Poland). *Remote Sens.* **2022**, *14*, 1281. [CrossRef]
37. *PN-EN ISO 7027-1:2016-09*; Water Quality—Determination of Turbidity (ISO_7027:1999). Polish Committee for Standardization EN_ISO_7027:1999: Warsaw, Poland, 2016.
38. *PN-86/C-05560/02*; Water and Sewage. Polish Committee for Standardization: Warsaw, Poland, 1986. Available online: https://www.ydylstandards.org.cn/static/down/pdf/PN%20C05560-02-1986_0000.pdf (accessed on 1 January 2022).
39. Kokaly, R.F.; Clark, R.N.; Swayze, G.A.; Livo, K.E.; Hoefen, T.M.; Pearson, N.C.; Wise, R.A.; Benzel, W.M.; Lowers, H.A.; Driscoll, R.L.; et al. *USGS Spectral Library Version 7*; U.S. Geological Survey Data Series 1035; U.S. Geological Survey: Reston, VA, USA, 2017; 61p. [CrossRef]
40. Hachaj, P.S. *Analysis of Hydrodynamics of Barrier Reservoirs for the Needs of Water Management: Model and Its Application*; Cracow University of Technology: Kraków, Poland, 2019; ISBN 9788365991614.

applied sciences

MDPI

Article

Evaluation of Seasonal and Spatial Variations in Water Quality and Identification of Potential Sources of Pollution Using Multivariate Statistical Techniques for Lake Hawassa Watershed, Ethiopia

Semaria Moga Lencha [1,2,*], Mihret Dananto Ulsido [2,3] and Alemayehu Muluneh [2]

[1] Faculty of Agriculture and Environmental Sciences, University of Rostock, 18051 Rostock, Germany
[2] Faculty of Biosystems and Water Resource Engineering, Institute of Technology, Hawassa University, Hawassa P.O. Box 05, Ethiopia; mihret@gmail.com (M.D.U.); muluneh96@yahoo.com (A.M.)
[3] Center for Ethiopian Rift Valley studies, Hawassa University, Hawassa P.O. Box 05, Ethiopia
* Correspondence: semaria.lencha@uni-rostock.de or semaria@hu.edu.et; Tel.: +49-1521-121-2094

Abstract: The magnitude of pollution in Lake Hawassa has been exacerbated by population growth and economic development in the city of Hawassa, which is hydrologically closed and retains pollutants entering it. This study was therefore aimed at examining seasonal and spatial variations in the water quality of Lake Hawassa Watershed (LHW) and identifying possible sources of pollution using multivariate statistical techniques. Water and effluent samples from LHW were collected monthly for analysis of 19 physicochemical parameters during dry and wet seasons at 19 monitoring stations. Multivariate statistical techniques (MVST) were used to investigate the influences of an anthropogenic intervention on the physicochemical characteristics of water quality at monitoring stations. Through cluster analysis (CA), all 19 monitoring stations were spatially grouped into two statistically significant clusters for the dry and wet seasons based on pollution index, which were designated as moderately polluted (MP) and highly polluted (HP). According to the study results, rivers and Lake Hawassa were moderately polluted (MP), while point sources (industry, hospitals and hotels) were found to be highly polluted (HP). Discriminant analysis (DA) was used to identify the most critical parameters to study the spatial variations, and seven significant parameters were extracted (electrical conductivity (EC), dissolved oxygen (DO), chemical oxygen demand (COD), total nitrogen (TN), total phosphorous (TP), sodium ion (Na^+), and potassium ion (K^+) with the spatial variance to distinguish the pollution condition of the groups obtained using CA. Principal component analysis (PCA) was used to qualitatively determine the potential sources contributing to LHW pollution. In addition, three factors determining pollution levels during the dry and wet season were identified to explain 70.5% and 72.5% of the total variance, respectively. Various sources of pollution are prevalent in the LHW, including urban runoff, industrial discharges, diffused sources from agricultural land use, and livestock. A correlation matrix with seasonal variations was prepared for both seasons using physicochemical parameters. In conclusion, effective management of point and non-point source pollution is imperative to improve domestic, industrial, livestock, and agricultural runoff to reduce pollutants entering the Lake. In this regard, proper municipal and industrial wastewater treatment should be complemented, especially, by stringent management that requires a comprehensive application of technologies such as fertilizer management, ecological ditches, constructed wetlands, and buffer strips. Furthermore, application of indigenous aeration practices such as the use of drop structures at critical locations would help improve water quality in the lake watershed.

Keywords: monitoring; mitigations; spatial and temporal variabilities; principal component analysis; cluster analysis; discriminant analysis; water quality; pollution; correlation

Citation: Lencha, S.M.; Ulsido, M.D.; Muluneh, A. Evaluation of Seasonal and Spatial Variations in Water Quality and Identification of Potential Sources of Pollution Using Multivariate Statistical Techniques for Lake Hawassa Watershed, Ethiopia. *Appl. Sci.* **2021**, *11*, 8991. https://doi.org/10.3390/app11198991

Academic Editor: Jorge Rodríguez-Chueca

Received: 2 August 2021
Accepted: 23 September 2021
Published: 27 September 2021

1. Introduction

Studies have shown that urban, agricultural, and industrial discharges have a direct effect on surface water quality. Similarly, urban wastewaters cause fecal contamination of surface waters, and urban stormwater runoff, which contains large amounts of fecal microbes, also affects surface water quality [1]. Surface water bodies are vital natural resources that are vulnerable to pollution. The contaminants are chemical, physical, and biological constituents resulting from anthropogenic activities and are of greater environmental consideration [2]. Surface water bodies are extensively used as the major sources for domestic, non-domestic, industrial, and irrigation purposes. Therefore, monitoring and assessment of water bodies is imperative to obtain reliable information on water quality for effective management [3]. Anthropogenic uses of the waterbodies in the study basin can degrade the quality of surface water and impair its usability as potable water supply or for industry, agriculture, recreation, or other purposes. Hence, regular monitoring of water quality of rivers and lake is indispensable [4,5]. The most affected river stretches are those that flow through urbanized and exceedingly populated urban areas where there is no adequate sanitation. Upstream rural areas are mainly affected by pollutants from non-point sources such as agricultural runoff, whereas urban areas are polluted by point sources, sewage discharges, urban runoff, and pollutants from upstream areas [6,7].

Studies have shown that some lakes and wetlands around the world have disappeared or are showing changes in their ecosystem. Furthermore, factors such as intensive land use for urbanization and agriculture have had significant impact on the hydrology, ecology, and ecosystem services of lakes, which has eventually led to a decline in lake levels [8]. In addition, pollutants have long been a concern, as their accumulation can have serious effects on fauna, flora, and human health when the huge amount of urban and industrial wastewater reaches the shores [9].

Lake Hawassa is located near the city of Hawassa and is surrounded by agricultural land, industries and residential areas. Therefore, it is susceptible to a variety of pollutants that enter the lake directly or indirectly. On the other hand, the Lake Hawassa Watershed is experiencing rapid land cover change, and natural resources have overwhelmingly diminished. The lake is hydrologically closed and has no apparent outlet, so all pollutants entering the lake are retained. As a result, the lake faces numerous problems, and the water quality deteriorates over time, threatening biodiversity [10].

Significant industrialization, augmented with rapid urbanization and increasing economic development, has increased the extent of pollution [11]. The pollution is mainly from non-point sources caused by urban and agricultural runoff, overgrazing, deforestation, soil erosion, land development, and industrial effluents. This leads to numerous environmental concerns that have resulted in substantial hydrological disturbances. The main factories in the study area are a ceramics factory, a flourmill, a cement products factory, a Moha soft drink factory, a BGI (St. George Brewery factory), an Etabs soap factory, an industrial park in Hawassa, and other small-scale industries. They are virtually all concentrated along the main road, which is close to the shallow swamp, and discharge their effluents into the lake through streams. On the other hand, deforestation and irrigation of the land have caused the drying up of Lake Cheleleka by reducing the streamflow [12].

Various studies have been conducted to examine water quality in the LHW catchment and identify sources of pollution. Teshome [11] investigated the eastern catchment of Lake Hawasa Watershed to assess the seasonal water quality and its suitability for the designated uses. The findings revealed that the rivers in the eastern part of Lake Hawassa Watershed are suitable for agriculture and livestock but unpleasant for aquatic life, and the lake is hypereutrophic.

Amare [13] investigated the primary sources of non-point source pollution and their relative contribution in Lake Hawassa Watershed using the Annualized Agricultural Non-Point Source (AnnAGNPS) model. The pollutant-loading model revealed non-point source pollutants originating from agricultural lands and associated with deleterious anthropogenic activities responsible for the water quality impairment of Lake Hawassa.

These point sources have been determined to be the source of numerous pollutants in the lake ecosystem if the effluent control system put in place is unsuitable [14].

Kebede [15] studied the impact of land cover changes on water quality and streamflow in Lake Hawassa Watershed and concluded that water quality in the upper watershed of the three rivers was better than the lower sections of the catchment with respect to the parameters studied, which might be correlated to the observed land use.

A study conducted by Lencha et al. [16] at Lake Hawassa revealed that most of the population, including the inner part of the city, are using latrines. Larger buildings have conventional flushing systems but without any wastewater treatment. Furthermore, industrial and commercial point sources are known to discharge their effluents into streams or rivers that end up in the Lake. In addition, Hawassa Industrial Park and Referral Hospital discharge their effluents directly into the lake. This is a threat to the people who rely on rivers, streams, and the lake for domestic and other purposes and to the survival of aquatic life as well.

To sum up, some studies regarding the water quality have been conducted in either the eastern or the western catchment of Lake Hawassa, while others have been carried out only at Lake Hawassa. Nonetheless, there is no sufficient water quality study to connect agricultural and urban land use with the watershed pollution level to identify the sources of pollution. The previous studies mainly relied on random monitoring and data from literature and focused only on a few water quality parameters, which cannot reflect the whole picture of water quality in the watershed. Additionally, some previous studies also obtained contradictory findings. On the other hand, urbanization, industrialization, commercial activities, and population growth are increasing rapidly, which could increase sewage and effluents production. Through monitoring data, consistent data analysis, and homogenization of parameters, this study aimed to (1) statistically analyze multiple-parameter data by using principal component analysis (PCA), cluster analysis (CA), and discriminant analysis (DA); (2) investigate the broad-spectrum variation in the parameters of LHW; and (3) cluster monitoring stations with similar characteristics and identify potential sources of pollution in LHW.

2. Materials and Methods

2.1. Study Area

Lake Hawassa Watershed (LHW) is located 275 km from the capital Addis Ababa, in the capital of Sidama regional state, on the main road leading to Nairobi, Kenya via Moyale. LHW has a total area of 1431 km^2 and lies between 6°45′ to 7°15′ N latitude and 38°15′ to 38°45′ E longitude (Figure 1). LHW comprises five sub-watersheds [17].

The watershed is known for its flat plains and dissevered undulating landscape with elevation ranging from 1571 to 2962 m above sea level [18]. The area comprises mountains and low-lying areas, with a wide flat wetland called Cheleleka. Perennial rivers and streams on the north and northeast sides of the catchment and runoff on the east wall feed Cheleleka. The sub-basin of Tikur-Wuha consists of only a tributary called Tikur-Wuha that flows into Lake Hawassa. In this lake system, no surface water flows out from the lake except by evaporation and abstraction, so the catchment can be considered hydrologically closed [15]. The climate of the Hawassa sub-basin is sub-humid and distinctly seasonal. The months from April to October are wet and humid, and the main rainy season is between July and September, with a mean annual precipitation of about 955 mm. The mean minimum precipitation is 17.8 mm in December (dry season) and the mean maximum precipitation is 119.8 mm in August (rainy season) [19].

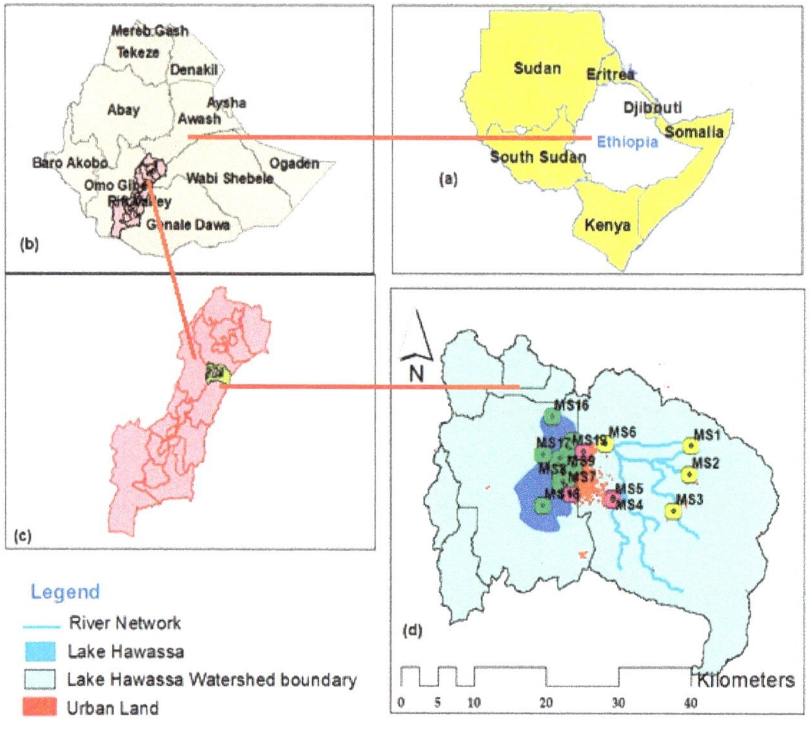

Figure 1. Study area map and monitoring station locations (**a**) = countries sharing boundaries with Ethiopia, (**b**) = major river basins in Ethiopia, (**c**) = Rift Valley lake basin and (**d**) = Lake Hawassa sub-basin and monitoring stations).

2.2. Sampling and Monitoring Parameters

The monitoring sites and sampling strategy were planned to cover a wide range of factors contributing to the water quality of the river, taking into account tributaries and point sources whose effluents end up in the lake and have a substantial impact on the water quality of the lake. The criteria for selecting monitoring points were hydrological, with confluence of sub-basins having distinct characteristics and land use types, with the intention of transferring parameters to unmonitored sub-basins. Furthermore, factors such as availability of point and non-point sources, land use type, and urban and wastewater drains were considered in the selection of monitoring sites.

Hence, a total of nineteen (19) monitoring stations were selected (Table 1 and Figure 1). Four (4) monitoring sites were selected purposively at the Wesha, Hallow, Wedessa, and Tikur-Wuha river mouths of the respective sub-watersheds.

Eleven (11) monitoring sites were distributed evenly along the entire course of Lake Hawassa for water quality monitoring. Three (3) monitoring sites were selected near the industrial disposal site, and one (1) was at the health care center.

The monitoring sites in the Tikur-Wuha catchment were Wesha River (MS1), Hallo River (MS2), and Wedessa River (MS3), which are located in the upstream part of Lake Hawassa, where agricultural runoff from the catchment flows directly or through its tributaries into the Cheleleka wetland. The three rivers were purposively selected based on their size and spatial location to represent their respective sub-basins. Monitoring station 6 (MS6) is a critical area with mostly fresh water where factories discharge their effluent into the Tikur-Wuha River, and the river eventually flows into Lake Hawassa. This is an area where river inputs to the lake are high.

Table 1. Monitoring stations in Lake Hawassa Watershed.

No	Monitoring Stations	Site Code	Location
1	Wesha River	MS1	LHW upstream
2	Hallo River	MS2	LHW upstream
3	Wedessa River	MS3	LHW upstream
4	BGI effluent discharge site	MS4	LHW middle
5	Pepsi factory oxidation pond	MS5	LHW middle
6	Tikur-Wuha River	MS6	LHW middle
7	Amora-Gedel (fish market)	MS7	Eastern side of LH
8	Amora-Gedel (Gudumale)	MS8	Eastern side of LH
9	Nearby Lewi resort	MS9	Eastern side of LH
10	Fikerhayk center (FH)	MS10	Center of LH
11	Fikerhayk (meznegna)	MS11	Eastern side of LH
12	Center of LH (Towards HR)	MS12	Center of LH
13	Nearby Haile resort	MS13	Eastern side of LH
14	Tikur-Wuha site	MS14	Eastern side of LH
15	Referral Hospital	MS15	Eastern side of LH
16	Ali-Girma site (opposite to HR)	MS16	Western side of LH
17	Sima Site (opposite to Mount Tabor)	MS17	Western side of LH
18	Dore-Bafana Betemengist	MS18	Southern part of LH
19	Hawassa Industrial Park	MS19	LHW middle

The site codes are indicated in Figure 1. FH designates Fikerhayk, HR labels Haile Resort, LHW designates Lake Hawassa Watershed, LH designates Lake Hawassa.

Monitoring sites for point sources were selected from available industries in the catchment that directly or indirectly feed Lake Hawassa. The selected sites were the St. George Brewery factory, BGI (MS4), and the Moha soft drink factory (MS5), whose effluents discharge into the Cheleleka wetland and eventually enter Lake Hawassa via Tikur-Wuha River, as well as the Referral Hospital (MS15) and Hawassa Industrial park (MS19), which discharge their effluents directly in to Lake Hawassa.

The monitoring stations for Lake Hawassa were selected based on the presence of major pollution sources in the lake, existence of point sources, health facilities, industrial effluent emission sites, availability of boating and recreational activities, presence of service rendering facilities such as Haile and Lewi resorts, fish market (Amora-Gedel and Gudumale), and also the central part of the lake where the disturbance is minimum.

For this purpose, eight (8) monitoring sites were selected in the eastern part (northeast to southeast) of the lake and designated as MS7, MS8, MS9, MS10, MS11, MS12, MS13, and MS14.

The other three (3) monitoring sites were located on the western (northwest to southwest) sides of the lake and were designated as MS16 for the local village Ali-Girma site (opposite Haile Resort), MS17 for Sima site that is opposite side of Mount Tabor, and MS18 for Dore-Bafana Betemengist site. In this part of the lake, although there is no point source pollution, there is enormous anthropogenic activity in the form of non-point source pollution from recreational activities, agricultural runoff, and animal waste.

The analyses of physicochemical water quality parameters at selected sites and periods were conducted from May 2020 to January 2021 to see seasonal variation. Sample collection for the wet season was event-based, i.e., samples were collected after rainfall events. The coordinates of each sampling stations was determined using GNSS.

Composite samples were collected in pre-cleaned 2L polyethylene plastic bottles (sterilized glass bottles were used for biochemical oxygen demand (BOD) and chemical oxygen demand (COD) analyses) for different parameters. The bottles were washed with concentrated nitric acid and distilled water before sample collection and thoroughly rinsed with sample water during collection to avoid possible contamination. The water samples were aseptically handled, labelled, preserved in sterile glass bottles, stored in the cooler (Mobicool v30 AC/DC, Germany) and ice box, and transported to the laboratory of Hawassa University Environmental Engineering, Addis Ababa City Government Environmental Protection, and Green Development Commission and Engineering Corporation of Oromiya for analysis.

The collection, handling, preservation, and treatment of the water samples followed the standard methods outlined for the examination of water and wastewater by the American Public Health Association guidelines [20] and all the parameters were presented with their respective analytical methods and instruments used for analysis in Table 2 below.

Table 2. Analytical methods and instruments used for analysis.

Parameter	Analytical Method and Instrument
pH, EC, TDS, and Temperature	Portable multi-parameter analyzer (Zoto, Germany)
Turbidity	Nephelometric (Hach, model 2100A)
DO	Modified Winkler
BOD	Manometric, BOD sensor
COD	Closed Reflux, colorimetric
SRP and TP	Spectrophotometrically by molybdovandate (Hach, model DR 3900)
TN	Spectrophotometrically by TNT Persulfate digestion (Hach, model DR 3900)
NO_2^- and TAN ($NH_3-N + NH_4-N$)	Spectrophotometrically by salicylate (Hach, model DR 3900)
NO_3^-	Photometric measurements, Wagtech Photometer 7100 at 520 nm wavelength
SS	Filtration by standard glass fiber filter
Mg^{2+}, Na^+, Ca^{2+}, and K^+	Atomic Absorption Spectrophotometer, AAS, model NOVAA400

Total ammonium nitrogen (TAN), electrical conductivity (EC), total dissolved solids (TDS), dissolved oxygen (DO), biochemical oxygen demand (BOD_5), chemical oxygen demand (COD), soluble reactive phosphorous (SRP), total phosphorous (TP), nitrate (NO_3^-), nitrite (NO_2^-), magnesium ion (Mg^{+2}), sodium ion (Na^+), potassium ion (K^+), calcium ion (Ca^{+2}), and suspended solids (SS).

Un-Ionized Ammonia Determination from Total Ammonium Nitrogen (TAN)

The un-ionized free ammonia was calculated by the mass action law in its logarithmic form (1). The pKa as function of temperature was taken from Emerson et al. [21]:

$$\% \text{ Un} - \text{ionized NH3} - \text{N} = \frac{1}{\left(1 + 10^{(\text{pK}_a - \text{pH})}\right)} \tag{1}$$

$$\text{pK}_a = \frac{0.09108+2729.92}{(\text{T}_k)} \tag{2}$$

where T_k is temperature in Kelvins (273 + °C).

3. Multivariate Statistical Techniques and Data Treatment

3.1. Multivariate Statistical Techniques

Multivariate statistical techniques (MVST) are a valuable tool to estimate efficiently the spatio-temporal variability in a watershed and the influences of human intervention on the characteristics of physicochemical parameters at monitoring stations [22]. In addition, MVST like cluster analysis (CA), discriminant analysis (DA), and PCA/factor analysis can be implemented to interpret complex databases to offer better visualization of water quality in the studied watershed [23]. The statistical techniques PCA, CA, and DA are vital to determine the primary relationships among the physicochemical parameters measured in experimental data standardized to the Z-scale to avoid inaccurate grouping because of the huge variability in the data dimensionality [5,24–26].

Principal component analysis (PCA), cluster analysis (CA) and discriminant Analysis (DA) were carried out to examine the seasonal variations, identify possible pollution sources, and analyze and interpret surface water quality data to draw meaningful information in China [2,7,27–30], South Asia—Bangladesh [31], the Middle East—Iran [3], India [23,32,33], South African [34], Ethiopia [22,35], South Asia Malaysia [36], the Middle East—Lebanon [6,37], Spain [38], and Serbia [39].

XLSTAT 2016 (Addinsoft, New York, USA), Microsoft Excel 2016, and "Statistical Package for the Social Sciences Software, IBM SPSS 25 for Windows" were employed to perform statistical analysis integrally.

3.2. Data Treatment and Multivariate Statistical Methods

PCA is sensitive to outliers, missing data, and poor linear correlation among variables due to insufficient assigned variables. Thus, the data treatment needs to be performed for missing data and outliers in the monitored water quality data before executing multivariate statistical analysis. There might be a real shift in the value of an observation that arises from non-random causes. In this study, outliers were detected according to Grubbs [40] test method using XLSTAT 2016. On the other hand, data collection and analysis were conducted with great prudence to minimize the amount of missing data. However, the incidence of missing data is inevitable and was handled by the multiple imputation of missing values technique using Markov Chain Monte Carlo (MCMC) [41].

The raw water quality parameters were standardized to a mean of 0 and variance of 1 using Z-scale transformation to examine the normality of the distribution of data sets and to ensure that the different variables were equally weighted in the statistical analyses [36]. The data were further checked for normality using Kaiser–Meyer–Olkin (KMO) and Bartlett's sphericity tests to determine if our measured variables may be factorized efficiently. KMO is the degree of sampling adequacy, which shows the percentage of variance that is likely attributable to the underlying factors. Generally, the KMO index ought to be greater than 0.5 for satisfactory factor analysis. When the KMO index is close to 1, the PCA of the variables is suitable; however, when it is close to 0, the PCA is not relevant. In this study, the KMO had a value of 0.68. Bartlett's test of sphericity shows whether the correlation matrix is an identity with variables that are unrelated. The significance level, which is 0 in this study (less than 0.05), indicates that there are significant relationships among the variables.

3.2.1. Principal Component (PCs)/Factor Analysis (FA)

PCA reduces the dimensionality of the data set by explaining the correlations amongst a large number of variables in terms of a smaller number of underlying factors without losing much information [42,43]. The original variables of PCs produce loadings that have correlation coefficients with PCs. The PCs' formula was taken from [33,36]:

$$\mathcal{Y}\mathrm{mn} = \mathcal{Z}\mathrm{m1}\mathcal{X}\mathrm{1n} + \mathcal{Z}\mathrm{m2}\mathcal{X}\mathrm{2n} + \mathcal{Z}\mathrm{m3}\mathcal{X}\mathrm{3n} \ldots + \mathcal{Z}\mathrm{mi}\mathcal{X}\mathrm{in} \tag{3}$$

where z is the component loading, y is the component score, x is the measured value of a variable, m is the component number, n is the sample number, and m is the total number of variables.

Meanwhile, FA attempts to extract a lower-dimensional linear structure from the data set and extracts the new group of variables known as varifactors (VFs) via rotation along the PCA axis. In FA, the basic concept is borrowed from [33,36]:

$$\mathcal{Y}\mathrm{mn} = \mathcal{Z}\mathrm{p1}P\mathrm{1m} + \mathcal{Z}\mathrm{p2}P\mathrm{2m} + \mathcal{Z}\mathrm{p3}P\mathrm{3m} + \ldots + \mathcal{Z}\mathrm{p}r P r\mathrm{m} + e\mathrm{pm} \tag{4}$$

where y is the measured value of the variable, z refers to the factor loading, p is the factor score, m is the sample number, n is the variable number, r is the total number of factors, and e is the residual term accounting for errors or other sources of variation.

In this study, PCA was employed for qualitative determination of pollution sources.

3.2.2. Discriminant Analysis

DA was used for discriminating between and among groups by applying discriminating variables. These variables measure characteristics regarding which the groups are expected to differ [44]. DA applies a linear equation of a regression analysis on raw data with prior knowledge of membership of objects to particular clusters and provides statistical classification of samples, expressed in the following equation [43,45]:

$$\mathrm{f(Gi)} = \mathrm{Ki} + \sum_{\mathrm{i}=1}^{\mathrm{n}} (\mathrm{Wij} * \mathrm{Pij}) \tag{5}$$

where Ki is a constant specific to each particular group, i is the number of groups (G), n is the number of parameters used in group classification, and Wij is the weight coefficient designated by DA for the specific parameter (Pij).

Independent variables are entered into DA either all together or stepwise, using both backward and forward approaches. In the first approach of variable entry, the discriminant function is calculated by engaging all the independent variables at once. This approach is used when there are a limited number of independent variables in the interest of discovering how well certain variables perform as discriminants in the absence of others. The stepwise method, on the other hand, involves entering the independent variables into the discriminant function (DF) one at a time. This stepwise input is based on the fact that variables with relative importance to the cluster variables with greater discriminant weights were entered first [46].

In this study, standard, forward, and backward stepwise approaches of DA were applied to each matrix of the primary data. In the forward stepwise mode, discriminant function analysis (DFA) variables were added stepwise until no significant change occurred, while in the backward stepwise mode, variables were removed starting from least significant until a significant change occurred. For this purpose, two groups obtained from CA were selected for spatial evaluations [35].

3.2.3. Pollution Index (PI)

Pollution index (PI) is a simple technique to examine surface water quality and was applied by Tiwan EPA. The parameters such as DO, BOD, SS, and NH_3-N employed to determine PI were classified into four index scores (Table 3) and computed using the equation formulated by [47,48]. In particular, PI refers to the arithmetic mean of the index values with respect to the water quality.

$$PI = \frac{1}{4}\sum_{K=1}^{4} Si \tag{6}$$

Table 3. Classification system for pollution index.

Item	Rank			
	Non-Polluted (Good)	**Slightly Polluted (LP)**	**Moderately Polluted (MP)**	**Highly Polluted (HP)**
DO (mg/L)	>6.5	4.6–6.5	2.0–4.5	<2.0
BOD_5 (mg/L)	<3	3.0–4.9	5.0–15.0	>15
SS (mg/L)	<20	20–49	50–100	>100
NH_3-N (mg/L)	<0.5	0.5–0.9	1.0–3.0	>3.0
Index score	1	3	6	10

PI classifies water quality into four categories: (0–2) for good or non-polluted, (2–3) for slightly polluted, (3–6) for moderately polluted, and (>6) for highly polluted. Anthropogenic activities have been associated with water quality degradation [47,49].

3.3. Cluster Analysis

Hierarchical agglomerative CA was carried out on the normalized data set using Ward's approach, where Euclidean distances were used as the degree of similarity among samples, and a distance was represented by the distinction among analytical values. In hierarchical clustering, sequentially higher clusters formed [23,45,50–52]. In cluster analysis, cases are classified into classes based on similarities between two samples, which are usually given by the Euclidean distance between analytical values of the two samples. The squared Euclidean distance can be calculated by [53]:

$$\text{Distance}\left(Q_i, Q_j\right) = \sum_{j=1}^{n}(X_{1i} - X_{2j})^2 \quad (7)$$

where Qi is the ith object, and X_{ij} is the value of the jth variable of the ith object.

The dendrogram provides a visual summary of the clustering process to classify a sample of entities into a smaller number of mutually exclusive groups on the basis of multivariate similarities among entities [33].

Therefore, CA, DA, PCA, and pollution index were applied in this study to identify the underlying interrelationship among the parameters and monitoring stations. CA was applied based on prior knowledge of monitoring stations and the results of DA and pollution index to accurately cluster monitoring stations. PCA was employed to qualitatively identify pollution sources and the type of contaminants contributing to pollution.

4. Results and Discussion

4.1. Correlation Matrix Evaluation and Seasonal Variation

Correlation coefficients are established to portray a correlation among variables and measure statistical significance between pairs of water quality variables [54,55]. Correlation analysis measures the proximity between the identified dependent and independent variables. Correlation coefficients that are close to −1 or +1 demonstrates a strong correlation between x and y, which have a linear correlation. The correlation between the parameters is referred to as strong from (+0.8 to 1.0) or (−0.8 to −1.0), moderate from (+0.5 to 0.8) or (−0.5 to −0.8) and weak from (+0.0 to 0.5) or (−0.0 to −0.5) [56]. In cases where the correlation coefficient between variables is zero, there could be no correlation with a degree of $p < 0.05$ between the two variables [57]. In this study, a correlation matrix was constructed for each dry and wet season using the physicochemical parameters. Pearson's correlation coefficient (r) is determined using correlation matrix to identify the highly correlated and interrelated water quality parameters. To test the significance of the pair of parameters, the *p*-value is determined.

In the wet season, strong positive correlations were observed between TDS values and EC, temperature, TP, TN, and Na^+ values (r = 0.992, r = 0.874, r = 0.850, r = 0.836; $p < 0.05$), and strong negative correlations between TDS and DO with −0.825 at $p < 0.05$. Moderate positive correlations were found between TDS and PO_4−P, BOD, COD, and K^+ values (r = 0.797, r = 0.698, r = 0.695, r = 0.523; $p < 0.05$), and low positive correlation between TDS and pH with r = 0.26; $p < 0.05$ (Table 4). Strong negative correlations were found between DO and EC, TDS, TP, and TN (r = −0.825, r = −0.850, r = −0.851, r =−0.806; $p < 0.05$), and moderate negative correlations were observed between DO and temperature, BOD, COD, and Na+ values (r = −0.526, r = −0.544, r = −0.692, r = −0.599; $p < 0.05$).

Table 4. Correlation matrix Pearson (r) and alpha (p) values for the wet season.

Parameters	TDS	EC	NH_3−N	NO_3−N	PO_4−P	DO	BOD	COD	TN	TP	Temp	Mg^{2+}	Ca^{2+}	Na^+	K^+
TDS	**1**														
EC	**0.992**	**1**													
NH_3−N	0.446	0.379	**1**												
NO_3−N	0.183	0.172	−0.030	**1**											
PO_4−P	**0.797**	**0.824**	0.416	−0.116	**1**										
DO	**−0.825**	**−0.850**	−0.216	−0.275	**−0.793**	**1**									
BOD	**0.698**	**0.719**	0.106	−0.173	**0.712**	**−0.526**	**1**								
COD	**0.695**	**0.714**	0.204	−0.111	**0.730**	**−0.544**	**0.965**	**1**							
TN	**0.874**	**0.855**	**0.481**	0.059	**0.825**	**−0.851**	**0.587**	**0.602**	**1**						
TP	**0.850**	**0.871**	0.249	0.255	**0.602**	**−0.806**	**0.485**	**0.482**	**0.736**	**1**					
Temperature	**0.860**	**0.864**	0.331	0.410	**0.594**	**−0.692**	0.454	0.447	**0.669**	**0.82**	**1**				
Mg^{2+}	−0.005	0.029	−0.317	0.070	−0.013	−0.085	0.224	0.159	0.046	0.09	−0.020	**1**			
Ca^{2+}	0.375	0.397	−0.085	−0.080	0.350	−0.394	**0.523**	**0.528**	0.429	0.24	0.137	0.401	**1**		
Na^+	**0.836**	**0.853**	0.314	0.268	**0.709**	**−0.599**	**0.619**	**0.632**	**0.572**	**0.68**	**0.849**	−0.062	0.19	**1**	
K^+	**0.523**	0.431	**0.531**	0.155	0.290	−0.429	0.149	0.190	**0.700**	0.34	0.320	−0.080	0.20	0.19	**1**

Values in bold are different from 0 with a significance level alpha = 0.05.

Strong positive correlations were observed between temperature and the values of EC, TDS, Na^+ and TP (r = 0.86, r = 0.864, r = 0.849, r = 0.821; $p < 0.05$), and a moderate positive correlation was observed between temperature and the values of TN and PO_4-P (r = 0.525, r = 0.669, r = 0.594; $p < 0.05$). There was also a moderate negative correlation between temperature and DO, with r = −0.692 at $p < 0.005$. There was a weak correlation between temperature and the values of COD and BOD (r = 0.447, r = 0.454; $p < 0.05$).

NH_3-N had a moderate positive correlation with K^+, with r = 0.531 at $p < 0.005$, and weak positive correlations with TN and temperature (r = 0.331, r = 0.481 at $p < 0.05$). NO_2-N correlated moderately positively with BOD and COD (r = 0.721, r = 0.664 at $p < 0.05$) and weakly positively with PO_4-P and Ca^{+2} (r = 0.449, r= 0.404 at $p < 0.05$).

A strong positive correlation was found between PO_4-P and TN, with r = 0.825 at $p < 0.005$, moderate positive correlations were found between PO_4-P and COD, BOD, TP, and temperature (r = 0.712, r = 0.709, r = 0.730, r = 0.602, r = 0.594; $p < 0.05$), and a moderate negative correlation was observed between PO_4-P and DO values (r = −0.793; $p < 0.05$). No statistically significant difference was found between pH and NO_3-N and the rest of the parameters of LHW ($p > 0.05$).

In the dry season, strong positive correlations were observed between TDS values and EC, TP, Na^+, $PO_4{}^-P$, and temperature values (r = 0.999, r = 0.814, r = 0.899, r =0.839, r = 0.933; $p < 0.05$), moderate positive correlations were observed between TDS and BOD, COD, K^+, and TN values (r = 0.686, r = 0.561, r = 0.645, r = 0.534; $p < 0.05$), and a strong negative correlation was found between TDS values and DO (r = −0.819 at $p < 0.05$).

Strong negative correlations were observed between the values of DO and TDS, EC, and Na^+ (r = −0.819, r = 0.817, r = −0.826; $p < 0.05$), moderate negative correlations were observed between DO and TN, TP, BOD, K^+, and temperature values (r = −0.577, r = −0.568, r = −0.687, r = −0.639 r = −0.729; $p < 0.05$), and a moderate negative correlation was observed between DO and $NO_3{}^-N$, with r = −0.464 at $p < 0.005$).

Strong positive correlations were found between temperature and EC and TDS (r = 0.839, r = 0.842; $p < 0.05$), and moderate positive correlations were found for temperature with TP and PO_4-P(r = 0.730, r = 0.532; $p < 0.05$). There was also a moderate negative correlation observed between temperature and DO, with r = −0.729 at $p < 0.005$. NH_3-N had a moderate positive correlation with COD, TP, temperature, and Na^+ (r = 0.476, r = 0.484, r = 0.550, r = 0.343; $p < 0.005$).

A strong positive correlation was found between PO_4-P and TP, with r = 0.921 at $p < 0.005$, moderate positive correlations were found for PO_4-P with BOD, COD, TP, Na^+, and temperature (r = 0.749, r = 0.647, r = 0.680, r = 0.76; $p < 0.05$), and a moderate negative correlation was found between PO_4-P and DO values r = −0.626; $p < 0.05$) (Table 5).

Table 5. Correlation matrix Pearson (r) and alpha (p) values for dry season.

Parameters	TDS	EC	NH_3-N	NO_3-N	PO_4-P	DO	BOD	COD	TN	TP	Tem	Mg^{2+}	Ca^{2+}	Na^+	K^+
TDS	**1**														
EC	**0.999**	**1**													
NH_3-N	0.433	0.419	**1**												
NO_3-N	0.208	0.212	−0.10	**1**											
PO_4-P	**0.814**	**0.815**	0.383	−0.04	**1**										
DO	**−0.82**	**−0.82**	−0.31	**−0.46**	**−0.63**	**1**									
BOD	**0.686**	**0.686**	0.450	−0.12	**0.749**	**−0.58**	**1**								
COD	**0.561**	**0.564**	**0.476**	−0.19	**0.647**	−0.41	**0.871**	**1**							
TN	**0.645**	**0.642**	0.410	0.184	**0.680**	**−0.57**	**0.520**	**0.619**	**1**						
TP	**0.899**	**0.899**	**0.484**	−0.03	**0.921**	**−0.69**	**0.804**	**0.683**	**0.535**	**1**					
Temperature	**0.839**	**0.842**	0.343	0.237	**0.532**	**−0.73**	0.436	0.344	0.291	**0.730**	**1**				
Mg^{2+}	−0.27	−0.27	0.25	0.13	0.01	0.305	0.13	0.20	0.16	−0.13	−0.42	**1**			
Ca^{2+}	0.385	0.392	−0.19	0.398	0.091	−0.33	0.235	0.324	0.208	0.17	0.455	−0.33	**1**		
Na^+	**0.933**	**0.931**	**0.550**	0.173	**0.760**	**−0.83**	**0.813**	**0.694**	**0.601**	**0.881**	**0.788**	−0.38	0.37	**1**	
K^+	**0.534**	**0.531**	0.182	0.419	0.261	**−0.64**	0.237	0.240	**0.701**	0.197	0.335	−0.39	0.42	**0.53**	**1**

Values in bold are different from 0 with a significance level alpha = 0.05.

The pH of rivers was 7.4 (7.1 to 7.6) in the dry season and 8.2 (7.5 to 8.7) in the wet season, and the pH of lake was 8.2 (7.3 to 8.9) in the dry season and 8.5 (7.5 to 9) in the wet season. The pH of point sources was 8.3 (7.1 to 9) in the dry season and 8.3 (8.1 to 8.7) in the wet season. The recommended pH as per the standard for drinking, irrigation, and aquatic life is 6.5–8.6, and the pH of LHW was within the accepted limit (Table 6). The EC (TDS) of rivers was 148mg/L (297 µS/cm) in dry seasons and 89 mg/L (179 µS/cm) in wet seasons, and EC (TDS) of lakes was 453 mg/L (877 µS/cm) in dry season and 421 mg/L (829 µS/cm) in wet seasons. The EC (TDS) of point sources was 1655 mg/L (3509 µS/cm) in dry season and 1395 mg/L (2809 µS/cm) in wet seasons. This shows that the EC (TDS) of rivers, lakes, and point sources increases significantly with increasing temperature (Table 6). The NO_3-N concentration of rivers was 0.5 mg/L, NO_3-N concentration of Lake Hawassa was 1.4 mg/L, and that of point sources was 1.5 mg/L for the dry season. In the wet season, the NO_3-N concentration was 0.7, 1.9, and 1.9 for rivers, Lake Hawassa, and point sources, respectively. The value of NO_3-N increases in the rainy season due to the contribution of agricultural runoff and use of fertilizers. The PO_4-P concentration of rivers was 6.5 mg/L, PO_4-P of Lake Hawassa was 3.3 mg/L, and that of point sources was 43.8 mg/L in dry season. In the wet season, the PO_4-P concentration was 7.4, 2.9, and 25.7 for rivers, Lake Hawassa, and point sources, respectively (Table 6). Similarly, Gebre-Mariam [58] reported that Ethiopian Rift Valley lakes generally have lower EC values in the rainy season than in the dry season, due to dilution by rain coupled with minimal evaporation rates during the rainy season.

Table 6. Descriptive statistics (mean and standard deviation) of the physicochemical characteristics of LHW collected during dry season.

Codes	SS	TDS	EC	pH	NH_3-N	NO_3-N	PO_4-P	DO	BOD	COD	TN	TP	Mg^{2+}	Ca^{2+}	Na^+	K+	Temperature
MS1	17.3	89	178	7.1	0.04	0.6	3.6	4.1	13.8	88.3	5.8	0.001	7.2	20	32.5	6.7	19.2
	(1.6)	(4)	(7)	(0.2)	(0.01)	(0.01)	(2)	(0.7)	(1.5)	(26.9)	(1.5)	(0)	(2.1)	(7.4)	(1.5)	(0.6)	(0.8)
MS2	27.3	100	200	7.6	0.16	0.4	10.2	3.5	23.7	107.5	7.2	0.5	54.0	9	26.2	8.1	17.7
	(5.8)	(15)	(30)	(0.5)	(0.07)	(0.04)	(6.7)	(1)	(7.2)	(32.5)	(1.8)	(0.5)	(16.4)	(8.4)	(3)	(0.6)	(1)
MS3	54.5	87	175	7.7	0.10	0.6	5.9	4.4	69.0	313.8	7.5	0.001	153.4	4.6	25.8	6.8	18.1
	(3)	(6)	(12)	(0.3)	(0.01)	(0.1)	(0.1)	(0.4)	(20.5)	(93.3)	(2.5)	(0)	(50.2)	(4.2)	(3)	(0.6)	(0.7)
MS4	58.0	1575	3825	7.1	7.60	2.8	18.7	1.5	63.3	263.7	23.8	15	11.4	50.4	501.1	19.8	33.8
	(10.4)	(59)	(108)	(0.4)	(1.49)	(0.5)	(2.9)	(0.1)	(10.8)	(84.9)	(5.9)	(3)	(0.1)	(5.6)	(83)	(0.3)	(0.3)
MS5	27.7	2349	4698	9.5	12.35	0.6	118.3	0.9	190	600	41.3	6.5	2.9	15.0	1078.1	19.3	29
	(1.5)	(193)	(385)	(0.6)	(5.15)	(0.05)	(40)	(0.1)	(1.3)	(241)	(16.7)	(1.6)	(1.8)	(10.6)	(178)	(0.8)	(0.6)
MS6	23.3	317	635	7.6	0.06	1	6.3	4	5.3	26.3	11.3	0.001	5.7	18.7	111.8	9.4	24.5
	(0.1)	(63)	(126)	(0.1)	(0.03)	(0.2)	(0.6)	(0.5)	(1.8)	(5.8)	(3.7)	(0)	(1)	(0.7)	(24)	(0.9)	(0.4)
MS7	10.6	388	765	8.8	0.37	0.9	2.5	4.5	5.9	116	6.8	0.8	5.1	16.2	221.9	20.1	22.8
	(1.4)	(7)	(25)	(0.003)	(0.08)	(0.02)	(0.5)	(0.5)	(0.3)	(88)	(1.2)	(0.2)	(0.7)	(1.1)	(15.9)	(0.3)	(1)
MS8	13.6	518	851	8.9	11.75	1	4.3	5.3	9.5	135	4.5	0.4	3.9	24.2	255.0	22.2	22.8
	(0.1)	(26)	(32)	(0.02)	(3.9)	(0.02)	(0.1)	(0.3)	(1.5)	(5)	(1.5)	(0.1)	(0.1)	(1.8)	(30.1)	(0.8)	(1)
MS9	9.0	392	748	8.7	0.38	2	3.0	4	9.9	45	3	0.001	12.8	22.2	191.9	20.0	22.7
	(1.2)	(3)	(24)	(0.1)	(0.11)	(1)	(0.1)	(0)	(0)	(0)	(0)	(0)	(1)	(1.9)	(5.4)	(0.3)	(1.8)
MS10	10.8	473	955	8.5	0.12	0.6	2.5	4.3	71.8	326	1.1	0.1	5.2	18.4	224.2	20.0	21.3
	(0.4)	(8)	(5)	(0.04)	(0.04)	(0.1)	(0.7)	(0.3)	(22.8)	(104)	(0.1)	(0.1)	(0.5)	(1.4)	(7.9)	(0.1)	(1.4)
MS11	13.5	463	880	8.6	3.71	3.1	2.0	3.3	9.0	96	4.5	0.001	5.4	20.9	205.1	20.7	23.1
	(0.2)	(3)	(20)	(0.04)	(1.23)	(1.8)	(0.3)	(0.1)	(1)	(20.5)	(1.5)	(0)	(0.1)	(0.1)	(4.9)	(0.3)	(1.3)
MS12	10.3	460	921	8.6	1.34	1	2.3	4.5	10.1	46	4.0	0.001	10.1	26	225.0	23.4	22.6
	(1.8)	(18)	(35)	(0.2)	(0.56)	(0)	(0.4)	(0)	(0.4)	(1.8)	(1.8)	(0)	(1.9)	(1.9)	(10.9)	(2.7)	(1.1)
MS13	12.5	411	807	8.5	0.15	3.1	3.1	4.0	47.3	255	6.9	0.5	13.5	40.9	280.8	19.0	23.2
	(3)	(9)	(33)	(0.2)	(0.03)	(2.1)	(0.6)	(0.5)	(8.8)	(55)	(2.1)	(0.1)	(2.5)	(7.1)	(29.2)	(0.7)	(1.3)
MS14	9.3	358	714	7.3	1.19	1.3	3.8	3.5	20.2	134	3.8	0.001	6.3	16.9	150.8	16.7	20.8
	(1.3)	(82)	(166)	(0.1)	(0.38)	(0.1)	(0.8)	(0.5)	(4)	(23.8)	(1.2)	(0)	(0.3)	(0.2)	(37.3)	(3.3)	(0.9)
MS15	24.2	1632	3266	8.3	24.97	1.6	36.7	1.5	63.5	290	49.5	5.6	13.7	33.7	420.2	44.7	23.9
	(0.9)	(39)	(78)	(0.005)	(7.06)	(0.8)	(6.8)	(0.03)	(9.1)	(40)	(15.5)	(1.9)	(2.1)	(2.9)	(41.3)	(3.3)	(0.8)
MS16	16.6	483	935	8.6	0.96	1.0	3	4.2	22.6	75.5	6.3	3.8	3.2	8.8	197.2	17.8	21.5
	(0.8)	(8)	(45)	(0.1)	(0.78)	(0.1)	(0.3)	(0.1)	(3.1)	(10.5)	(0.8)	(1.2)	(0.3)	(1.3)	(13.7)	(2.3)	(0.3)
MS17	14.3	479	935	8.6	3.17	1	2.7	4.2	48	160	5.3	0.001	14.1	33.8	159.0	18.0	22.0
	(0.2)	(1)	(25)	(0.01)	(0.04)	(0.01)	(0.1)	(0.1)	(3)	(10)	(0.2)	(0)	(1.7)	(1.3)	(16.3)	(2.3)	(0.5)
MS18	96.4	561	1133	8.7	0.86	1	7.8	4.3	55.5	185	12.3	0.8	16	34	243.7	19.1	23
	3.0	(34)	(48)	(0.1)	(0.21)	(0.01)	(0.7)	(0.3)	(1.5)	(5)	(2.7)	(0.2)	(1.7)	(2)	(11.2)	(1.2)	(0.3)
MS19	7.2	1065	2246	8.4	0.05	0.8	9.8	4.2	126	420	12.8	2	18.3	53.7	301.4	20.9	21.2
	(0.8)	(215)	(469)	(0.1)	(0.02)	(0.04)	(1.4)	(0.1)	(3)	(10)	(2.3)	(0.3)	(3.8)	(4.9)	(25.5)	(0.05)	(0.4)

All units in mg/L except pH (Dimensionless), Temperature (°C), EC (µS/cm) and Turbidity (NTU).

The TN (TP) of rivers was 8 (0.12) mg/L in dry seasons and 5(0.26) mg/L in wet season, and TN (TP) of lakes was 5.3 (0.2) mg/L) in dry season and 5.2 (0.6) mg/L in wet season. Hence, there is an obvious increase of TN in rivers and Lake Hawassa when temperature increases due to lower dilution and greater agricultural contribution from the upper stream by irrigation, whereas TP in rivers and Lake Hawassa increases in wet seasons due to greater agricultural, rural, and urban runoff. The TN (TP) from point sources was 31.8 (7.2) mg/L in dry season and 13.9 (5.4) mg/L in wet season. This shows that TN (TP) of point sources increases significantly with increasing temperature due to lower dilution. The NH_3-N of rivers was 0.2 mg/L, NH_3-N of Lake Hawassa was 0.83 mg/L, and that of point sources was 4.72 mg/L in dry season. In the wet season, the NH_3-N values were 0.03, 0.71, and 3.6 for rivers, Lake Hawassa, and point sources, respectively. The decreases in NH_3-N level in the rainy season might be due to dilution effect (Table 6).

The positive correlation between temperature and TN, TP, EC, TDS, NH_3-N, and PO_4-P indicates the increase in the concentration of nutrients as the temperature increases (dry period). It also confirms the major contributors of nutrients were the point sources that are releasing a relatively higher amount of pollutants than the agricultural and other sources, as this value lowers during the wet season due to dilution effect. However, the increase in nutrient (NO_3-N) concentration in rivers and Lake Hawassa in the wet season might be due to the increased contribution of agricultural runoff and use of fertilizers.

Sodium, calcium, magnesium, and potassium concentrations of the rivers were 49.1, 13.06, 55.1, and 7.74 mg/L in dry season and 28.9, 32.7, 10.1, and 5.7 mg/L in wet seasons. Sodium, calcium, magnesium and potassium concentrations of the lake were 214, 23.8, 8.7, and 19.7 mg/L in dry season and 178.9, 25.1, 7.3, and 17.2 mg/L in wet season. The sodium, calcium, magnesium, and potassium concentrations of the point sources were 575.2, 38.2, 11.5, and 26.2 mg/L, respectively, in the dry season and 375.2, 38.2, 9.5, and 50.1 mg/L. respectively in the wet season (Table 6). There was an observed decrease in ions when the temperature decreased in the study area. This can be ascribed to the discharge of industrial and domestic effluents, which contribute large amounts of alkaline ions to the river system, as the conductivity depends mainly on the ion concentration in surface water [52]. The natural range of sodium ions in water and soil is so low that their existence can show river pollution caused by human activities. Calcium is added to water from soil, industrial wastes, and natural resources. Magnesium is an essential nutrient required for numerous biochemical and physiological functions [59].

The TDS of water generally increases with the level of dissolved pollutants (such as nitrate, ammonium, and phosphate). Conductivity of ions in water depends on water temperature, and ions move faster when water is warm. Hence, conductivity apparently increases when water has a higher temperature [60]. In addition, Taylor et al. [61] pointed out a strong relationship between these variables or ions, such as nitrate, ammonium, and phosphate, and stated that high concentrations of EC indicate high concentrations of soluble salts. There are strong correlations between EC/TDS, as evidenced by an increase in conductivity as the concentration of all dissolved constituents increases [62] Table 6.

The BOD (COD) of rivers was 19.7 (96.5) mg/L in dry seasons and 6.9 (89.4) mg/L in the wet season, and the BOD and COD of lakes was 28.1 (133.3) mg/L in dry season and was 19.1 (112.9) mg/L in wet season. The BOD and COD concentrations for point sources were 116.2 (398.6) mg/L in dry season and 111.6 (353.7) mg/L in wet season (Table 6). The DO of rivers was 3.5 mg/L in dry season and 6 mg/L in wet season, and the DO of lakes was 4.2 mg/L in dry season and 4.4 mg/L in wet season. The DO of point sources was 2 mg/L in dry season and 2.3 mg/L in the wet season (Table 6).

The DO of the rivers in the dry seasons and Lake Hawassa were well below the standard value. This indicates that the discharge of industrial and domestic effluents has resulted in serious organic pollution of these rivers, as the decrease of DO was mainly caused by the decomposition of organic compounds. Moreover, an extremely low DO content usually indicates the degradation of an aquatic system [63].

The DO showed a negative correlation with most parameters in both dry and rainy seasons, revealing the value of DO decreases with the increase in other water quality parameters. This could explain the temporal variations, as more oxygen was available for reaction with the pollutants, especially metals and organic pollutants, during dry seasons. Additionally, the characteristics of temporal variation in water quality of LHW were affected by DO. DO was strongly correlated with organic matters, nutrients, and metals, and thus seasonal variation should be considered when DO is used as an indicator to evaluate surface water quality. Low dissolved oxygen (DO) is primarily the result of excessive algal growth caused by nutrients. As the algae die and decompose, this process consumes dissolved oxygen. This may result in insufficient dissolved oxygen for fish and other aquatic life. Temperature was significantly correlated with water quality parameters such as EC, TDS, TP, PO_4-P, and DO in both seasons. Temperature had significant negative correlation with DO in the dry and wet seasons, indicating that when water temperature increases, the metabolic rate of microorganisms also increases, and the amount of DO in the water decreases. This might be because faster biodegradation of organic matter during dry seasons can effectively improve water quality. The solubility of oxygen was inversely related to temperature, as the water becomes warmer and more easily saturated with oxygen, hence holds less DO during the dry season. Singh et al. [32] observed the inverse relationship between temperature and DO in natural processes, as water can hold less DO with increasing temperature.

4.2. Pollution Index (PI)

The mean pollution index of the rivers in the lake watershed was 4.5 in dry and 3.3 in wet season, indicating a moderately polluted condition of rivers. Lake Hawassa PI was 5 in both dry and rainy season, indicating that the quality of the lake was moderately polluted. Anthropogenic activities were causing deterioration of the water quality of the rivers and Lake Hawassa, and the overall status of the water quality is moderately polluted. The PI for the point sources was measured for comparison purposes, and it was found to be highly polluted, having a PI index of 6.8 and 7.3 for the wet and dry seasons, respectively (Table 7).

Table 7. Average concentrations of monitoring stations for rivers, Lake Hawassa, and point sources (PS) observed in both dry and wet seasons.

Parameters	Seasons	Rivers	Lake Hawassa	PS
DO (mg/L)	dry seasons	4.2	4.2	1.7
	wet seasons	6	4.3	2.1
BOD_5 (mg/L)	dry seasons	19.7	28.1	116.2
	wet seasons	6.9	19.1	111.6
SS (mg/L)	dry seasons	30.6	19.7	29.3
	wet seasons	51.1	20.9	28.1
NH_3-N (mg/L)	dry seasons	0.2	0.8	1.2
	wet seasons	0.002	0.71	14.4
PI	dry seasons	4.5	5	7.3
	wet seasons	3.3	5	6.8
Rank	dry seasons	MP	MP	HP
	wet seasons	MP	MP	HP

4.3. Cluster Analysis

Spatial and Temporal Similarities

Cluster analysis was applied to find out if the monitoring stations had similar characteristics in terms of water quality parameters. It was implemented with the water quality data set to group comparable monitoring sites (spatial variability) spread over the watershed. Results from CA display high homogeneity within clusters and high heterogeneity between clusters [64]. Hierarchical agglomerative CA was carried out with the normalized data set employing Ward's method, using Euclidean distances as a measure of similarity. In this approach, the analysis of variance method is used to evaluate the distances between clusters, attempting to reduce the sum of squares of all clusters that can be made at each step. In this method, the clusters are grouped sequentially, beginning with the most comparable pair of objects and establishing better clusters one after the other, demonstrated through a dendrogram [2,65].

The dendrogram presents a visual summary of the clustering processes and provides the map of the groups with a dramatic reduction in the dimensionality of the original records [2,5,32,43,44]. The CA grouped all 19 monitoring stations into two statistically significant clusters for the dry and wet seasons in LHW, and the dendrogram displays the grouping of stations for the wet and dry seasons, as demonstrated in Figure 2. Regarding the clustering for the dry and wet seasons, monitoring stations from most of the watershed upstream, from the eastern and western sides of the lake, and from the center of Lake Hawassa have been grouped in Cluster 1. Stations in these clusters typically consist of rivers and Lake Hawassa and are categorized as moderately polluted. The monitoring stations in these clusters are MS1-MS3, MS6-MS14, and MS16-MS18, which can be labeled as "moderate anthropogenic effect". This cluster received pollution from point sources and non-point sources, consisting of animal waste and runoff. It is characterized by moderate anthropogenic impact and labelled as moderately polluted.

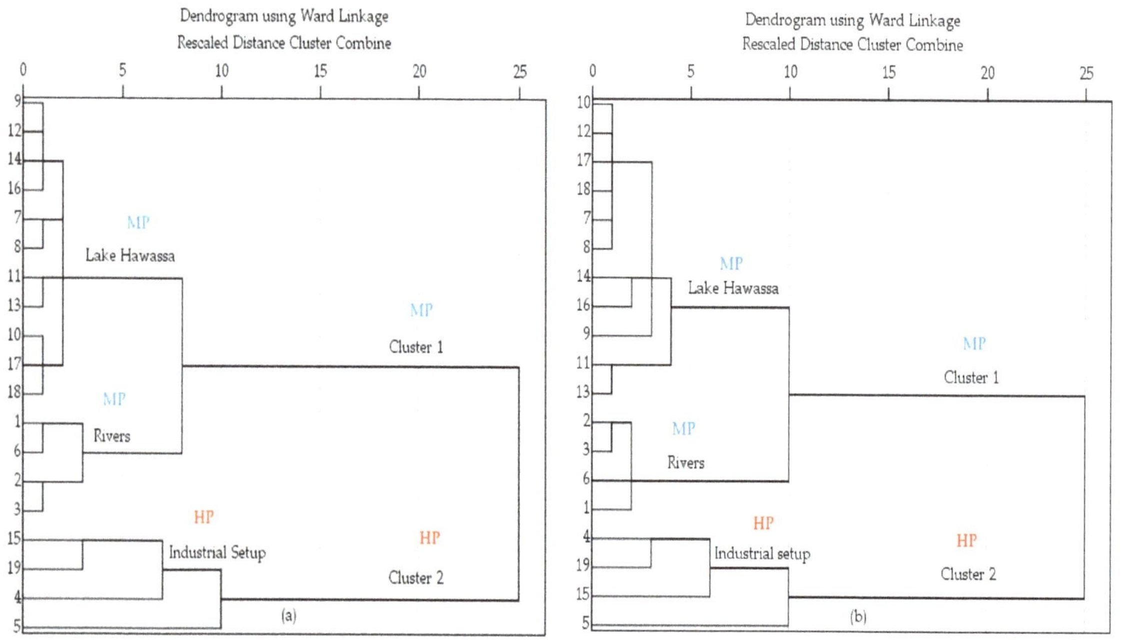

Figure 2. Dendrogram for LHW based on Ward's method showing the clustering of 19 monitoring stations into two significant clusters for both dry (**a**) and wet (**b**) seasons.

The pollution sources for monitoring stations MS1-MS3 were mainly anthropogenic activities from non-point pollution sources such as agricultural and sewage pollution, whereas pollution sources for monitoring stations MS6 (Tikur-Wuha river) and Lake Hawassa (MS7–MS14, MS16–MS18) were mainly industrial pollution, dispersed point sources, agricultural pollution, urban runoff, and sewage pollution.

Owing to their relative sources, all stations in this cluster were rivers and lakes, suggesting that clustering is reasonable for both dry and wet seasons.

The spatial trend of water quality was generally driven by anthropogenic activities from point and non-point sources of pollution, especially anthropogenic activities with respect to pollutant loading and land use.

Cluster 2 includes four monitoring stations in the middle part of the LHW and groups monitoring stations in this cluster as MS4, MS5, MS15, and MS19. Four point sources, specifically BGI, Pepsi Factory, Referral Hospital, and Industrial Park monitoring stations, were assigned to this cluster. Consequently, this cluster is characterized by comparatively heavy pollution.

4.4. Discriminant Analysis

Discriminant analysis (DA) was used to evaluate the spatial variations in water quality and to distinguish the most critical parameters in relation to variations between clusters. Both the standard and stepwise modes were applied to the primary data by dividing them into wet and dry seasons, and the two spatial groups resulting from CA were used in DA. In this case, the WQ parameters were treated as independent variables, while the clusters were considered as dependent variables. The confusion matrixes (CM) showed that 100%, 100%, and 100% of the data points were correctly classified in the standard, forward stepwise, and backward stepwise modes for both dry and wet seasons, respectively (Table 8).

Table 8. Classification matrix for standard, forward stepwise, and backward stepwise DA of spatial variation in LHW for both dry and wet seasons, showing percentage of correct assignation for discriminating parameters.

Monitoring Stations	% Correct	Stations Assigned by DA	
		C1	C2
Standard DA mode for dry season			
C1	100	15	0
C2	100	0	4
Total	100	15	4
Standard DA mode for wet season			
C1	100	15	0
C2	100	0	4
Total	100	15	4
Forward stepwise DA mode for dry season			
C1	100	15	0
C2	100	0	4
Total	100	15	4
Forward stepwise DA mode for wet season			
C1	94	15	0
C2	85	0	4
Total	84.5	15	4
Backward stepwise DA mode for dry season			
C1	100	15	0
C1	100	0	4
Total	100	15	4
Backward stepwise DA mode for wet season			
C1	100	15	0
C2	75	0	4
Total	87.5	15	4

C1: Includes stations (MS1-MS3, MS6-MS14, and MS16-MS18). C2: Includes stations (MS4, MS5, MS15, and MS19).

The standard DA method builds DFs using eighteen parameters, while only three and seven parameters were the critical parameters useful to make distinction within the two pollution groups for both the forward stepwise modes and backward stepwise modes, respectively, for both dry and wet seasons. In forward stepwise mode, most of the parameters such as turbidity, TDS, pH, NH_3-N, NO_3-N, PO_4-P, DO, COD, NO_2-N, TN, TP, temperature, Mg^{2+}, Ca^{2+}, and K^+ were insignificant variables leading to less variation, and they were deleted in the further process. However, in the forward stepwise DA mode, the three significant variables that were useful to make distinctions within the two pollution groups with 100% correct assignation were EC, BOD, and Na^+. The backward stepwise mode deleted the least significant and identified seven significant variables: EC, DO, COD, TN, TP, Na^+ and K^+. These seven parameters, which were 100% correctly assigned, were the critical parameters useful to make distinctions within the two pollution groups. This implies that the expected spatial variation in water quality can be explained sufficiently using variables EC, DO, COD, TN, TP, Na^+, and K^+. Wilks' lambda shows that the discriminant distribution is skewed towards high concentrations.

On the other hand, the standard DA functions was constructed using eighteen parameters, of which three and four parameters were used for forward stepwise mode and backward stepwise mode, respectively, for wet season. In forward stepwise mode, the pollutants that were found to be insignificant variables and had less variation in terms of their spatial distribution were deleted in the further process. However, in the backward stepwise DA mode, the three significant variables that were useful to make distinctions within the two pollution groups with 84.5% correct assignment were EC, Na^+, and COD. The backward stepwise mode deleted the least significant and identified two significant variables: EC and Ca^{+2}. These two parameters were the critical parameters useful to make distinctions within the two pollution groups with 87.5% correct assignation (Table 8). This implies the spatial water quality variation can be sufficiently explained by using variables EC, Na^+, COD, and Ca^{2+}, with Wilks' lambda value showing discriminatory distribution is skewed toward high concentration, as shown in Figure 3.

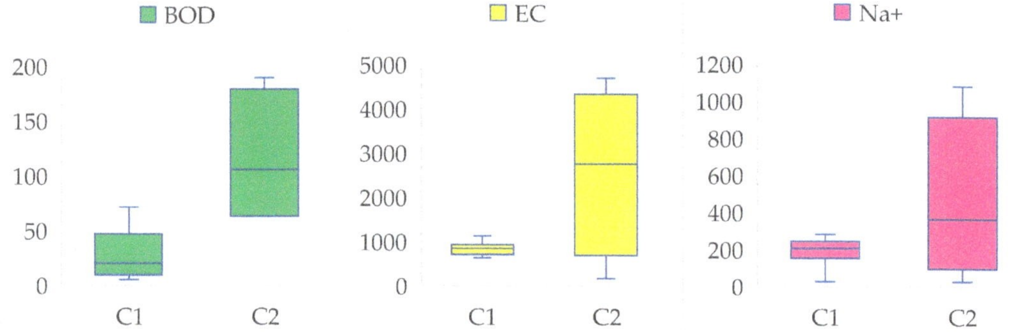

Figure 3. Box plot of the most discriminating parameters, BOD (mg/L), EC (µS/cm) and Na^+ (mg/L) and Wilks' lambda showing skewedness of discriminatory distribution toward high concentration.

4.5. Pollution Source Identification of Monitored Variables

Principal Component Analysis

PCA was applied to the normalized data and was able to identify three principal components (PCs) using the Kaiser criterion [66] based on loading higher than 0.5. The scree plot graphs are used widely to identify the number of PCs to be retained to understand the underlying data structure [26]. Based on the scree plot and the eigenvalues >1 criterion, three factors were chosen as principal factors. The variables with eigenvalues lower than 1 were removed due to their low significance [67].

In this study, the scree plot (Figure 4) shows the sorted eigenvalues from large to small as a function of the number of PCs. This figure shows a pronounced change in slope after the third eigenvalue; three components were retained (Table 9). After the third PC (Figure 4a,b), beginning with the upward curve, the remaining components were circumvented. It was used to classify the number of PCs to be retained in order to figure out the underlying data structure [25]. Consequently, a new set of data is obtained that may explain the variation of data set having fewer variables.

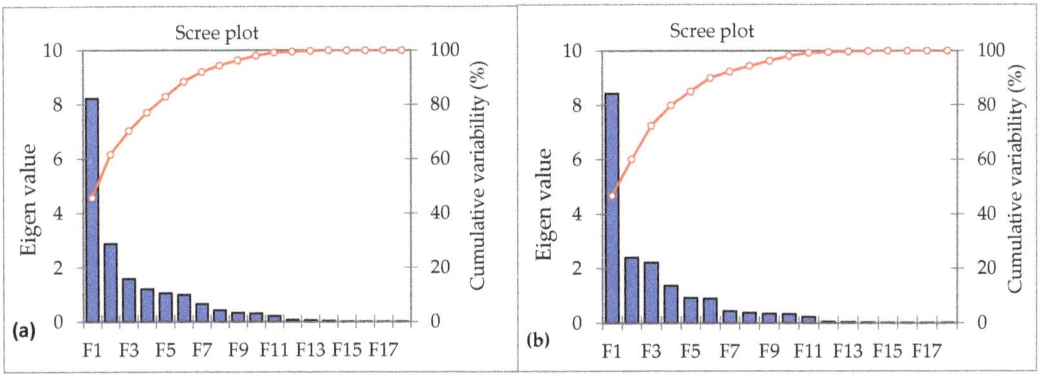

Figure 4. Factor loadings derived from scree plot and eigenvalue for LHW and three factors are retained for dry (**a**) and wet (**b**) seasons.

Table 9. Matrix of factor loadings calculated based on water quality parameters measured in the period from May to January in the Lake Hawassa Watershed and factor loadings of variables on the first three PCs extracted by using eigenvalue for both wet (a) and dry (b) seasons.

Parameters	F1 (a)	F2 (a)	F3 (a)	F1 (b)	F2 (b)	F3 (b)
Turbidity	0.282	−0.420 [c]	**0.452** [c]	−0.032	**−0.781** [a]	−0.320 [c]
TDS	**0.974** [a]	0.136	0.044	**0.962** [a]	0.020	−0.084
EC	**0.978** [a]	0.078	0.079	**0.961** [a]	0.018	−0.098
pH	0.285	0.324 [c]	**−0.710** [b]	0.056	−0.178	**0.775** [a]
NH_3-N	0.416 [c]	**0.516** [b]	−0.313 [c]	0.521 [b]	−0.244	**0.700** [c]
NO_2-N	0.428 [c]	−0.475 [c]	**−0.620** [b]	−0.088	**−0.531** [b]	−0.064
NO_3-N	0.131	0.398 [c]	**0.507** [b]	0.195	**0.599** [b]	−0.168
PO_4-P	**0.871** [a]	−0.035	−0.174	**0.830** [a]	−0.414 [c]	−0.200
DO	**−0.842** [a]	−0.055	−0.365 [c]	**−0.847** [a]	−0.246	0.186
BOD	**0.784** [a]	−0.461 [c]	−0.297	**0.796** [a]	−0.394 [c]	0.015
COD	**0.793** [a]	−0.388 [c]	−0.302 [c]	**0.721** [b]	−0.320 [c]	0.135
TN	**0.898** [a]	0.064	0.101	**0.724** [b]	−0.015	0.047
TP	**0.812** [a]	0.139	0.436 [c]	**0.897** [a]	−0.333 [c]	−0.105
Temp	**0.825** [a]	0.290	0.194	**0.783** [a]	0.246	−0.143
Mg^{2+}	0.077	**−0.654** [b]	0.389 [c]	−0.350 [c]	**−0.567** [b]	−0.380 [c]
Ca^{2+}	0.449 [c]	**−0.627** [b]	0.103	0.401 [c]	**0.524** [b]	−0.246
Na^+	**0.832** [a]	0.205	−0.116	**0.973** [a]	0.001	0.076
K^+	**0.477** [c]	0.335 [c]	0.035	**0.572** [b]	0.522 [b]	0.106
Eigenvalue	8.4	2.4	2.2	8.2	2.9	1.6
Variability (%)	**46.8**	**13.4**	**12.3**	**45.7**	**16**	**8.8**
Cumulative %	46.8	60.2	**72.5**	45.7	61.7	**70.5**

[a] strongly correlated factor loadings, [b] moderately correlated factor loadings, [c] weakly correlated factor loadings.

Moreover, scree plots are used to visually evaluate which components or factors elucidate the maximum variability in the data.

The PCA results, which include the loadings (participation of the original variable in the new one), are summarized in Table 9. The FA in LHW extracted three factors by retaining the PCs through varimax rotation that explained 72.5% of the total variance for the wet season. An eigenvalue offers a degree of the importance of the factor, and factors having the highest eigenvalues are the most significant. Eigenvalues of 1.0 or more are considered significant. Liu et al. [26] additionally categorized the factor loadings as 'strong', 'moderate', and 'weak', corresponding to absolute loading values of >0.75, 0.75–0.50, and 0.50–0.30, respectively.

The first factor (F1), accounting for 46.8% of the total variance, showed strong positive loadings of TDS, EC, PO_4-P, BOD, COD, TP, TN, Na^+, and temperature with factor loadings of 0.974, 0.978, 871, 0.811, 0.784, 0.793, 0.898, 0.812, 0.825, and 0.832, respectively; a weak positive loading of K^+ (0.477); and strong negative loading of DO (−0.842) (Table 9). High positive loadings of temperature and high negative loading of DO might suggest the impact of seasonal variation, and temperature is inversely related to DO. The strong and moderate positive loading of BOD and COD signify biodegradation of organic matter and are negatively affected by DO of water bodies. F1 stands clearly for pollution by BOD or COD, and nutrients and oxygen depletion is a consequence. When the temperature of water bodies decreases, the biodegradation of organic matter decreases, and the solubility of oxygen in the water increases. Similar reports of high concentrations of BOD and COD exist elsewhere [42,44,45]. Similarly, the strong negative DO loading indicates the utilization of DO under anaerobic conditions in rivers and lakes for the degradation of organic matter. F1 showed strongly positive loadings for both COD and BOD, while the loading for DO was strongly negative. This indicates a group of purely organic pollution indicator parameters from industrial effluents, domestic discharges, and livestock affecting water bodies [23,27,51].

High nutrient loadings of factors such as TN and TP represent pollution from point and non-point sources from industrial setup, agriculture areas, domestic sewage, and urban runoff. The high loading of metals demonstrates the influences of industrial effluents and agriculture activities. Phosphorus and nitrogen can originate from point sources such as sewage pollution, industrial facilities and livestock, as well as from non-point sources, mainly from agricultural activities, runoff from rural and urban areas, soil erosion, and livestock. These results are consistent with findings of other reports elsewhere [27,68]. Consequently, the component is more likely to be explained by the combination of domestic pollution and industrial factors. These factors are characteristic of the monitoring stations in the upper catchment (MS1 and MS2), in the middle section including point sources (MS5 and MS15), along Tikur-Wuha River (MS6), and on the eastern side of Lake Hawassa (MS7, MS9, MS12, MS13, and MS14), where domestic and industrial effluents and agricultural runoff are predominant.

The strongly positive loadings of Na^+ and weak positive loadings of K^+ are likely due to industrial effluents discharged into the river Tikur-Wuha and Lake Hawassa. Reports also indicate that the sources of Na^+ and K^+ might be domestic sources, fertilizers, and residential waste in addition to industrial effluents [69]. During field observation, it was found that the major industries are discharging their treated and untreated effluents directly into the Tikur-Wuha River and the lake during the rainy period when the flow rate is high, resulting in high dilution, but during the dry period, the dilution effect is lower and consequent pollution is higher.

On the other hand, the strong loadings of TN and TP in F1 suggest higher contribution from point sources in industry and non-point sources such as agricultural land use, urban drainage, and residential areas during the rainy season. In general, these factors are symbolic of a blended source of contamination, encompassing industrial discharges, urban runoff, and agricultural land use. The results are in agreement with those of other studies [5,24,67,69]. Hence, they can be considered as the contamination index for surface water [44,45].

The second factor (F2) explained 13.4% of the total variance. It had a moderately negative loading of Mg^{2+} and Ca^{2+} (−0.654, −0.627) and a moderately positive loading of NH_3–N (0.516). This factor's moderately negative loading of Mg^{2+} and Ca^{2+} is likely to originate from industrial wastewater discharged into the Tikur-Wuha River and Lake Hawassa, usually from carbonate minerals, which are naturally present in the soils of the Lake Hawassa watershed. This factor is more pronounced at monitoring stations affected by point sources, agricultural lands, and rural and urban runoff, such as MS3 in the upper catchment, MS19 in the middle section (point source), and MS8, MS11, MS16, and MS18 monitoring stations on both eastern and western sides of Lake Hawassa.

A moderately positive loading of NH_3-N (0.7) indicates biodegradation of organic matter. This variable is primarily from runoff, with high loading of solids and wastes from point sources of pollution from domestic and industrial areas. Furthermore, NH_3-N is triggered by the decomposition of organic matter, indicating the discharge of domestic sewage to surface water. Studies elsewhere have showed comparable results [42,44,45,69,70].

The third factor (F3), explaining 12.3% of the total variance, had a moderately negative loading for pH (−0.710), suggesting the dominance of physical reactions by aquatic plants and natural weathering of the basin, possibly due to industrial impact from different sources [22]. It had weak positive loading of turbidity (0.452), moderate negative loading of NO_2-N (−0.620), and moderate positive loading of NO_3-N (0.507). NO_3–N may additionally have derived from agricultural areas in the region, where inorganic nitrogen fertilizers are in common use and the role of domestic waste is strong, and hence, this component can be best explained by a "nutrient" factor representing influences from non-point sources such as agricultural runoff and the domestic pollution factor. The reports of Yilma et al. [35] in Ethiopia and Zhang et al. [27] elsewhere were comparable with this result. This factor is typical of the monitoring stations in the middle section including point sources and eastern and western sides of Lake Hawassa (MS4, MS10, and MS17), where domestic sewage, industrial effluents, and agricultural runoff are predominant.

The FA in LHW extracted three factors by retaining the PCs through varimax rotation that explained 70.5% of the total variance for the dry season. The first factor (F1), accounting for 45.7% of the total variance, showed strong positive loadings of TDS, EC, PO_4-P, BOD, DO, TP, Na^+, and temperature, having factor loadings of 0.962, 0.961, 0.830, 0.796, 0.897, 0.783, and 0.973, respectively; moderate positive loadings of K+, COD, and TN (0.572, 0.721, 0.724); and strong negative loadings of DO (−0.847). Strong positive loadings of temperature and strong negative loadings of DO might suggest the impact of seasonal variations. The strong and moderate positive loading of BOD and COD signify biodegradation of organic matters and negatively affect DO of water bodies. F1 stands clearly for pollution by BOD or COD, and nutrients and oxygen depletion is a consequence. High temperature increases biodegradation and reduces solubility of oxygen in the water. This PC was correlated with COD and BOD_5, indicating a group of purely organic pollution indicator parameters from uncontrolled domestic discharges caused by rapid urbanization and industrial effluents. Biodegradation of organic matter causes concentrations of BOD and dissolved oxygen in water [23,27,51].

A high loading of nutrients represents pollution from industrial setup and domestic wastewater. High loading of metals demonstrates the influences of industrial discharges. Phosphorus and nitrogen may originate from point sources such as sewage pollution, agricultural runoff in the upper stream due to irrigation, industrial facilities, and livestock. Consequently, this component is more likely to be explained by the combination of domestic pollution factors and industrial factors. Strongly positive loading of Na^+ and moderate positive loadings of K^+ are likely to originate from industrial effluents discharged directly into the Tikur-Wuha River and Lake Hawassa. These results are also supported by similar findings obtained elsewhere [27,69].

This factor is more pronounced at monitoring stations in the upper catchment (MS1 and MS3), monitoring stations in the middle section including point sources (MS4, MS5, MS15 and MS19), Tikur-Wuha River (MS6), and monitoring stations from both eastern and western sides of Lake Hawassa (MS9, MS10, MS14, MS16, and MS17), where domestic sewage, industrial effluents, and agricultural activities are predominant. The major industries discharge their treated and untreated effluents directly into Tikur-Wuha River and the lake during the dry period when the flow is low, which might lead to higher pollution. On the other hand, the strong loadings of TN and TP at F1 suggest a higher contribution of point sources from industrial facilities and agricultural runoff in the upper stream due to irrigation. Generally, these factors suggest a blended source of contamination encompassing municipal and industrial point source and livestock. This result is also confirmed by other studies [5,23,33,67,69]. Hence, it can be considered to be the contamination index for surface water [44,45].

The second factor (F2) explained 16% of the total variance and had a strong negative loading of turbidity (−0.781), a moderate negative loading of NO_2−N and Mg^{+2} (−0.567, −0.531), and a moderate positive loading of NO_3−N and Ca^{+2} (0.599, 0.524). NO_3-N could be mainly from point sources, and the role of domestic waste is also strong. Hence, this component can be explained by the "nutrient" factor, which represents influences from non-point sources such as the domestic pollution factor [24,27,32,35,66,69]. A moderately positive loading of K^+ and a moderately negative loading of Mg^{2+} in this factor likely originate from industrial discharges into the Tikur−Wuha River and Lake Hawassa. This PC is more influenced by industrial discharges, and monitoring stations from the LHW, where industry is predominant, are more pronounced. This factor is more pronounced in monitoring stations in the upper catchment (MS2) and the monitoring stations in the eastern and western sides of Lake Hawassa (MS11, MS12, MS13, and MS18), where domestic, industrial, and agricultural activities are predominant in the upper stream due to irrigation.

The third factor (F3), explaining 8.8% of the total variance, had a strong positive loading of pH (0.775), suggesting the dominance of physical reactions by aquatic plants and natural weathering of the basin, and attributed to industrial impact from different sources [22]. A moderate positive loading of NH_3−N (0.7) indicates the biodegradation of organic matter causing concentrations of waterborne factors such as NH_3−N. This variable originated primarily from wastes from point sources of pollution from domestic and industrial areas. Furthermore, NH_3−N is triggered by organic matter decomposition, indicating the discharge of domestic sewage to surface water. Reports elsewhere support the findings of this study [42,44,45,70]. This factor is more pronounced in monitoring stations on the eastern side of Lake Hawassa (MS7 and MS8), where domestic sewage, industrial effluents, and agricultural activities are prevalent.

The bi-plot of PCs on key parameters TDS, EC, PO_4−P, DO, BOD, COD, TN, TP, temperature, Na^+, K^+, Turbidity, NO_2−N, NO_3−N, Mg^{2+}, and Ca^{2+} that characterize monitoring stations from rivers in the upper and middle catchment, point sources in the middle catchment, and the eastern and western sides of Lake Hawassa are presented in Figure 5a,b for dry and wet seasons. In fact, the average values of EC, TDS, BOD, COD, Na^+, K^+, Mg^{2+}, Ca^{2+}, and NH_3−N of point sources were exceedingly higher than that of rivers in the upper and middle catchment (MS1–MS3, and MS6) and Lake Hawassa (MS7-MS14, MS16 and MS18) in Table 6. In addition, NO_3−N, NO_2−N, TN, TP, and PO_4−P were the main parameters characterizing the stated monitoring sites in both seasons. These stations predominantly include rural areas, urban and peri-urban areas, and industrial sites from which domestic sewage, urban runoff, and effluents are discharged into the lake. Furthermore, the influence of agricultural activities in the upper catchment and Tikur-Wuha River feeding the lake was evident. The results of this investigation were comparable to the findings of the studies conducted by Tibebe et al. [71] and Meshesha et al. [72] on Lake Ziway. In particular, higher EC and TDS values were recorded for similar monitoring stations in both seasons (Table 6). In an aquatic environment, EC is used to categorize the pollution status of surface waters, and an increase in conductivity indicates the presence of dissolved ions that can affect aquatic life and water quality [73].

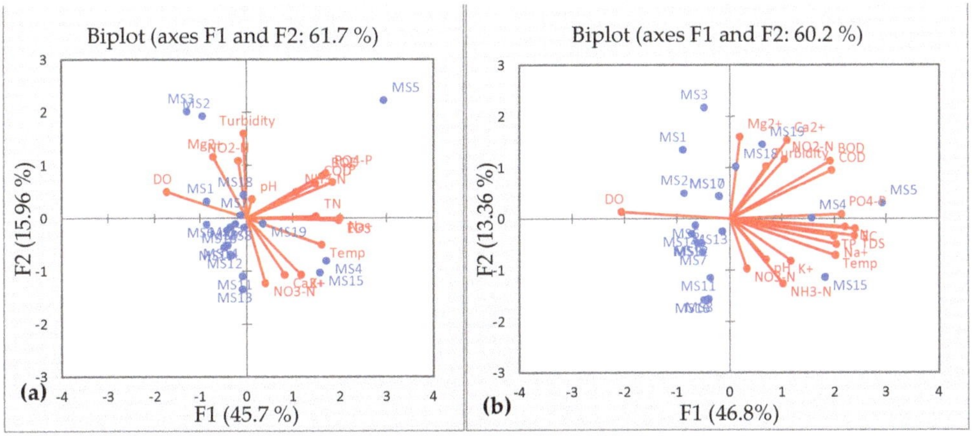

Figure 5. PCA biplots (**a**,**b**) suggest the projection of the monitoring sites (blue dots) and the variable loadings of the primary components (F1 and F2). The biplots additionally display the relationship between highly correlated variables and monitoring stations for dry (**a**) and wet (**b**) seasons. High and low values indicate strong positive and negative correlation, respectively, while values close to 0 imply weak correlation between F1 and F2 and the respective parameter.

4.6. Total Nitrogen to Total Phosphorus (TN:TP) Ratio

The TN:TP ratio in lakes and reservoirs is a key element, as it gives an idea of which of these nutrients is either in excess or limiting to growth, and it was used to estimate the nutrient limitation in the lake. According to Smith [74], blue-green algae (cyanobacteria) has a capacity to dominate in the lake section when the TN:TP ratio was less than 29, and it tends to be rare in the lake when TN:TP > 29. On the other hand, Fisher et al. [75] used a more conservative ratio of TN:TP. According to them, the ratio > 20 is designated as the phosphorus limitation and nitrogen limitation when the ratio is <10, while a TN:TP ratio of 10 to 16 demonstrates either phosphorus or nitrogen (or both) are limiting for growth. The estimated ratio for Lake Hawassa was 31, which is higher than 20 and 30, revealing cyanobacteria dominance in the lake section, which is rare. The TN:TP ratio > 20 in Lake Hawassa indicated that phytoplankton growth in the lake might be phosphorous deficient.

5. Conclusions

Multivariate statistical techniques help researchers to scrutinize the relationships between parameters in a broader fashion by applying different approaches such as cluster analysis, correlation, factor analysis, discriminant analysis, and multiple regressions to determine the association between dependent and independent variables. They reduce the dimensionality of data so that the whole picture can be visualized more easily than looking at specific cases allows. Furthermore, multivariate techniques provide powerful significance testing compared to univariate techniques. Despite their various merits, the results of multivariate statistical modeling are not easy to interpret and require a large data set to get meaningful results due to the high standard errors. In particular, PCA/FA is likely to lose information if PCs or factors are not chosen judiciously.

This study was conducted to evaluate seasonal and spatial variations in water quality and to identify potential sources of pollution using multivariate statistical techniques for the Lake Hawassa Watershed. The results of this study show that the condition of Lake Hawassa Watershed was classified into moderately and highly polluted categories in both dry and wet seasons. In data-limited developing countries such as Ethiopia, it is especially clumsy to identify possible sources of pollution due to certain contaminants, as this requires frequently monitored water quality data, which are often not available. To address this serious problem, this study applied MVST. Multivariate statistics were used to perform temporal and spatial assessment of surface water quality to reduce the number of monitoring stations and chemical parameters in LHW. In this study, we used Pearson correlation, PCA/FA, CA, and DA to evaluate spatial and temporal variance in surface water quality.

CA grouped the monitoring stations into two statistically significant clusters for the dry and wet seasons, labelled MP and HP, using PI. Accordingly, this resulted in a dendrogram with two clusters for the dry and wet seasons. The findings of the study revealed that rivers in the upstream and middle portion of the lake watershed and Lake Hawassa were moderately polluted (MP), while point sources (industries, hospitals, and hotels) in the middle of the LHW were found to be highly polluted (HP).

DA was used to identify the most critical parameters to investigate the spatial variations and extracted seven significant parameters: EC, DO, COD, TN, TP, Na+, and K+, with spatial variance to distinguish the pollution statuses of the groups obtained using CA.

PCA/FA techniques helped to identify the potential sources of water quality degradation. This study comprehensively analyzed the water quality of LHW and identified three significant sources responsible for pollution of Lake Hawassa Watershed in dry and wet seasons affecting the water quality. Accordingly, the pollution is due to mixed sources including point sources such as municipal and industrial effluents, natural processes, livestock, urban runoff, and non-point sources from agricultural activities.

Poor industrial effluent management combined with non-point sources from agriculture and urban runoff contribute significantly to the pollution of Lake Hawassa. Discharge of industrial effluents into the surface water system is the largest point source of anthropogenic pollution. Diffuse sources that contribute enormously to LHW come from agricultural activities, i.e., intensive farming and livestock (F1, F2, and F3).

We conclude that effective management of point and non-point source pollution is imperative to improve domestic, industrial, livestock, and agricultural runoff to reduce pollutant inputs into the lake. A stringent management that requires a comprehensive application of technologies such as fertilizer management, ecological ditches, constructed wetlands, and buffer strips should complement proper municipal and industrial wastewater treatment set-up.

Furthermore, application of indigenous aeration practices such as the use of drop structures at critical locations would help improve water quality in the lake watershed.

Author Contributions: Conceptualization and improvement of the methodology, S.M.L., M.D.U. and A.M.; Data collection, analysis, and interpretation, S.M.L.; Writing of the original manuscript, S.M.L.; Supervision, follow-up of the work, reviewing and modifying of the manuscript, A.M. and M.D.U. All authors have read and agreed to the published version of the manuscript.

Funding: This research was part of the DAAD-EECBP Home Grown PhD Scholarship Program at EECBP Homegrown PhD Program, 2019 (57472170). The Open Access Department, University of Rostock, has funded the APC.

Institutional Review Board Statement: Not applicable.

Informed Consent Statement: Not applicable.

Data Availability Statement: The data presented in this study are available on request from the first author. The data are not publicly available, as they are experimental.

Acknowledgments: We are grateful to the German Academic Exchange Service (DAAD) for offering a stipend for the first author in the course of the study.

Conflicts of Interest: The authors claim no conflict of interest in connection with the work submitted.

References

1. Cherif, E.K.; Salmoun, F.; Mesas-Carrascosa, F.J. Determination of Bathing Water Quality Using Thermal Images Landsat 8 on the West Coast of Tangier: Preliminary Results. *Remote Sens.* **2019**, *11*, 972. [CrossRef]
2. Fan, X.; Cui, B.; Zhao, H.; Zhang, Z., Zhang, H. Assessment of river water quality in Pearl River Delta using multivariate statistical techniques. *Procedia Environ. Sci.* **2010**, *2*, 1220–1234. [CrossRef]

3. Noori, R.; Sabahi, M.; Karbassi, A.; Baghvand, A.; Zadeh, H.T. Multivariate statistical analysis of surface water quality based on correlations and variations in the data set. *Desalination* **2010**, *260*, 129–136. [CrossRef]
4. Wang, Q.; Li, S.; Jia, P.; Qi, C.; Ding, F. A Review of Surface Water Quality Models. *Sci. World J.* **2013**, *2013*, 1–7. [CrossRef]
5. Simeonov, V.; Stratis, J.; Samara, C.; Zachariadis, G.; Voutsa, D.; Anthemidis, A.; Sofoniou, M.; Kouimtzis, T. Assessment of the surface water quality in Northern Greece. *Water Res.* **2003**, *37*, 4119–4124. [CrossRef]
6. Massoud, M.A.; El-Fadel, M.; Scrimshaw, M.D.; Lester, J.N. Factors influencing development of management strategies for the Abou Ali River in Lebanon. I: Spatial variation and land use. *Sci. Total Environ.* **2006**, *362*, 15–30. [CrossRef]
7. Wang, X.-L.; Lu, Y.-L.; Han, J.-Y.; He, G.-Z.; Wang, T.-Y. Identification of anthropogenic influences on water quality of rivers in Taihu watershed. *J. Environ. Sci.* **2007**, *19*, 475–481. [CrossRef]
8. Abebe, Y.D.; Geheb, K. Wetlands of Ethiopia. In *Proceedings of a Seminar on the Resources and Status of Ethiopia's Wetlands*; IUCN, The World Conservation Union: Gland, Switzerland, 2003.
9. Cherif, E.K.; Salmoun, F.; El Yemlahi, A.; Magalhaes, J.M. Monitoring Tangier (Morocco) coastal waters for As, Fe and P concentrations using ESA Sentinels-2 and 3 data: An exploratory study. *Reg. Stud. Mar. Sci.* **2019**, *32*, 100882. [CrossRef]
10. Amare, T.A.; Yimer, G.T.; Workagegn, K.B. Assessment of Metal concentration in water, sediment and macrophyte plant collected from Lake Hawassa, Ethiopia. *Environ. Anal. Toxicol.* **2014**, *05*, 1–7. [CrossRef]
11. Teshome, F.B. Seasonal water quality index and suitability of the water body to designated uses at the eastern catchment of Lake Hawassa. *Environ. Sci. Pollut. Res.* **2019**, *27*, 279–290. [CrossRef]
12. Zigde, H.; Tsegaye, M.E. Evaluation of the current water quality of Lake Hawassa, Ethiopia. *Int. J. Water Resour. Environ. Eng.* **2019**, *11*, 120–128.
13. Amare, D. *Assessment of Non-Point Source Pollution in Lake Awassa Watershed Using the Annualized Agricultural Non-Point Source (AnnAGNPS) Model*; Addis Ababa University: Addis Ababa, Ethiopia, 2008.
14. Drevnick, P.E.; Engstrom, D.; Driscoll, C.T.; Swain, E.; Balogh, S.J.; Kamman, N.C.; Long, D.T.; Muir, D.; Parsons, M.J.; Rolfhus, K.R.; et al. Spatial and temporal patterns of mercury accumulation in lacustrine sediments across the Laurentian Great Lakes region. *Environ. Pollut.* **2012**, *161*, 252–260. [CrossRef] [PubMed]
15. Kebede, W.; Tefera, M.; Habitamu, T.; Alemayehu, T. Impact of Land Cover Change on Water Quality and Stream Flow in Lake Hawassa Watershed of Ethiopia. *Agric. Sci.* **2014**, *5*, 647–659. [CrossRef]
16. Lencha, S.M.; Tränckner, J.; Dananto, M. Assessing the Water Quality of Lake Hawassa Ethiopia—Trophic State and Suitability for Anthropogenic Uses—Applying Common Water Quality Indices. *Int. J. Environ. Res. Public Health* **2021**, *18*, 8904. [CrossRef]
17. Nigussie, K.; Chandravanshi, B.S.; Wondimu, T. Correlation among trace metals in Tilapia (Oreochromis niloticus), sediment and water samples of lakes Awassa and Ziway, Ethiopia. *Int. J. Biol. Chem. Sci.* **2011**, *4*. [CrossRef]
18. Wondrade, N.; Dick, Ø.B.; Tveite, H. GIS based mapping of land cover changes utilizing multi-temporal remotely sensed image data in Lake Hawassa Watershed, Ethiopia. *Environ. Monit. Assess.* **2014**, *186*, 1765–1780. [CrossRef]
19. Abiye, T.A. Environmental resources and recent impacts in the Awassa collapsed caldera, Main Ethiopian Rift. *Quat. Int.* **2008**, *189*, 152–162. [CrossRef]
20. *APHA Standard Methods for the Examination of Water and Wastewater*, 23rd ed.; Baird, R.B.; Eaton, A.D.; Rice, E.W.; Brigewater, L.L. (Eds.) American Public Health Association (APHA): Washington, DC, USA; American Water Works Association (AWWA): Washington, DC, USA; Water Environment Federati, Water Environment Federation: Washington, DC, USA, 2017; ISBN 9780123821652.
21. Emerson, K.; Russo, R.C.; Lund, R.E.; Thurston, R.V. Aqueous Ammonia Equilibrium Calculations: Effect of pH and Temperature. *J. Fish. Res. Board Can.* **1975**, *32*, 2379–2383. [CrossRef]
22. Angello, Z.A.; Tränckner, J.; Behailu, B.M. Spatio-Temporal Evaluation and Quantification of Pollutant Source Contribution in Little Akaki River, Ethiopia: Conjunctive Application of Factor Analysis and Multivariate Receptor Model. *Pol. J. Environ. Stud.* **2021**, *30*, 23–34. [CrossRef]
23. Singh, K.P.; Malik, A.; Sinha, S. Water quality assessment and apportionment of pollution sources of Gomti river (India) using multivariate statistical techniques—A case study. *Anal. Chim. Acta* **2005**, *538*, 355–374. [CrossRef]
24. Sharma, S.; Reddy, A.S.; Dalwani, R.R. Ecological water quality index development and evaluation of water quality of the Satluj river. *Indian J. Environ. Prot.* **2015**, *35*, 477–489.
25. Palma, P.; Alvarenga, P.; Palma, V.L.; Fernandes, R.M.; Soares, V.M.; Amadeu, M.A.; Barbosa, I.R. Assessment of anthropogenic sources of water pollution using multivariate statistical techniques: A case study of the Alqueva's reservoir, Portugal. *Environ. Monit. Assess.* **2010**, *165*, 539–552. [CrossRef]
26. Liu, C.-W.; Lin, K.-H.; Kuo, Y.-M. Application of factor analysis in the assessment of groundwater quality in a blackfoot disease area in Taiwan. *Sci. Total. Environ.* **2003**, *313*, 77–89. [CrossRef]
27. Zhang, Q.; Li, Z.; Zeng, G.; Li, J.; Fang, Y.; Yuan, Q.; Wang, Y.; Ye, F. Assessment of surface water quality using multivariate statistical techniques in red soil hilly region: A case study of Xiangjiang watershed, China. *Environ. Monit. Assess.* **2009**, *152*, 123–131. [CrossRef] [PubMed]
28. Zhao, J.; Fu, G.; Lei, K.; Li, Y. Multivariate analysis of surface water quality in the Three Gorges area of China and implications for water management. *J. Environ. Sci.* **2011**, *23*, 1460–1471. [CrossRef]
29. Chen, P.; Li, L.; Zhang, H. Spatio-Temporal Variations and Source Apportionment of Water Pollution in Danjiangkou Reservoir Basin, Central China. *Water* **2015**, *7*, 2591–2611. [CrossRef]

30. Ma, X.; Wang, L.; Yang, H.; Li, N.; Gong, C. Spatiotemporal Analysis of Water Quality Using Multivariate Statistical Techniques and the Water Quality Identification Index for the Qinhuai River Basin, East China. *Water* **2020**, *12*, 2764. [CrossRef]
31. Mahmud, R.; Inoue, N.; Sen, R. Assessment of Irrigation Water Quality by Using Principal Component Analysis in an Arsenic Affected Area of Bangladesh. *J. Soil Nat.* **2007**, *1*, 8–17.
32. Singh, K.P.; Malik, A.; Mohan, D.; Sinha, S. Multivariate statistical techniques for the evaluation of spatial and temporal variations in water quality of Gomti River (India)—a case study. *Water Res.* **2004**, *38*, 3980–3992. [CrossRef]
33. Sharma, M.; Kansal, A.; Jain, S.; Sharma, P. Application of Multivariate Statistical Techniques in Determining the Spatial Temporal Water Quality Variation of Ganga and Yamuna Rivers Present in Uttarakhand State, India. *Water Qual. Expos. Health* **2015**, *7*, 567–581. [CrossRef]
34. Banda, T.D.; Kumarasamy, M. Application of Multivariate Statistical Analysis in the Development of a Surrogate Water Quality Index (WQI) for South African Watersheds. *Water* **2020**, *12*, 1584. [CrossRef]
35. Yilma, M.; Kiflie, Z.; Windsperger, A.; Gessese, N. Assessment and interpretation of river water quality in Little Akaki River using multivariate statistical techniques. *Int. J. Environ. Sci. Technol.* **2018**, *16*, 3707–3720. [CrossRef]
36. Mohd, N.F.; Samsudin, M.S.; Mohamad, I.; Awaluddin, M.R.; Mansor, A.; Hafizan, J.; Ramli, N. River water quality modeling using combined principle component analysis (PCA) and multiple linear regressions (MLR): A case study at Klang River, Malaysia. *World Appl. Sci. J.* **2011**, *14*, 73–82.
37. Kazi, T.; Arain, M.; Jamali, M.; Jalbani, N.; Afridi, H.; Sarfraz, R.; Baig, J.A.; Shah, A.Q. Assessment of water quality of polluted lake using multivariate statistical techniques: A case study. *Ecotoxicol. Environ. Saf.* **2009**, *72*, 301–309. [CrossRef] [PubMed]
38. Helena, B.; Pardo, R.; Vega, M.; Barrado, E.; Fernandez, J.M.; Fernand, L. Temporal evolution of groundwater composition in an alluvial aquifer (Pisuerga River, Spain) by principal component analysis. *Water Res.* **2000**, *34*, 807–816. [CrossRef]
39. Sojka, M.; Siepak, M.; Ziola-frankowska, A.; Frankowski, M. Application of multivariate statistical techniques to evaluation of water quality in the Mala Welna River (Western Poland). *Environ. Moni. Assess.* **2008**, *147*, 159–170. [CrossRef]
40. Grubbs, F.E.; Beck, G. Extension of Sample Sizes and Percentage Points for Significance Tests of Outlying Observations. *Technometrics* **1972**, *14*, 847–854. [CrossRef]
41. Suhaimi, N.; Ghazali, N.A.; Nasir, M.Y.; Mokhtar, M.I.; Ramli, N.A. Markov Chain Monte Carlo Method for Handling Missing Data in Air Quality Datasets. *Malays. J. Anal. Sci.* **2017**, *21*, 552–559.
42. Vega, M.; Pardo, R.; Barrado, E.; Debán, L. Assessment of seasonal and polluting effects on the quality of river water by exploratory data analysis. *Water Res.* **1998**, *32*, 3581–3592. [CrossRef]
43. Alberto, W.D.; del Pilar, D.M.; Valeria, A.M.; Fabiana, P.S.; Cecilia, H.A.; Ángeles, B.M.d.l. Pattern Recognition Techniques for the Evaluation of Spatial and Temporal Variations in Water Quality. A Case Study: Suquía River Basin (Córdoba–Argentina). *Water Res.* **2001**, *35*, 2881–2894. [CrossRef]
44. Shrestha, S.; Kazama, F. Assessment of surface water quality using multivariate statistical techniques: A case study of the Fuji river basin, Japan. *Environ. Model. Softw.* **2007**, *22*, 464–475. [CrossRef]
45. Wu, E.M.-Y.; Kuo, S.-L. Applying a Multivariate Statistical Analysis Model to Evaluate the Water Quality of a Watershed. *Water Environ. Res.* **2012**, *84*, 2075–2085. [CrossRef]
46. Swanson, R.A.; Holton, E.F. *Research in Organizations Foundations and Methods of Inquiry*; Holton, E.F., Ed.; Berrett-Koehler Organization: San Francisco, CA, USA, 2005; pp. 115–142.
47. Liou, S.-M.; Lo, S.-L.; Wang, S.-H. A Generalized Water Quality Index for Taiwan. *Environ. Monit. Assess.* **2004**, *96*, 35–52. [CrossRef]
48. Chen, Y.-C.; Yeh, H.-C.; Wei, C. Estimation of River Pollution Index in a Tidal Stream Using Kriging Analysis. *Int. J. Environ. Res. Public Health* **2012**, *9*, 3085–3100. [CrossRef] [PubMed]
49. Mahesh Kumar, M.K.; Mahesh, M.K.; Sushmitha, B.R. CCME Water Quality Index and Assessment of Physico- Chemical Parameters of Chikkakere, Periyapatna, Mysore District, Karnataka State, India. *Int. J. Innov. Res. Sci. Eng. Technol.* **2014**, *3*. [CrossRef]
50. Varol, M.; Şen, B. Assessment of surface water quality using multivariate statistical techniques: A case study of Behrimaz Stream, Turkey. *Environ. Monit. Assess.* **2008**, *159*, 543–553. [CrossRef]
51. Zhou, F.; Liu, Y.; Guo, H. Application of Multivariate Statistical Methods to Water Quality Assessment of the Watercourses in Northwestern New Territories, Hong Kong. *Environ. Monit. Assess.* **2006**, *132*, 1–13. [CrossRef] [PubMed]
52. Parinet, B.; Lhote, A.; Legube, B. Principal component analysis: An appropriate tool for water quality evaluation and management—Application to a tropical lake system. *Ecol. Model.* **2004**, *178*, 295–311. [CrossRef]
53. Rencher, A. *Methods of Multivariate Analysis Second Edition*, 2nd ed.; A Wiley-Interscience Publication: New York, NY, USA, 2002; pp. 156–504. ISBN 0471418897.
54. Gummadi, S.; Swarnalatha, G.; Vishnuvardhan, Z.; Harika, D. Statistical Analysis of the Groundwater Samples from Bapatla Mandal, Guntur District, Andhra Pradesh, India. *IOSR J. Environ. Sci. Toxicol. Food Technol.* **2014**, *8*, 27–32. [CrossRef]
55. Karakuş, C.B. Evaluation of water quality of Kızılırmak River (Sivas/Turkey) using geo-statistical and multivariable statistical approaches [Internet]. *Environ. Dev. Sustain.* **2020**, *22*, 4735–4769. [CrossRef]
56. Shroff, P.; Vashi, R.T.; Champa neri, V.A.; Patel, K.K. Correlation Study among water quality parameters of groundwater of Valsad District of south Gujarat (India). *J. Fundam. Appl. Sci.* **2015**, *151*, 1–10. [CrossRef]

57. Kumar, M.; Ramanathan, A.; Rao, M.S.; Kumar, B. Identification and evaluation of hydrogeochemical processes in the groundwater environment of Delhi, India. *Environ. Earth Sci.* **2006**, *50*, 1025–1039. [CrossRef]
58. Gebre-mariam, Z. The effects of wet and dry seasons on concentrations of solutes and phytoplankton biomass in seven Ethiopian rift-valley lakes. *Limnologica* **2002**, *179*, 169–179. Available online: http://www.sciencedirect.com/science/article/B7GX1-4GWP957-6/2/4d9fcc9b986796cd8cfb4a1b3193ad03 (accessed on 1 December 2002).
59. Yilmaz, E.; Koç, C. Physically and Chemically Evaluation for the Water Quality Criteria in a Farm on Akcay. *J. Water Resour. Prot.* **2014**, *06*, 63–67. [CrossRef]
60. Poisson, A. Conductivity/salinity/temperature relationships of diluted and concentrated standard seawater. *Mar. Geodesy* **1982**, *5*, 359–361. [CrossRef]
61. Taylor, M.; Elliott, H.A.; Navitsky, L.O. Relationship between total dissolved solids and electrical conductivity. *Water Sci. Technol.* **2018**, 1–7. [CrossRef]
62. Dhanasekarapandian, M.; Chandran, S.; Devi, D.S.; Kumar, V. Spatial and temporal variation of groundwater quality and its suitability for irrigation and drinking purpose using GIS and WQI in an urban fringe. *J. Afr. Earth Sci.* **2016**, *124*, 270–288. [CrossRef]
63. Boyle, T.P.; Fraleigh, H.D. Natural and anthropogenic factors affecting the structure of the benthic macroinvertebrate community in an effluent-dominated reach of the Santa Cruz River, AZ. *Ecol. Indic.* **2003**, *3*, 93–117. [CrossRef]
64. Johnson, R.A. *Applied Multivariate Statistical Analysis*, 6th ed.; Recter, P., Ryan, D., Behrens, L.M., Eds.; Pearson Education, Inc.: San Antonio, TX, USA, 2007; ISBN 0131877151.
65. McKenna, J. An enhanced cluster analysis program with bootstrap significance testing for ecological community analysis. *Environ. Model. Softw.* **2003**, *18*, 205–220. [CrossRef]
66. Kasier, H.F. The application of electronic computers to factor analysis. *Educ. Psychol. Meas.* **1960**, *20*, 141–151. [CrossRef]
67. Barakat, A.; El Baghdadi, M.; Rais, J.; Aghezzaf, B.; Slassi, M. Assessment of spatial and seasonal water quality variation of Oum Er Rbia River (Morocco) using multivariate statistical techniques. *Int. Soil Water Conserv. Res.* **2016**, *4*, 284–292. [CrossRef]
68. Boyacioglu, H.; Boyacioglu, H. Water pollution sources assessment by multivariate statistical methods in the Tahtali Basin, Turkey. *Environ. Geol.* **2008**, *54*, 275–282. [CrossRef]
69. Su, S.; Zhi, J.; Lou, L.; Huang, F.; Chen, X.; Wu, J. Spatio-temporal patterns and source apportionment of pollution in Qiantang River (China) using neural-based modeling and multivariate statistical techniques. *Phys. Chem. Earth Parts A/B/C* **2011**, *36*, 379–386. [CrossRef]
70. Gazzaz, N.M.; Yusoff, M.K.; Ramli, M.F.; Aris, A.Z.; Juahir, H. Characterization of spatial patterns in river water quality using chemometric pattern recognition techniques. *Mar. Pollut. Bull.* **2012**, *64*, 688–698. [CrossRef] [PubMed]
71. Tibebe, D.; Beshah, F.Z.; Lemma, B.; Kassa, Y.; Bhaskarwar, A.N. External Nutrient Load and Determination of the Trophic Status of Lake Ziway. *CSVTU Int. J. Biotechnol. Bioinform. Biomed.* **2018**, *3*, 1–16. [CrossRef]
72. Meshesha, D.T.; Tsunekawa, A.; Tsubo, M. Continuing land degradation: Cause-effect in Ethiopia's Central Rift Valley. *Land Degrad. Dev.* **2012**, *23*, 130–143. [CrossRef]
73. Badillo-Camacho, J.; Reynaga-Delgado, I.; Barcelo-Quintal, I.; Valle, P.F. Water Quality Assessment of a Tropical Mexican Lake Using Multivariate Statistical Techniques. *J. Environ. Prot.* **2015**, *6*, 215–224. [CrossRef]
74. Smith, V.H. Low Nitrogen to Phosphorus Ratios Favor Dominance by Blue-Green Algae in Lake Phytoplankton. *Science* **1983**, *221*, 669–672. [CrossRef] [PubMed]
75. Fisher, M.M.; Reddy, K.R.; James, R.T. Internal Nutrient Loads from Sediments in a Shallow, Subtropical Lake. *Lake Reserv. Manag.* **2005**, *21*, 338–349. [CrossRef]

applied sciences

MDPI

Article

Evaluation of the Hydrochemical and Water Quality Characteristics of an Aquifer Located in an Urbanized Area

Juan G. Loaiza [1], Yaneth Bustos-Terrones [2], Victoria Bustos-Terrones [3], Sergio Alberto Monjardín-Armenta [4], Alberto Quevedo-Castro [1], Rogelio Estrada-Vazquez [1] and Jesús Gabriel Rangel-Peraza [1,*]

1 División de Estudios de Posgrado e Investigación, TecNM-Instituto Tecnológico de Culiacán, Juan de Dios Batíz 310. Col. Guadalupe, Culiacan 80220, Mexico; d18170809@culiacan.tecnm.mx (J.G.L.); d08170173@itculiacan.edu.mx (A.Q.-C.); d13170726@culiacan.tecnm.mx (R.E.-V.)

2 División de Estudios de Posgrado e Investigación, CONACYT-TecNM-Instituto Tecnológico de Culiacán, Juan de Dios Batíz 310. Col. Guadalupe, Culiacan 80220, Mexico; yabustoste@conacyt.mx

3 Academic Department of Engineering in Environmental Technology, Polytechnic University of the State of Morelos, Boulevard Cuauhnáhuac, No. 556. Col. Lomas del Texcal, Jiutepec 62550, Mexico; vbustos@upemor.edu.mx

4 Faculty of Earth and Space Sciences, Autonomous University of Sinaloa, Circuito Interior Oriente, Cd Universitaria, Culiacan 80040, Mexico; sa.monjardin12@info.uas.edu.mx

* Correspondence: jesus.rp@culiacan.tecnm.mx

Abstract: Groundwater is an important source of fresh water in the world. However, the excessive extraction and increasing pollution represent a major challenge for water sustainability in Mexico. Nowadays, since water quality changes in aquifers are not noticeable, aquifer monitoring and assessment are imperious. In this study, the water quality of the Cuernavaca aquifer was evaluated using a database of 23 parameters in 4 sampling points from 2012 to 2019. The spatial behavior of water quality variables was described by using interpolation. The temporal evaluation of groundwater quality was carried out through time series. Water quality indices (WQI) were obtained in this aquifer and the WQI values suggest that the groundwater could be considered as good quality for potable use and of medium-high quality for irrigation. The chemical characteristics of the groundwater were also evaluated using Gibb, Piper, and Schoeller diagrams. Finally, with a total of 34 samples of each parameter in each sampling site, a multivariate statistical analysis was performed using a Pearson correlation and hierarchical cluster analysis. This analysis showed a correlation between hydrochemical features and groundwater quality parameters, where nitrates presented the highest number of significant correlations with other parameters. These results may be useful for the authorities to adopt planning methods to improve the sustainable development of the aquifer.

Keywords: Cuernavaca aquifer; hydrochemistry; water quality index; time series analysis; spatial analysis

Citation: Loaiza, J.G.; Bustos-Terrones, Y.; Bustos-Terrones, V.; Monjardín-Armenta, S.A.; Quevedo-Castro, A.; Estrada-Vazquez, R.; Rangel-Peraza, J.G. Evaluation of the Hydrochemical and Water Quality Characteristics of an Aquifer Located in an Urbanized Area. *Appl. Sci.* **2022**, *12*, 6879. https://doi.org/10.3390/app12146879

Academic Editors: Amit Kumar, Santosh Subhash Palmate and Rituraj Shukla

Received: 18 June 2022
Accepted: 5 July 2022
Published: 7 July 2022

1. Introduction

Groundwater is one of the most important natural resources and plays an important role in ecosystems [1]. It is widely used for domestic, industrial, and agricultural activities; hence, its demand is constantly increasing [2–5]. Population growth, accidental spills, surface leaching, runoff, and the extensive use of fertilizers in irrigated areas are considered the main causes of groundwater alteration [6,7]. Furthermore, agricultural activities modify groundwater conditions with nutrients and pesticides coming from leachate infiltration into the soil. Therefore, the use of fertilizers, pesticides, and herbicides in agriculture are major threats to aquifers [8–10]. In addition, the deterioration of an aquifer can be also related to natural causes such as floods, droughts, and salinization [11–13]. Once the aquifer is altered, it is complex and expensive to reverse the damage [12,14]. Therefore, groundwater quality must be monitored regularly to prevent aquifer alterations.

Many studies have been proposed for the assessment of aquifer vulnerability. Bannenberg et al. [6] evaluated the hydrological regime and hydrochemical features of the Flamouria aquifer in Edessa, Greece. They found that groundwater quality was not suitable for irrigation use since the high alkalinity and total dissolved solids found in groundwater could generate excessive salinization of the soil. Kumar et al. [8] conducted a hydrochemical study to assess the water quality suitability for drinking and irrigation purposes. As a result, they found a high concentration of some ions, such as As, Fe, and Mn, in an aquifer located in the Central Ganga Basin. Loh et al. [15] evaluated the suitability of an aquifer in Ghana for domestic and irrigation purposes. They used conventional hydrochemical and mass balance models to reveal relationships between water parameters and the main influence on the chemistry of the aquifer under study. As a result, they found that the groundwater in the area is permissible for agricultural irrigation.

Zakaria et al. [16] evaluated the hydrochemistry of groundwater in the Anayari catchment to identify the hydrogeochemical processes that are responsible for the main ions in groundwater. Their results showed good quality for irrigation without prior treatment. Wisitthammasri et al. [17] studied the water quality and hydrochemical characteristics of an aquifer in Thailand using multivariate statistical analysis to identify preliminary ion sources. This multivariate analysis evidenced the ion exchange between Ca^{2+} and Na^{+} from the weathering of silicates and calcite. Some commonly used multivariate statistical techniques, such as Pearson correlations and Hierarchical Cluster Analysis (HCA), have been used to illustrate the relationship between many groundwater variables and describe the relationship between them [18]. These studies have been carried out to develop appropriate groundwater management strategies and policies to protect aquifers. According to Elumalai et al. [4], multivariate statistical analyses are important because they provide essential information on groundwater quality and the processes responsible for its alteration.

Another tool for groundwater water quality assessment is the water quality index (WQI). This tool has been widely used by several researchers [8,14,19] since it simplifies the interpretation of water quality behavior. El Osta et al. [19] used this technique to classify the suitability of groundwater. As a result, they found that only a low percentage of the samples were classified as good to excellent to be used, while the rest of the samples were inadequate and required treatment to be used as drinking water. Another effective tool for assessing groundwater quality and its variability is that recommended by Kumar et al. [8] who evaluated the groundwater quality based on the Geographic Information System (GIS) through the Groundwater Quality Index (GQI) in an aquifer in southern India. They mention that this method is reliable for groundwater quality assessment and serves as a useful tool for decision-makers for efficient groundwater quality monitoring and management mainly in agricultural areas which have a great influence on groundwater recharge and quality.

In recent years, population growth and the increase of agricultural areas and indus-trial activities have intensified water demand, threatening the sustainable use of ground-water in the Cuernavaca aquifer. Despite this situation has been locally evidenced, no formal studies have been carried out to demonstrate the effect of these activities on groundwater resources. The novelty of this study lies in describing the hydrological and hydrochemical conditions of this aquifer to identify its vulnerability. The geohydrological features of the Cuernavaca aquifer are described using Gibb, Piper, and Schoeller diagrams. Groundwater quality evaluation is carried out based on time series analysis, water quality distribution maps, and water quality indices. This study performs a multivariate and correlation analysis to identify possible pollution sources and proposes better water management strategies for this aquifer.

2. Materials and Methods

2.1. Study Area

The study area is in the state of Morelos, Mexico with an approximate area of 820 km^2 (Figure 1). Mean annual precipitation, evapotranspiration, and air temperature are 1278 mm, 874.7 mm, and 19.4 °C, respectively. The highest rainfall values are observed from July to September, which corresponds to the summer season, and less significant precipitations are registered in winter from October to January mainly caused by cold fronts. The Cuernavaca aquifer is a free, heterogeneous, and anisotropic aquifer with surface geology that is represented by lithological units mainly of sedimentary and volcanic origin [20] and does not show any significant structural complications. The static water level of this aquifer varies from 20 to 100 m.

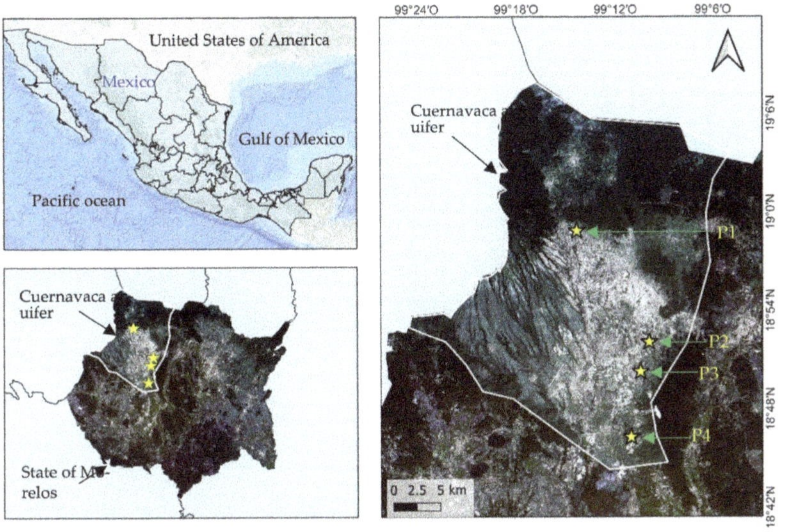

Figure 1. Location of the Cuernavaca aquifer and location of sampling wells.

2.2. Data Collection

In Mexico, the National Water Commission (CONAGUA) is a federal agency responsible for monitoring, surveillance, and management of aquifers [21]. The analysis of the groundwater samples was carried out by an accredited laboratory [22] which applied a methodology based on the standard methods (APHA) [23]. The data used in this study were obtained by this federal agency. For economic and strategic reasons, four sampling wells were monitored, which were in sites with intense anthropogenic activity. These wells are used for water consumption. The extraction of groundwater for consumption purposes is carried out using pumping systems.

The hydrochemical and water quality parameters considered in this study were: bicarbonates (HCO^{3-}), fecal coliforms (FC), total organic carbon (TOC), ammonium (NH_3), nitrites (NO^{2-}), nitrates (NO^{3-}), organic nitrogen (ON), total nitrogen (TN), total phosphorus (TP), total dissolved solids (TDS), electrical conductivity (EC), pH, chlorides (Cl^-), fluorides (F), silicon oxides (SiO_2), potassium (K^+), manganese (Mn), sodium (Na^+), sulfates ($SO_4{}^{2-}$), calcium (Ca^{2+}), magnesium (Mg^{2+}), total hardness (TH), water temperature (WT).

Quality control (QC) and quality assurance (QA). The sampling wells were monitored by CONAGUA. An accredited laboratory carried out the analysis of 23 parameters sampled every six months in the 2012–2019 period. The water sampling was carried out according to Mexican regulations. Based on these regulations, chemical products of analytical grade were required for the preparation of the standard solutions and reagents. In addition, replicates were performed to ensure the reliability of the results and comply with quality control required by the General Directorate of Standards and the Federal Law on Metrology and Normalization.

2.3. Spatial and Temporal Assessment of Groundwater Quality

The spatial evaluation of the groundwater quality of the Cuernavaca aquifer was carried out through interpolation of the measured parameters as suggested by other studies [24,25]. The inverse distance weighting (IDW) interpolation method was used to describe the spatial distribution of the groundwater quality values through the study area by using the QGIS 3.18 software. The weights used in the IDW method were calculated according to the weighting strategy proposed by Bartier [25]. These weight values are determined based on the distance between the sampling points according to Equation (1).

$$z_{x,y} = \frac{\sum_{i=1}^{n} z_i d_{x,y,i}^{-\beta}}{\sum_{i=1}^{n} d_{x,y,i}^{-\beta}} \tag{1}$$

where $z_{x,\,y}$ is the water quality parameter to be estimated; z_i represent the measured value for the sampling point; $d_{x,\,y,\,i}$ is the distance between $z_{x,\,y}$ and z_i; and β is a user-defined coefficient (the software default value of 2 was used for the β coefficient).

Temporal evaluation of groundwater quality was performed using time series analysis to determine possible groundwater quality temporal trends using biannual sampling data from 2012 to 2019. A temporal analysis was carried out by describing the groundwater quality variations over time. Finally, groundwater quality data were compared to World Health Organization (WHO) and local guidelines.

2.4. *Water Quality Assessment*

2.4.1. Drinking Water Quality Index

The drinking water quality index (DWQI) is frequently used to determine the suitability of groundwaters. In this study, the determination of DWQI was performed according to Equations (2)–(5) [14,26,27].

$$W_i = \frac{w_i}{\sum_i^n w_i} \tag{2}$$

$$Q_i = \frac{e_i - v_i}{b_i - v_i} * 100 \tag{3}$$

$$SI = W_i * Q_i \tag{4}$$

$$DWQI = \sum_{i=1}^{n} SI \tag{5}$$

where W_i is the relative weight; w_i is the weight assigned to each parameter according to its relative importance for drinking water (the maximum weight of "5" has been assigned for the highest importance and the minimum weight of "2" for the lowest importance); "n" is the number of groundwater parameters; Q_i: is the rating according to the distribution of the "ith" parameter. e_i: is the concentration of each parameter; v_i: is the optimum value of the parameter ("0" is considered as optimum value for all parameters, except pH which is "7"); b_i is the guideline value [28] for each parameter; *SI*: is the sub-index of "ith" parameter. According to some researchers [14,29,30], the optimum values and weights for the parameters of the DWQI are: pH (b_i = 8.5, w_i = 4, W_i = 0.13), TDS (mg/L, b_i = 500, w_i = 4, W_i = 0.13), total hardness (mg/L, b_i = 300, w_i = 3, W_i = 0.10), calcium (mg/L, b_i = 75, w_i = 3, W_i = 0.10), magnesium (mg/L, b_i = 30, w_i = 3, W_i = 0.10), nitrates (mg/L, b_i = 45, w_i = 4, W_i = 0.13), chlorides (mg/L, b_i = 250, w_i = 2, W_i = 0.06), sulfates (mg/L, b_i = 200, w_i = 2, W_i = 0.06), fluorides (mg/L, b_i = 1, w_i = 4, W_i = 0.13), and total alkalinity (mg/L, b_i = 200, w_i = 2, W_i = 0.06). Based on the results of Equation (4), the aquifer water was then classified into different categories: DWQI < 50 (excellent), DWQI = 50–100 (good), DWQI = 100–150 (moderate), DWQI = 150–200 (poor) and DWQI $\geq$ 200 (extremely poor).

2.4.2. Hydrochemical Characteristics

The chemical composition of groundwater is highly variable. Hence, the hydrochemical classification and groundwater chemical composition evolution were determined by using the Gibb, Piper, and Schoeller plots [15,31]. Then, the suitability of the groundwater for irrigation was evaluated by using the groundwater indices shown in Table 1.

2.4.3. Multivariate Statistical Analysis

Multivariate statistical techniques such as Pearson correlation and hierarchical cluster analysis (HCA) were used to figure out the relationship between the water quality variables. This multivariate statistical analysis was used to identify the factors and possible sources that could explain the behavior of the groundwater quality of the aquifer [32–35]. In addition, a dendrogram was performed using the ward conglomeration method with a Euclidean distance metric [7,15].

Table 1. Groundwater indices based on hydrochemical features.

Indices	Acronym	Equation	References
Sodium adsorption ratio	SAR	$SAR = \frac{Na^{+}}{\sqrt{\frac{Ca^{2+}+Mg^{2+}}{2}}}$	Abdelaziz et al. [34]
Soluble sodium percentage	SSP	$SSP = \frac{Na^{+}}{Ca^{2+}+Mg^{2+}+K^{+}} * 100$	Tefera et al. [31]
Sodium percentage	%Na	$\%Na = \frac{(Na^{+}+K)}{(Ca^{2+}+Mg^{2+}Na^{+}+K^{+})} * 100$	Abbasnia et al. [36]
Residual Sodium Carbonate	RSC	$RSC = (HCO_3^{-}) - (Ca^{2+} + Mg^{2+})$	Zakaria et al. [16]
Magnesium Hazard	MH	$MH = \frac{Mg^{2+}}{Mg^{2+}+Ca^{2+}} * 100$	Hossain et al. [37]
Permeability index	PI	$PI = \frac{Na^{+}+K^{+}+\sqrt{HCO_3^{-}}}{Ca^{2+}+Mg^{2+}+Na^{+}+K^{+}} * 100$	Kumar et al. [8]
Kelly Ratio	KR	$KR = \frac{Na^{+}}{Mg^{2+}+Ca^{2+}}$	Acharia et al. [27]
Total Hardness	TH	$TH = (Ca + Mg) * 50$	Tefera et al. [31]

3. Results and Discussion

3.1. Descriptive Analysis of Groundwater Quality Parameters

The total dissolved solids reflect the behavior of the salt concentration of the aquifer. These solids were found in a range of 75–688 mg/L and a mean value of 316 mg/L was registered. This value is low compared to that reported by Tefera et al. [31], who found concentrations up to 2777.6 mg/L. According to WHO [28], groundwaters with TDS values higher than 500 mg/L could be considered unsuitable for drinking water supply. The total hardness was found in a range of 24.6–456.8 mg/L. However, the mean value (179.2 mg/L) is below the concentration of 300 mg/L suggested by WHO [28] for drinking water. This value is also below the total hardness found by Kumar et al. [8], who presented concentrations greater than 292 mg/L in an unconfined aquifer located in the Central Ganga Basin, India.

The electrical conductivity of the Cuernavaca aquifer was between 90 and 991 µS/cm with a mean conductivity of 409.8 µS/cm. A high variation of electrical conductivity was observed in this aquifer, where the lowest conductivity values were found in sampling well one. Anthropogenic activities, such as agriculture, and rainwater filtration could be the reason for this variation. Jama et al. [38] presented concentrations up to 11,950 µS/cm in the unconfined Doukkala Aquifer located in a large agricultural region in Morocco. The groundwater of the Cuernavaca aquifer is slightly alkaline since its pH is in the range of 6.2–8.4 (the water is considered alkaline when pH > 8 and acidic when pH < 6). This pH range is within the drinking water standards of the WHO (6.5–8.5).

Nitrogen and phosphorus were below the permissible limits proposed by local standards. The TN and TP concentrations found were between 0.012–7.02 and 0.001–0.39 mg/L, respectively. Nitrogen concentrations are not usually frequent in natural soils, they occur due to the contact of the soil cover with nitrated fertilizers, animal waste, domestic effluents, and septic tanks [14]. The total organic carbon was found in a range of 0.07–2.57 mg/L. The presence of organic matter in the Cuernavaca aquifer could be related to the infiltration of the organic matter produced naturally by plants and animals due to excretion and decomposition. This situation is corroborated since fecal coliforms were found in the aquifer, with a mean value of 276.6 CFU/100 mL. The presence of fecal coliforms in groundwater could indicate pollution from anthropogenic sources since the sampling wells are in an urban area with a large population. Table 2 presents the mean values of the water quality parameters measured in the Cuernavaca aquifer from 2012 to 2019.

Table 2. Range, standard deviation and mean values for water quality parameters in the Cuernavaca aquifer from 2012 to 2019.

Parameters	Abbreviation	Minimum	Maximum	Mean	Standard Deviation
Bicarbonates (mg/L)	HCO^{3-}	48.40	294.90	145.12	79.75
Fecal coliforms (MPN/100 mL)	FC	1.00	2909.00	276.62	719.15
Total organic carbon (mg/L)	TOC	0.08	2.87	0.92	0.72
Ammonium (mg/L)	NH_3	0.00	0.61	0.07	0.12
Nitrites (mg/L)	NO^{2-}	0.00	0.04	0.01	0.01
Nitrates (mg/L)	NO^{3-}	0.00	6.87	3.24	2.13
Organic Nitrogen (mg/L)	ON	0.00	1.55	0.24	0.29
Total nitrogen (mg/L)	TN	0.01	7.03	3.56	2.20
Total phosphorus (mg/L)	TP	0.00	0.40	0.15	0.09
Total dissolved solids (mg/L)	TDS	64.64	688.00	316.01	195.16
Electrical conductivity (μS/cm)	EC	90.00	991.00	409.79	266.25
pH	PH	6.20	8.40	7.29	0.45
Chlorides (mg/L)	Cl^-	8.44	78.25	23.89	20.62
Fluorides (mg/L)	F	0.04	0.97	0.32	0.19
Silicon oxides (mg/L)	SiO_2	30.42	91.29	67.10	13.56
Potassium (mg/L)	K^+	1.32	7.72	3.68	1.60
Manganese (mg/L)	Mn	0.00	0.48	0.01	0.08
Sodium (mg/L)	Na^+	1.92	40.01	18.59	9.43
Sulfates (mg/L)	$SO_4{}^{2-}$	0.82	226.38	48.41	49.98
Calcium (mg/L)	Ca^{2+}	3.87	121.10	41.93	37.57
Magnesium (mg/L)	Mg^{2+}	3.82	50.47	19.25	13.07
Total hardness (mg/L)	TH	24.60	456.80	184.46	141.14
Water temperature (°C)	WT	17.25	22.49	20.24	21.68

The concentrations of some mineral compounds such as calcium and magnesium cause the precipitation of these salts. In the Cuernavaca aquifer, calcium was found in concentrations from 3.88 to 121.1 mg/L and a mean value of 41.9 mg/L, while magnesium was found from 3.8 to 50.5 mg/L with a mean value of 19.2 mg/L. The presence of concentrations of these salts (Mg^{2+} and Ca^{2+}) is due to the geological features of the aquifer. Sodium and potassium were found in a range of 1.92–37.5 and 1.3–6.6 respectively. The mean values for all major cations were within the maximum permissible limit [28].

Bicarbonates were within a range of 48.4–295 mg/L and a mean value of 145.1 mg/L was calculated. It is noteworthy that carbonates were not found in the samples. Sulfates in the Cuernavaca aquifer are between 0.8 and 136 mg/L, which are below the $SO_4{}^{2-}$ concentrations reported in other studies [39] and the guidelines recommended by the WHO [28]. Moreover, the chlorides presented a concentration between 8.4 and 78.2 mg/L, while nitrates showed a maximum concentration up to 6.2 mg/L, with a mean value of 3.2 mg/L. Both anions' mean values were also below the WHO maximum allowable values. Adimalla and Qian [14] suggest that nitrates could be found in groundwaters due to anthropogenic activity. They reported NO_3 concentrations up to 198.17 mg/L in groundwater under the influence of agriculture activities in Nanganur, India. In this study, a high variation in NO_3 concentrations was found between sampling sites, where the highest concentration was found in sampling site 2. Cadmium, chromium, mercury, lead, zinc, and arsenic were also analyzed in this study; however, the concentrations found could be negligible because low concentrations were observed (cadmium < 0.0002 mg/L; chromium < 0.00088 mg/L; mercury < 0.00009 mg/L, lead < 0.00154 mg/L, zinc < 0.002 mg/L and arsenic < 0.00139 mg/L). According to these results, the influence of geogenic sources was evidenced, where leaching and weathering of rocks and the use of pesticides and fertilizers could be recognized as the main driving factors for the hydrochemical and water quality of the aquifer.

3.2. Spatial and Temporal Variations of Measuring Indicators

A total of 23 water quality parameters were analyzed at four sampling wells. These sampling wells are located within the urban area of the city of Cuernavaca, which has different elevations as shown in Figure 2.

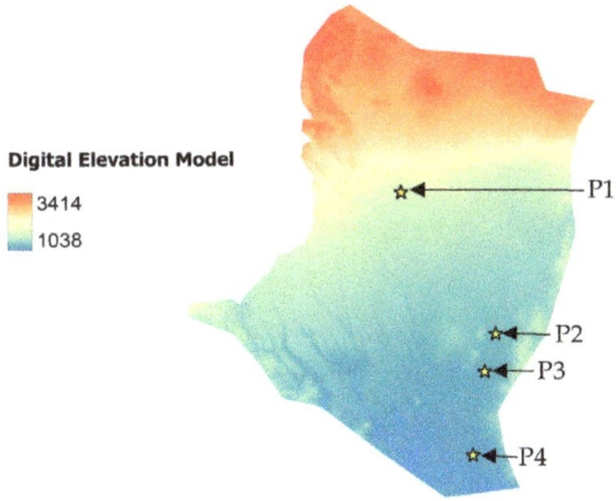

Figure 2. Elevations of the Cuernavaca aquifer.

The land use and soil classifications in the study area are shown in Figures 3 and 4, respectively. These figures demonstrate that P1 is in a wooded area with little human settlement, close to the annual rainfed agricultural area. The dominant soil type in this area is Luvic phaeozem which is characterized by organic matter and scarce carbonates. This sample site is next to an oak-pine forest land-use zone. Sample sites P2, P3, and P4 have similar characteristics because they are in irrigated agricultural areas close to the urban area. These sites are in a Pelic vertisol soil characterized by high mineral content.

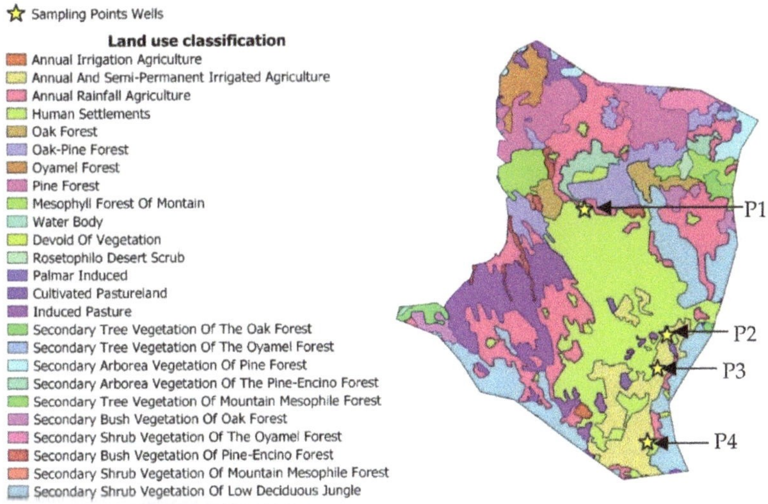

Figure 3. Land use classification of the Cuernavaca aquifer.

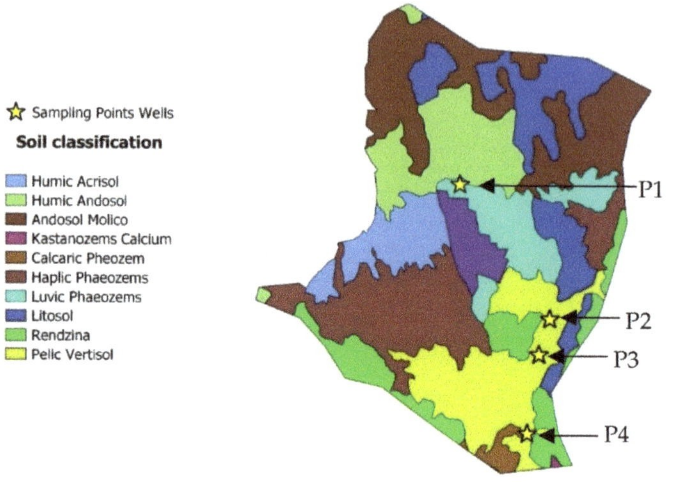

Figure 4. Soil classification of the Cuernavaca aquifer.

Figure 5 presents the spatial interpolation of the water quality parameters in the Cuernavaca aquifer. The highest values for all the parameters analyzed were observed in sampling point two (P2), while sampling point one (P1) presented the lowest values. This situation could be related to the soil type in the area. Since the highest elevation is observed in P1, the rest of the sample points located in lower elevation areas could be influenced by the erosion, transport, and deposition of contaminants.

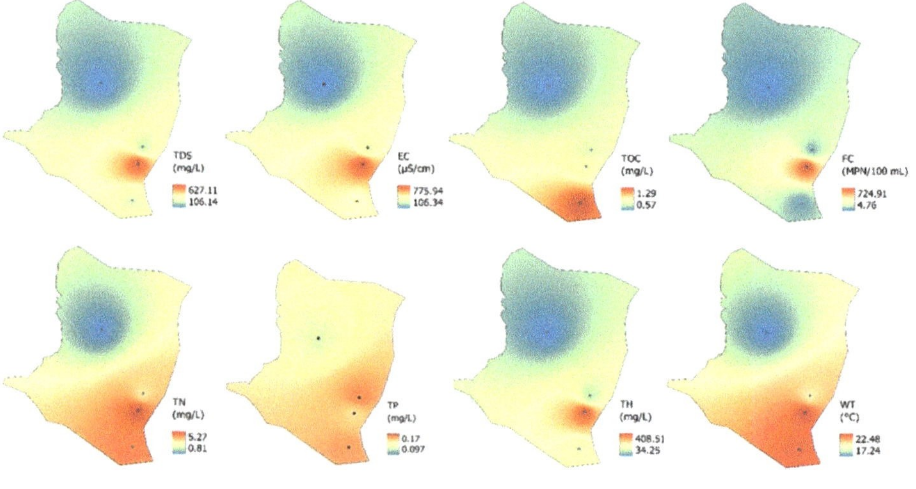

Figure 5. Spatial behavior of physicochemical parameters in the Cuernavaca aquifer.

Figure 6 presents the spatial behavior of the major ions. The presence of ions in the sampling wells is due to interactions with the geological material of the aquifer, natural processes of rock dissolution, and ion leaching. This figure shows that higher concentrations of ions were found at the P2. At this site, groundwater is not suitable for domestic use according to WHO [28] guidelines. The spatial distribution of these chemical elements highlights the vulnerability of the aquifer, especially at P2. Since the concentrations of ions at P1, P3, and P4 sites are similar, they could be considered reference values for the major ions in the Cuernavaca aquifer. It is noteworthy that the concentration of ions in all the groundwater samples was found to be within the WHO desirable limits for agricultural irrigation [6,28].

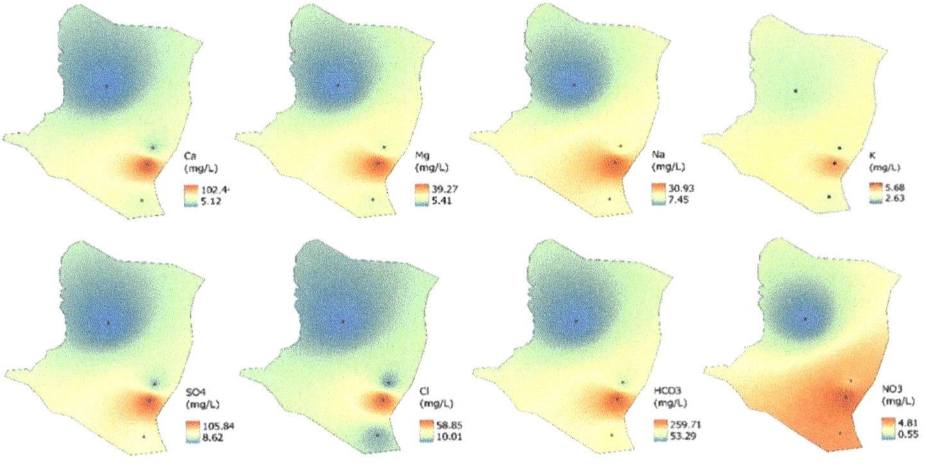

Figure 6. Spatial behavior of major ions in the Cuernavaca aquifer.

Figure 7 presents the temporal variation of the water quality parameters from 2012 to 2019. No trends, seasonal or cyclic patterns were found in the groundwater quality data. The time series also demonstrated that P2 showed higher values in almost all parameters. Lower concentrations of the physicochemical and major ions are observed at P1 since this site is at a higher elevation where the runoff of anthropogenic contaminants is significantly low. Similar values are presented by Adimalla et al. [14], who evaluated the groundwater of Nanganur county in India. Based on these results, they consider that groundwater quality does not represent health risks for drinking water use and only recommend groundwater defluoridation.

Table 3 presents the ANOVA statistical analysis of the groundwater quality parameters. This table showed that 15 parameters had a statistically significant variation from a spatial point of view. However, only 4 groundwater quality parameters showed a temporal significant variation.

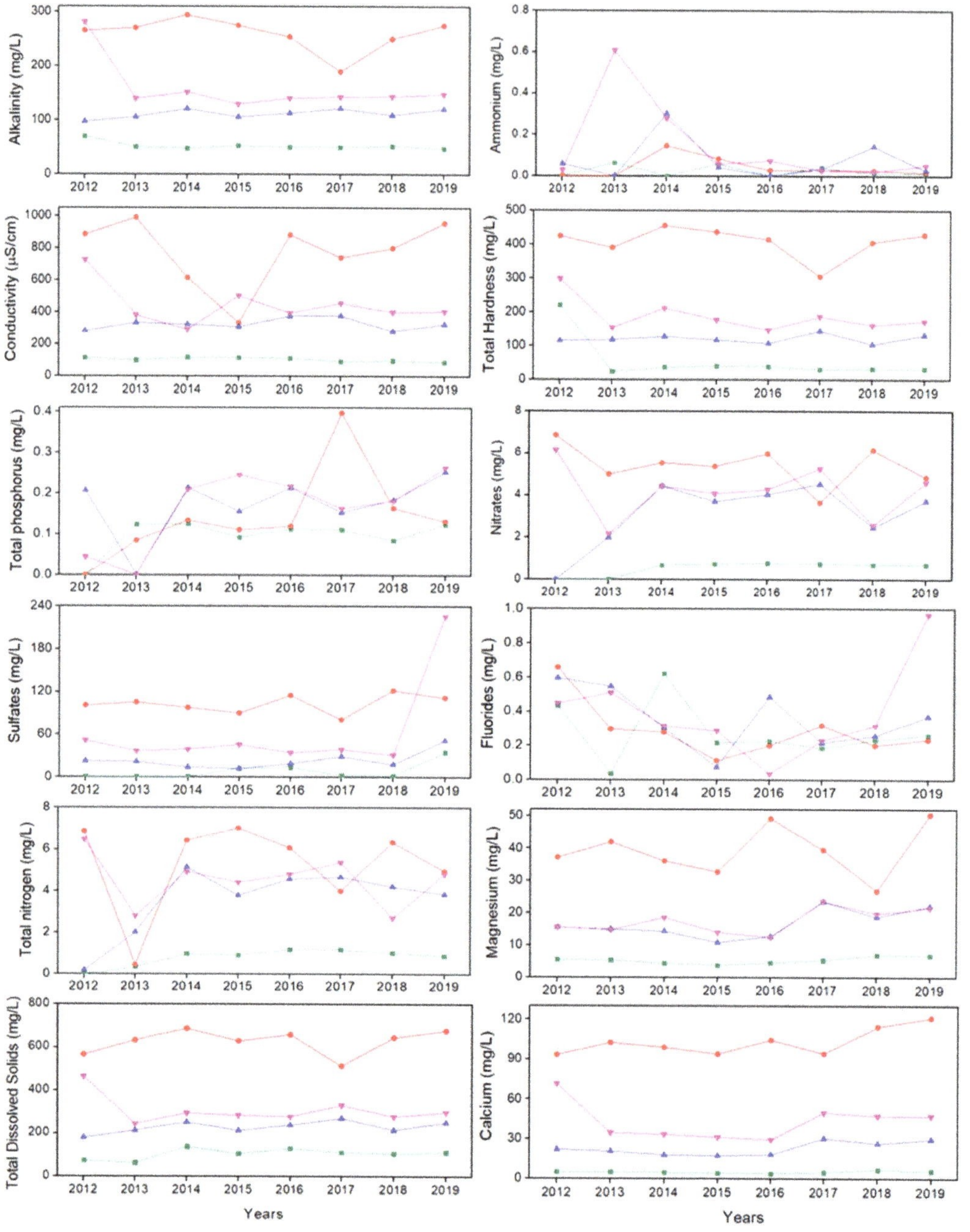

Figure 7. Variation of water quality parameters over time (2012–2019) at P1 (-■-), P2 (-●-), P3 (-▲-) y P4 (-▼-).

Table 3. Spatial and temporal statistical analysis (ANOVA) of the water quality parameters measured in the Cuernavaca aquifer.

Parameter	Site (Spatial)	Year (Temporal)
	p-Value	*p*-Value
Bicarbonates (mg/L)	0.0000 *	0.7058
Fecal coliforms NMP/100 mL	0.2229	0.8342
Total organic carbon (mg/L)	0.3044	0.1937
Ammonium (mg/L)	0.3219	0.4469
Nitrites (mg/L)	0.7001	0.4622
Nitrates (mg/L)	0.0000 *	0.0911
Organic Nitrogen (mg/L)	0.1192	0.1272
Total nitrogen (mg/L)	0.0000 *	0.1061
Total phosphorus (mg/L)	0.2540	0.0283
Total dissolved solids (mg/L)	0.0000 *	0.5686
Electrical conductivity µS/cm	0.0000 *	0.4360
pH	0.0002 *	0.0034
Chlorides (mg/L)	0.0000 *	0.4176
Fluorides (mg/L)	0.3992	0.0844
Silicon oxides (mg/L)	0.0315 *	0.0001 *
Potassium (mg/L)	0.0001 *	0.2912
Manganese (mg/L)	0.5720	0.8024
Sodium (mg/L)	0.0000 *	0.0143 *
Sulfates (mg/L)	0.0000 *	0.0134 *
Calcium (mg/L)	0.0000 *	0.2466
Magnesium (mg/L)	0.0000 *	0.0058 *
Total hardness (mg/L)	0.0000 *	0.3337
Water temperature °C	0.0000 *	0.8973

* p-value $\leq$ 0.05) is statistically significant.

3.3. Multivariate Statistical Analysis

Figure 8 shows the Pearson correlations between the groundwater quality parameters. Pearson correlation coefficient (r) ranges from −1 to +1 and measures the strength of the linear relationship between parameters [9]. A high negative correlation is found when r is close to −1 but r values close to +1 indicate a high positive correlation. A Pearson correlation (r) close to 0 indicates that there is no linear relationship between the two variables.

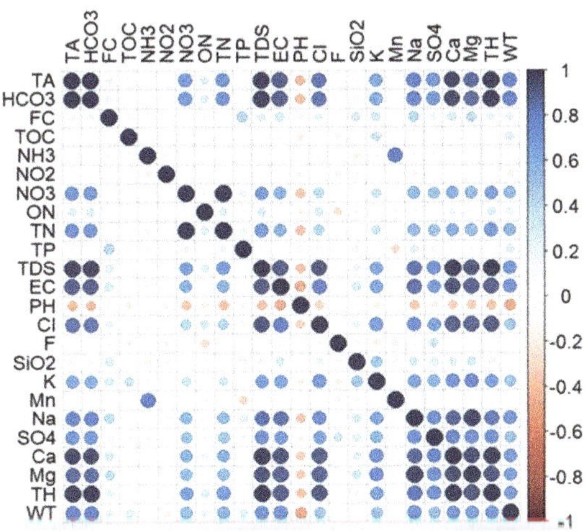

Figure 8. Pearson correlation coefficients of the water quality parameters of the Cuernavaca aquifer.

Nitrates presented the highest number of correlations with other parameters. This parameter is correlated with TN, TDS, EC, PH, Cl^-, K^+, Na^+, SO_4, Ca^{2+}, Mg^{2+}, TH, and WT. TA is related to HCO_3^-, NO_3^-, TN, TDS, EC, Cl^-, Na^+, Ca^{2+}, Mg^{2+}, TH, and WT. Bicarbonates showed a high correlation with NO_3^-, TN, TDS, EC, Cl^-, Na^+, Ca^{2+}, Mg^{2+}, TH, and WT. Total nitrogen is associated with TDS, EC, Na^+, SO_4^{2-}, Ca^{2+}, Mg^{2+}, TH, and WT. Total dissolved solids are highly related to ions, TH, EC, and WT. The electrical conductivity attributes a higher correlation with Cl^-, K^+, Na^+, SO_4^{2-}, Ca^{2+}, Mg^{2+}, TH, and WT. Chlorides are significantly related to K^+, Na^+, SO_4^{2-}, Ca^{2+}, Mg^{2+}, and TH. Potassium is related to other ions such as Na^+, SO_4, Ca^{2+}, Mg^{2+}, and TH. In turn, sodium is related to SO_4^{2-}, Ca^{2+}, Mg^{2+}, and TH). Sulfates are related to Ca^{2+}, Mg^{2+}, and TH, and calcium shows a correlation with Mg and TH. This method has been used for the evaluation of groundwater quality. Strong correlations between major ions are also reported by Miao et al. [40]. This situation evidenced that the groundwater quality of a coastal city in China was affected by various factors, such as dissolution and water evaporation.

Since a high amount of groundwater quality parameters were correlated with each other, a hierarchical cluster analysis was carried out (Figure 9). Hierarchical cluster analysis was used to further unearth the main chemical processes controlling groundwater chemistry in the aquifer [15,34]. This analysis included the 23 analyzed parameters and 34 water samples at different times of the year. The dendrogram formed the main cluster which in turn formed two groups. The first group includes only fluorides and sulfates. The second main group is composed of the rest of the water quality analyzed parameters. Several subgroups are evidenced, such as those formed by TDS, EC, and TH, TA, and HCO_3^-, and SiO_2, K, Ca^{2+}, and Mg^{2+}. These results corroborated the relationship between the observed parameters in the Pearson correlations. The relationship between the groundwater parameters indicated a common source. Due to the nature of these subgroups, groundwater quality is derived from geogenic sources, mainly carbonate mineral solutions [32]. Abdelaziz et al. [34] also noted that the dendrogram can be used to classify the groundwater quality parameters and found great similarities with the grouping carried out by the principal components analysis.

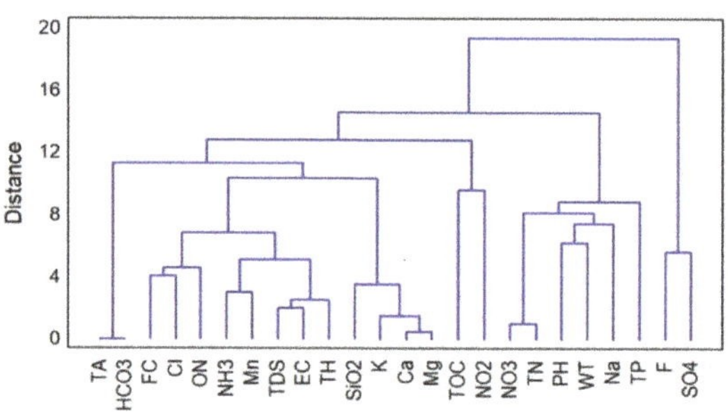

Figure 9. Hierarchical cluster analysis for groundwater quality parameters monitored in the Cuernavaca aquifer from 2012 to 2019.

3.4. Drinking Water Quality Index

Table 4 shows the DWQI obtained in the four sampling sites of the Cuernavaca aquifer. DWQI range from 11.2 to 78.2 were obtained from 2012 to 2019, where 70% of the samples showed an excellent groundwater quality, mainly in the P1, P3, and P4 sampling sites, as shown in Figure 10. P1 showed the best water quality, possibly because this sampling site is at the highest elevation and close to a protected green area. In contrast, P2 showed a high variation of groundwater quality because it is in a highly-populated area. Similar DWQI results are reported by Ahmed et al. [41], who mentioned that the DWQI ranged from 1.86 to 82.25 for water samples from different sampling sites of an aquifer in India.

Table 4. Classification of the water quality index and percentage of the values of the Cuernavaca aquifer samples.

DWQI	Classification of Water	Samples	% of Samples
<50	Excellent water	24	70.5
50–100	Good water	10	29.5
100–200	Poor water	0	0
200–300	Very poor water	0	0
>300	Unsuitable for drinking	0	0

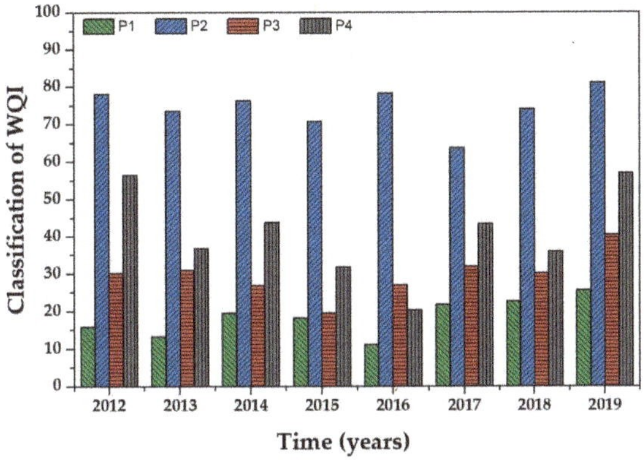

Figure 10. Variation of the water quality index in the sampling wells of the Cuernavaca aquifer.

3.5. Hydrochemical Characteristics

Hydrochemical analysis was carried out to characterize the Cuernavaca aquifer's groundwater. A high content of salts in groundwater could lead to the salinization of the soils and crop yield losses due to dehydration of plants [38,42]. The concentrations of salts in the Cuernavaca aquifer showed the following behavior:

$$HCO_3^- > Ca^{2+} > Na^+ > Mg^{2+} > Cl^- > SO_4^{2-} > K^+ > NO_3^-$$

Figure 11a shows the Piper triangular diagram. In this figure, the mean values of 34 samples at each sampling point were used. This diagram is a graphical representation of groundwater chemistry, where the relative concentrations of cations and anions are shown by separate ternary plots. In the lower-left ternary plot (cation diagram), a dominance of Mg^{2+} and $Na^+ + K^+$ is observed. This dominance could be related to progressive evaporation and ion exchange processes [43]. The lower-right ternary plot (anion diagram) indicated that the groundwater chemistry of the Cuernavaca aquifer is highly influenced by Calcium-bicarbonate type and Bicarbonate type [44]. These results are consistent with the results suggested by other researchers [45]. Figure 11b shows the Schoeller diagram which exhibits a similar behavior of cations and anions in the multiple samples from different wells. This diagram demonstrated that the highest equivalent concentrations of the ions were present in sampling site P2, where Ca^{2+} and HCO_3^- showed the highest equivalent concentrations of cations and anions, respectively. Similar results were presented by Tefera et al. [31] in Tana basin in Ethiopia. However, Abotalib et al. [46] obtained opposite results to those presented in this study for an aquifer located in hyperarid deserts in central Egypt.

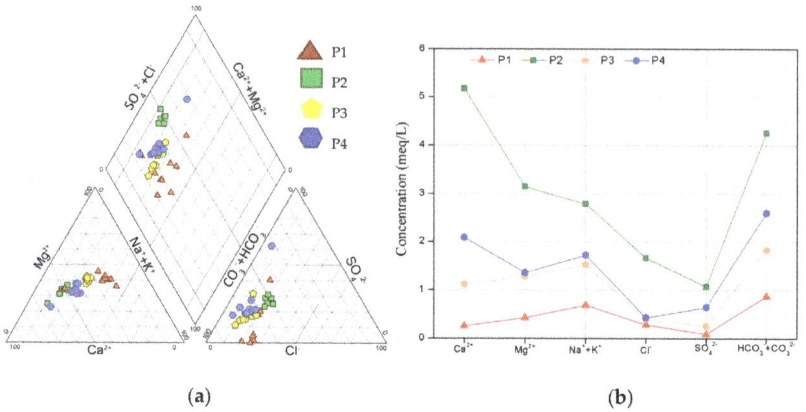

Figure 11. Piper (**a**), and Schoeller (**b**) diagrams for groundwater chemistry composition at P1 (-▲-), P2 (-■-), P3 (-⬟-) y P4 (-⬢-).

El Osta et al. [19] suggest that the groundwater chemistry of an aquifer is the result of evaporation, weathering, and rock-water interaction. In this study, the Gibbs diagram (Figure 12) showed that the cations and anions in the aquifer are primarily controlled by rock–water interaction. The dissolution of the rock in the aquifer is evidenced since a high content of chlorides and sulfates is observed. Therefore, this process regulates groundwater chemistry and quality. Likewise, this diagram suggests that the P2 site could be controlled by evaporation. This process produces dissolved solutes in groundwater and soil in areas with little depth [15,19,47,48].

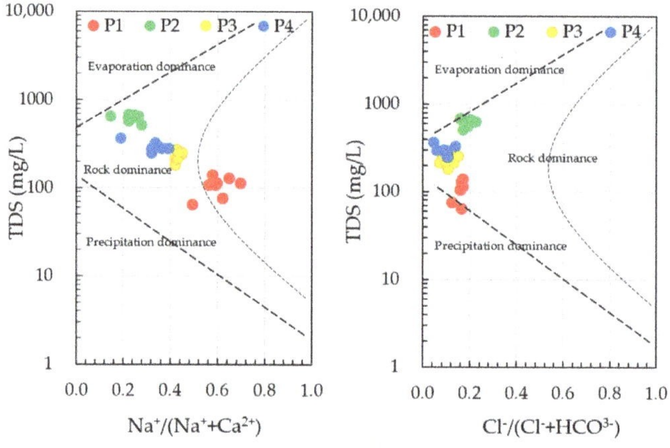

Figure 12. Gibbs diagram showing the source of cations and anions in the Cuernavaca aquifer at P1 (-•-), P2 (-•-), P3 (-•-) y P4 (-•-) sites.

3.6. Groundwater Indices Based on Hydrochemical Features

Figure 13 presents the classification of water quality for irrigation purposes. Some groundwater indices obtained in this study suggest that the Cuernavaca aquifer has a good quality for irrigation purposes. For example, the SAR index classified the groundwater as excellent, which indicates that there is no risk of sodium for irrigation. The RSC index also showed that all samples presented a good quality of water for irrigation. Based on the KR index, the groundwater of the Cuernavaca aquifer was adequate in most of the samples (84%).

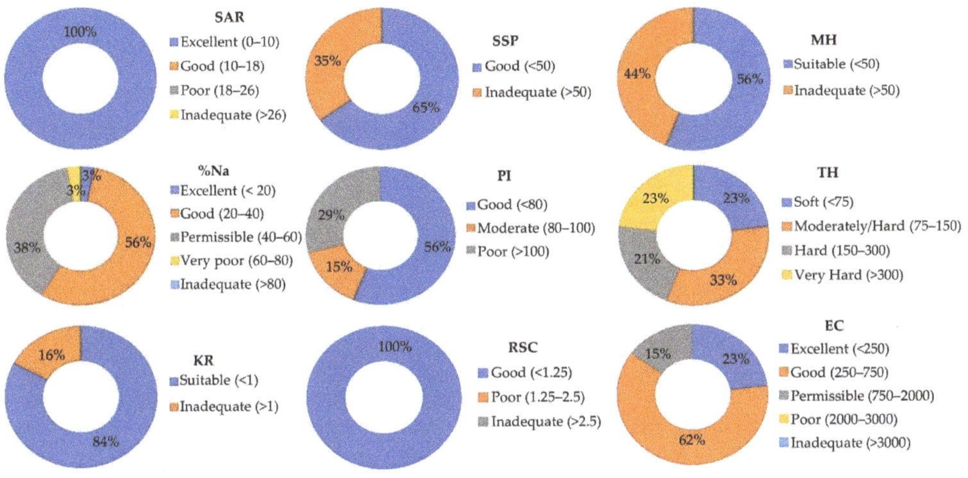

Figure 13. Classification of water samples for irrigation purposes in the Cuernavaca aquifer.

However, other groundwater indices suggest that groundwater quality is inadequate for irrigation. The SSP index showed that 65% of the samples have good water quality, mainly at P2 and P4. However, most of the samples at P1 and P3 showed an inadequate quality. Tefera et al. [31] reported that 53.3% of samples analyzed in alluvial aquifers in the Upper Blue Nile Basin, Ethiopia, could be considered good quality but 46.7% of the samples could be considered unsuitable. The MH index demonstrated that 56% of the samples have adequate quality at P2 and P4, but inadequate at P1 and P3. The high sodium levels found at these sampling sites could be related to weathering of Na-containing basaltic rocks. However, the content of calcium (41.93 mg/L) and magnesium (19.25 mg/L) in the groundwater of the aquifer maintains an equilibrium state. The groundwater of the Cuernavaca aquifer could be considered good quality according to the %Na. Most of the samples (56%) are within 20–40% Na. However, high sodium percentages were recorded in 38% of the samples, mainly at P1. The presence of high levels of sodium could reduce soil permeability. Similar results were obtained when using the PI index. Good water quality was observed in 56% of the samples, but it is noteworthy that 29% of the samples were within the poor-quality range (PI > 100). This groundwater quality index is related to the texture and structure of the soil. Since a high content of ions such as sodium, magnesium, calcium, and bicarbonates were found in the aquifer, the PI index also suggests that the use of groundwater for irrigation could affect the soil permeability [19]. Moreover, the high levels of bicarbonates over calcium and magnesium make groundwater unsuitable for irrigation uses.

The hardness of groundwater varied from soft to very hard. This variation is related to urbanization since soft groundwater was found at P1, characterized by the presence of agricultural areas, with low population density and small settlements, while very hard groundwater was located at P2 which is characterized by a mineralized subsoil. Hardness levels found in this study could be considered normal according to that suggested by Udeshani [49], who reports similar TH values in the groundwater of Sri Lanka.

The electrical conductivity in the Cuernavaca aquifer was found between 90 and 991 μS/cm. Despite this high variation, most of the samples were in a good quality range according to Tutmez's [50] classification (EC level between 0 and 750 mS/cm). The levels of electrical conductivity have been increasing during the last years. This increase could be also related to the loss of vegetation cover due to urbanization [51]. However, electrical conductivity in groundwater showed a satisfactory quality classification because the presence of ions (Ca^{2+}, Mg^{2+}, Na^{+}, K^{+}, Cl^{-}, HCO^{3-}, $SO4^{2+}$, and NO^{3-}) is within the permissible limits according to the standards of the World Health Organization

(WHO). Determining groundwater suitability is important to understand the potential negative impacts of the high content of ions on crop production and mitigate groundwater contamination problems to improve healthy crop production [31].

4. Conclusions

- The hydrological and hydrochemical conditions of the Cuernavaca aquifer were evaluated through the application of water quality indices and statistical techniques. This study provides an approach to the spatial and temporal behavior of an urbanized aquifer and assesses its vulnerability due to population growth.
- This study identified spatial variations between the sampling sites and evidenced the influence of urbanization on groundwater chemistry and quality in the Cuernavaca aquifer. The spatial variation of the chemical elements highlights the vulnerability of the aquifer, especially at P2.
- The time series analysis demonstrated no trend, seasonal, or cyclic patterns in the groundwater quality data. The multivariate statistical analysis showed a high number of correlations between the groundwater parameters. These parameters were grouped based on hierarchical cluster analysis which revealed the main chemical processes controlling groundwater chemistry in the aquifer.
- Most of the parameters (physicochemical and ions) measured in the Cuernavaca aquifer were within the standards allowed by the WHO for irrigation purposes. This situation was confirmed by the water quality indexes since the groundwater of the aquifer was classified as good quality. However, the presence of fecal coliforms, organic matter, and the high content of ions such as Ca^{2+}, Mg^{2+}, and HCO_3^- is an important situation that must be addressed to reduce the vulnerability of the aquifer.
- This study provides an approach to describe the behavior of the hydrochemical features and water quality of the Cuernavaca aquifer and points out the main driving factors for the deterioration of groundwater quality in an aquifer located in an urban area.

Author Contributions: Conceptualization, J.G.L.; formal analysis, S.A.M.-A.; investigation, R.E.-V. and Y.B.-T.; methodology, J.G.L. and V.B.-T.; project administration, V.B.-T.; supervision, Y.B.-T.; validation, J.G.R.-P.; visualization, S.A.M.-A. and A.Q.-C.; writing—original draft, J.G.L. and Y.B.-T.; writing—review and editing, J.G.R.-P. All authors have read and agreed to the published version of the manuscript.

Funding: This research received no external funding.

Institutional Review Board Statement: Not applicable.

Informed Consent Statement: Not applicable.

Data Availability Statement: The data that support the findings of this study are available on request from the corresponding author.

Acknowledgments: The authors kindly acknowledge the National Water Commission for providing data and the Polytechnic University of Morelos for their support.

Conflicts of Interest: The authors declare no conflict of interest.

References

1. El-Bindary, M.A.; El-Desouky, M.G.; El-Bindary, A.A. Adsorption of industrial dye from aqueous solutions onto thermally treated green adsorbent: A complete batch system evaluation. *J. Mol. Liq.* **2021**, *346*, 117082. [CrossRef]
2. Tapia-Villaseñor, E.M.; Shamir, E.; Cruz-Ayala, M.-B.; Megdal, S.B. Assessing Groundwater Withdrawal Sustainability in the Mexican Portion of the Transboundary Santa Cruz River Aquifer. *Water* **2022**, *14*, 233. [CrossRef]
3. Loaiza, J.G.; Rangel-Peraza, J.G.; Sanhouse-García, A.J.; Monjardín-Armenta, S.A.; Mora-Félix, Z.D.; Bustos-Terrones, Y.A. Assessment of Water Quality in A Tropical Reservoir in Mexico: Seasonal, Spatial and Multivariable Analysis. *Int. J. Environ. Res. Public Health* **2021**, *18*, 7456. [CrossRef] [PubMed]
4. Elumalai, V.; Nethononda, V.G.; Manivannan, V.; Rajmohan, N.; Li, P.; Elango, L. Groundwater quality assessment and application of multivariate statistical analysis in Luvuvhu catchment, Limpopo, South Africa. *J. Afr. Earth Sci.* **2020**, *171*, 103967. [CrossRef]

5. Mora, A.; Torres-Martínez, J.A.; Moreau, C.; Bertrand, G.; Mahlknecht, J. Mapping salinization and trace element abundance (including As and other metalloids) in the groundwater of north-central Mexico using a double-clustering approach. *Water Res.* **2021**, *205*, 117709. [CrossRef]
6. Bannenberg, M.; Ntona, M.M.; Busico, G.; Kalaitzidou, K.; Mitrakas, M.; Vargemezis, G.; Fikos, I.; Kazakis, N.; Voudouris, K. Hydrogeological and Hydrochemical Regime Evaluation in Flamouria Basin in Edessa (Northern Greece). *Environments* **2020**, *7*, 105. [CrossRef]
7. Ha, K.; Lee, E.; An, H.; Kim, S.; Park, C.; Kim, G.-B.; Ko, K.-S. Evaluation of Seasonal Groundwater Quality Changes Associated with Groundwater Pumping and Level Fluctuations in an Agricultural Area, Korea. *Water* **2020**, *13*, 51. [CrossRef]
8. Kumar, S.; Kumar, M.; Chandola, V.K.; Kumar, V.; Saini, R.K.; Pant, N.; Kumari, N.; Srivastava, A.; Singh, S.; Singh, R.; et al. Groundwater Quality Issues and Challenges for Drinking and Irrigation Uses in Central Ganga Basin Dominated with Rice-Wheat Cropping System. *Water* **2021**, *13*, 2344. [CrossRef]
9. Goyal, V.; Singh, O.; Singh, R.; Chhoden, K.; Kumar, J.; Yadav, S.; Singh, N.; Shrivastava, N.; Carvalho, L. Ecological health and water quality of village ponds in the subtropics limiting their use for water supply and groundwater recharge. *J. Environ. Manag.* **2020**, *277*, 111450. [CrossRef]
10. Tuqan, N.; Haie, N.; Ahmad, M.T. Assessment of the Agricultural Water Use in Jericho Governorate Using Sefficiency. *Sustainability* **2020**, *12*, 3634. [CrossRef]
11. Haque, C.N.; Haque, M.; Jana, H.; Basu, D.; Karak, S. Socio Economic Factors Responsible for Groundwater Consumption in Purba Bardhaman District of West Bengal. *Int. J. Environ. Clim. Chang.* **2021**, *11*, 187–194. [CrossRef]
12. Šperac, M.; Zima, J. Assessment of the Impact of Climate Extremes on the Groundwater of Eastern Croatia. *Water* **2022**, *14*, 254. [CrossRef]
13. Liu, L.; Qi, S.; Wang, W. Groundwater Quality in Agricultural Lands Near a Rapidly Urbanized Area, South China. *Int. J. Environ. Res. Public Health* **2021**, *18*, 1783. [CrossRef]
14. Adimalla, N.; Qian, H. Groundwater quality evaluation using water quality index (WQI) for drinking purposes and human health risk (HHR) assessment in an agricultural region of Nanganur, south India. *Ecotoxicol. Environ. Saf.* **2019**, *176*, 153–161. [CrossRef]
15. Loh, Y.S.A.; Akurugu, B.A.; Manu, E.; Aliou, A.S. Assessment of groundwater quality and the main controls on its hydro-chemistry in some Voltaian and basement aquifers, northern Ghana. *Groundw. Sustain. Dev.* **2020**, *10*, 100296. [CrossRef]
16. Zakaria, N.; Anornu, G.; Adomako, D.; Owusu-Nimo, F.; Gibrilla, A. Evolution of groundwater hydrogeochemistry and assessment of groundwater quality in the Anayari catchment. *Groundw. Sustain. Dev.* **2020**, *12*, 100489. [CrossRef]
17. Wisitthammasri, W.; Chotpantarat, S.; Thitimakorn, T. Multivariate statistical analysis of the hydrochemical characteristics of a volcano sedimentary aquifer in Saraburi Province, Thailand. *J. Hydrol. Reg. Stud.* **2020**, *32*, 100745. [CrossRef]
18. Badeenezhad, A.; Tabatabaee, H.R.; Nikbakht, H.-A.; Radfard, M.; Abbasnia, A.; Baghapour, M.A.; Alhamd, M. Estimation of the groundwater quality index and investigation of the affecting factors their changes in Shiraz drinking groundwater, Iran. *Groundw. Sustain. Dev.* **2020**, *11*, 100435. [CrossRef]
19. El Osta, M.; Masoud, M.; Alqarawy, A.; Elsayed, S.; Gad, M. Groundwater Suitability for Drinking and Irrigation Using Water Quality Indices and Multivariate Modeling in Makkah Al-Mukarramah Province, Saudi Arabia. *Water* **2022**, *14*, 483. [CrossRef]
20. Comisión Nacional del Agua (CONAGUA). 2020. Available online: https://sigagis.conagua.gob.mx/gas1/Edos_Acuiferos_18/morelos/DR_1701.pdf (accessed on 18 December 2021).
21. Comisión Nacional del Agua (CONAGUA). Estadísticas del Agua en México 2018. Available online: https://sina.conagua.gob.mx/publicaciones/EAM_2018.pdf (accessed on 4 December 2021).
22. Comisión Nacional del Agua (CONAGUA). Calidad del Agua en México. Available online: https://www.gob.mx/conagua/articulos/calidad-del-agua (accessed on 8 December 2021).
23. APHA. *Standard Methods for the Examination of Water and Wastewater*, 22nd ed.; American Public Health Association: Washington, DC, USA, 2012.
24. Chen, X.; Zhang, K.; Chao, L.; Liu, Z.; Du, Y.; Xu, Q. Quantifying natural recharge characteristics of shallow aquifers in groundwater overexploitation zone of North China. *Water Sci. Eng.* **2021**, *14*, 184–192. [CrossRef]
25. Bartier, P.M.; Keller, C. Multivariate interpolation to incorporate thematic surface data using inverse distance weighting (IDW). *Comput. Geosci.* **1996**, *22*, 795–799. [CrossRef]
26. El Baba, M.; Kayastha, P.; Huysmans, M.; De Smedt, F. Evaluation of the Groundwater Quality Using the Water Quality Index and Geostatistical Analysis in the Dier al-Balah Governorate, Gaza Strip, Palestine. *Water* **2020**, *12*, 262. [CrossRef]
27. Acharya, S.; Sharma, S.; Khandegar, V. Assessment of groundwater quality by water quality indices for irrigation and drinking in South West Delhi, India. *Data Brief* **2018**, *18*, 2019–2028. [CrossRef]
28. WHO. *Guidelines for Drinking Water Quality*, 4th ed.; World Health Organization: Geneva, Switzerland, 2011.
29. Gabr, M.E.; Soussa, H.; Fattouh, E. Groundwater quality evaluation for drinking and irrigation uses in Dayrout city Upper Egypt. *Ain Shams Eng. J.* **2020**, *12*, 327–340. [CrossRef]
30. Saikrishna, K.; Purushotham, D.; Sunitha, V.; Reddy, Y.S.; Linga, D.; Kumar, B.K. Data for the evaluation of groundwater quality using water quality index and regression analysis in parts of Nalgonda district, Telangana, Southern India. *Data Brief* **2020**, *32*, 106235. [CrossRef]

31. Tefera, A.K.; Wassie, A.B.; Sinshaw, B.G.; Defersha, D.T.; Takele, T.A.; Atanaw, S.B.; Tesfaye, A.T.; Getu, E.; Fenta, H.M.; Atinkut, H.B.; et al. Groundwater quality evaluation of the alluvial aquifers using GIS and water quality indices in the Upper Blue Nile Basin, Ethiopia. *Groundw. Sustain. Dev.* **2021**, *14*, 100636. [CrossRef]
32. Hasan, M.N.; Siddique, M.A.B.; Reza, A.S.; Khan, R.; Akbor, M.A.; Elius, I.B.; Hasan, M. Vulnerability assessment of seawater intrusion in coastal aquifers of southern Bangladesh: Water quality appraisals. *Environ. Nanotechnol. Monit. Manag.* **2021**, *16*, 100498. [CrossRef]
33. Doza, B.; Islam, A.T.; Ahmed, F.; Das, S.; Saha, N.; Rahman, M.S. Characterization of groundwater quality using water evaluation indices, multivariate statistics and geostatistics in central Bangladesh. *Water Sci.* **2016**, *30*, 19–40. [CrossRef]
34. Abdelaziz, S.; Gad, M.I.; El Tahan, A.H.M. Groundwater quality index based on PCA: Wadi El-Natrun, Egypt. *J. Afr. Earth Sci.* **2020**, *172*, 103964. [CrossRef]
35. Gaikwad, S.K.; Kadam, A.K.; Ramgir, R.R.; Kashikar, A.S.; Wagh, V.; Kandekar, A.M.; Gaikwad, S.P.; Madale, R.B.; Pawar, N.J.; Kamble, K.D. Assessment of the groundwater geochemistry from a part of west coast of India using statistical methods and water quality index. *HydroResearch* **2020**, *3*, 48–60. [CrossRef]
36. Abbasnia, A.; Yousefi, N.; Mahvi, A.H.; Nabizadeh, R.; Radfard, M.; Yousefi, M.; Alimohammadi, M. Evaluation of ground-water quality using water quality index and its suitability for assessing water for drinking and irrigation purposes: Case study of Sistan and Baluchistan province (Iran). *Hum. Ecol. Risk Assess.* **2019**, *25*, 988–1005. [CrossRef]
37. Hossain, M.; Patra, P.K.; Begum, S.N.; Rahaman, C.H. Spatial and sensitivity analysis of integrated groundwater quality index towards irrigational suitability investigation. *Appl. Geochem.* **2020**, *123*, 104782. [CrossRef]
38. Jamaa, H.; El Achheb, A.; Namr, K.I. Spatial variation of groundwater quality and assessment of water table fluctuations in Plio-Quaternary aquifer formations in Doukkala Plain, Morocco. *Groundw. Sustain. Dev.* **2020**, *11*, 100398. [CrossRef]
39. Trigui, M.R.; Trabelsi, R.; Zouari, K.; Agoun, A. Implication of hydrogeological and hydrodynamic setting on water quality of the Complex Terminal Aquifer in Kebili (southern Tunisia): The use of geochemical indicators and modelling. *J. Afr. Earth Sci.* **2021**, *176*, 104121. [CrossRef]
40. Miao, Q.; Li, X.; Xu, Y.; Liu, C.; Xie, R.; Lv, Z. Chemical characteristics of groundwater and source identification in a coastal city. *PLoS ONE* **2021**, *16*, e0256360. [CrossRef]
41. Ahmed, S.; Khurshid, S.; Madan, R.; Abu Amarah, B.A.; Naushad, M. Water quality assessment of shallow aquifer based on Canadian Council of Ministers of the environment index and its impact on irrigation of Mathura District, Uttar Pradesh. *J. King Saud Univ. Sci.* **2019**, *32*, 1218–1225. [CrossRef]
42. Pradhan, R.M.; Behera, A.K.; Kumar, S.; Kumar, P.; Biswal, T.K. Recharge and Geochemical Evolution of Groundwater in Fractured Basement Aquifers (NW India): Insights from Environmental Isotopes (δ18O, δ2H, and 3H) and Hydrogeochemical Studies. *Water* **2022**, *14*, 315. [CrossRef]
43. Dorjderem, B.; Torres-Martínez, J.; Mahlknecht, J. Intensive long-term pumping in the Principal-Lagunera Region aquifer (Mexico) causing heavy impact on groundwater quality. *Energy Rep.* **2020**, *6*, 862–867. [CrossRef]
44. Badmus, G.O.; Akinyemi, O.D.; Gbadebo, A.M.; Oyedepo, J.A. Hydrochemical analysis of groundwater quality along the coastal aquifers in part of Ogun Waterside, Ogun State, southwestern Nigeria. *Heliyon* **2020**, *6*, e05661. [CrossRef]
45. Dragon, K. Identification of groundwater conditions in the recharge zone of regionally extended aquifer system with use of water chemistry and isotopes (Lwowek region, Poland). *J. Hydrol. Reg. Stud.* **2021**, *34*, 100787. [CrossRef]
46. Abotalib, A.Z.; Heggy, E.; El Bastawesy, M.; Ismail, E.; Gad, A.; Attwa, M. Groundwater mounding: A diagnostic feature for mapping aquifer connectivity in hyper-arid deserts. *Sci. Total Environ.* **2021**, *801*, 149760. [CrossRef]
47. Adimalla, N.; Qian, H.; Li, P. Entropy water quality index and probabilistic health risk assessment from geochemistry of groundwaters in hard rock terrain of Nanganur County, South India. *Geochemistry* **2019**, *80*, 125544. [CrossRef]
48. Gad, M.; Elsayed, S.; Moghanm, F.; Almarshadi, M.; Alshammari, A.; Khedher, K.; Eid, E.; Hussein, H. Combining Water Quality Indices and Multivariate Modeling to Assess Surface Water Quality in the Northern Nile Delta, Egypt. *Water* **2020**, *12*, 2142. [CrossRef]
49. Udeshani, W.; Dissanayake, H.; Gunatilake, S.; Chandrajith, R. Assessment of groundwater quality using water quality index (WQI): A case study of a hard rock terrain in Sri Lanka. *Groundw. Sustain. Dev.* **2020**, *11*, 100421. [CrossRef]
50. Tutmez, B.; Hatipoglu, Z.; Kaymak, U. Modelling electrical conductivity of groundwater using an adaptive neuro-fuzzy inference system. *Comput. Geosci.* **2006**, *32*, 421–433. [CrossRef]
51. Lü, X.; Han, Z.; Li, H.; Zheng, Y.; Liu, J. Influence of Urbanization on Groundwater Chemistry at Lanzhou Valley Basin in China. *Minerals* **2022**, *12*, 385. [CrossRef]

applied sciences

MDPI

Article

Groundwater Quality Monitoring Using In-Situ Measurements and Hybrid Machine Learning with Empirical Bayesian Kriging Interpolation Method

Delia B. Senoro [1,2,3,4,*], Kevin Lawrence M. de Jesus [2,3,4], Leonel C. Mendoza [4,5], Enya Marie D. Apostol [4,5], Katherine S. Escalona [4,6] and Eduardo B. Chan [7]

1 School of Civil, Environmental and Geological Engineering, Mapua University, Intramuros, Manila 1002, Philippines
2 School of Graduate Studies, Mapua University, Intramuros, Manila 1002, Philippines; klmdejesus@mymail.mapua.edu.ph
3 School of Chemical, Biological, Materials Engineering and Sciences, Mapua University, Intramuros, Manila 1002, Philippines
4 Resiliency and Sustainable Development Center, Yuchengco Innovation Center, Mapua University, Manila 1002, Philippines; lcmendoza@minscat.edu.ph (L.C.M.); emddapostol@minscat.edu.ph (E.M.D.A.); kpsescalona@minscat.edu.ph (K.S.E.)
5 College of Teacher Education, Mindoro State University, Calapan 5200, Oriental Mindoro, Philippines
6 College of Arts and Sciences, Mindoro State University, Victoria 5205, Oriental Mindoro, Philippines
7 Dyson College of Arts and Sciences, Pace University, New York, NY 10038, USA; echan@pace.edu
* Correspondence: dbsenoro@mapua.edu.ph; Tel.: +63-2-8251-6622

Featured Application: In-Situ and Hybrid Machine Learning—Geostatistical Interpolation method for groundwater quality monitoring applications.

Abstract: This article discusses the assessment of groundwater quality using a hybrid technique that would aid in the convenience of groundwater (GW) quality monitoring. Twenty eight (28) GW samples representing 62 barangays in Calapan City, Oriental Mindoro, Philippines were analyzed for their physicochemical characteristics and heavy metal (HM) concentrations. The 28 GW samples were collected at suburban sites identified by the coordinates produced by Global Positioning System Montana 680. The analysis of heavy metal concentrations was conducted onsite using portable handheld X-Ray Fluorescence (pXRF) Spectrometry. Hybrid machine learning—geostatistical interpolation (MLGI) method, specific to neural network particle swarm optimization with Empirical Bayesian Kriging (NN-PSO+EBK), was employed for data integration, GW quality spatial assessment and monitoring. Spatial map of metals concentration was produced using the NN-PSO-EBK. Another, spot map was created for observed metals concentration and was compared to the spatial maps. Results showed that the created maps recorded significant results based on its MSEs with values such as 1.404×10^{-4}, 5.42×10^{-5}, 6.26×10^{-4}, 3.7×10^{-6}, 4.141×10^{-4} for Ba, Cu, Fe, Mn, Zn, respectively. Also, cross-validation of the observed and predicted values resulted to R values range within 0.934–0.994 which means almost accurate. Based on these results, it can be stated that the technique is efficient for groundwater quality monitoring. Utilization of this technique could be useful in regular and efficient GW quality monitoring.

Keywords: groundwater; heavy metals; physicochemical parameters; in-situ; machine learning; geostatistical analysis

Citation: Senoro, D.B.; de Jesus, K.L.M.; Mendoza, L.C.; Apostol, E.M.D.; Escalona, K.S.; Chan, E.B. Groundwater Quality Monitoring Using In-Situ Measurements and Hybrid Machine Learning with Empirical Bayesian Kriging Interpolation Method. *Appl. Sci.* **2022**, *12*, 132. https://doi.org/10.3390/app12010132

Academic Editors: Amit Kumar, Santosh Subhash Palmate and Rituraj Shukla

Received: 5 December 2021
Accepted: 17 December 2021
Published: 23 December 2021

1. Introduction

Water quality is associated with ecosystem preservation, economic growth and social development [1]. Groundwater (GW) quality is critical to the Philippines' overall water resource; hence, monitoring should be given attention. Population expansion and the acceleration of modernization as well as industrialization have resulted in an increased water demand [2]. It is inevitable that quality of both surface water and GW is compromised in areas where economy is in transition and where there is increasing urbanization, industrialization, and agricultural activities [3]. However, due to scarcity of elemental laboratory instruments in the Philippines, the preparation requirements of the laboratory station-based instruments for samples, and the travel time from sampling sites to laboratory stations become a challenge to GW quality monitoring especially to elemental concentration analysis.

The population of the Philippines has rice in its regular daily meals. Rice fields and techniques to produce high and good quality yields are among the agricultural areas and programs, respectively, that are being supported by the Philippine government. Calapan City, Oriental Mindoro province in the Philippines, with GW as primary water source, is among the top producers of rice in the country. Rice fields require highly flattened area and good water quality as among the criteria to increase yields with good rice quality. Having the Philippines within the tropical storm belt, Calapan City experienced regular flooding as shown in Figure 1. This condition become a challenge to the water quality for rice fields and for domestic supply.

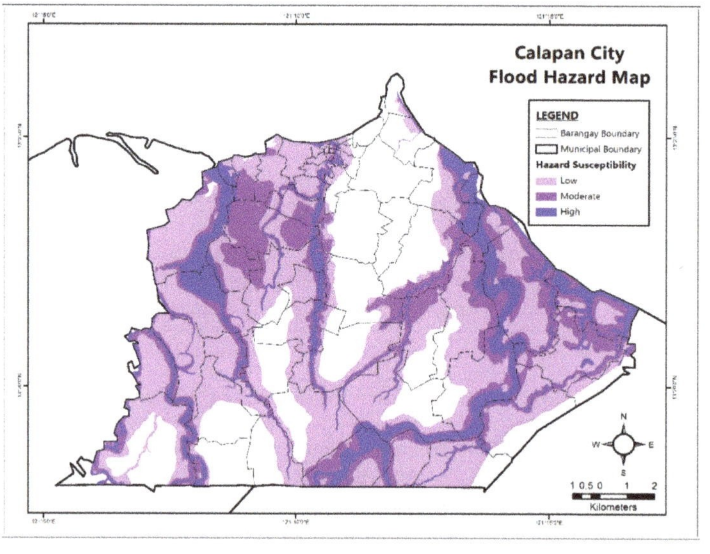

Figure 1. Flood Hazard Map of Calapan City [4].

The Philippines is rich in natural resources such as metals and non-metallic minerals [5] with tropical climate where annual rainfall is high. However, anthropogenic activities, due to economic development, population positive growth rate and urbanization, caused unintentional adverse effects to the environment. When the pristine environment is disturbed, and minerals are exposed to oxygen and water, chemical reactions happen. Similar condition takes place during weathering. Having rich in mineral resources, high annual rainfall, large flattened agricultural areas, and GW as primary source of water supply, GW quality monitoring is important. Water quality that has elevated concentration of metals such as arsenic (As), copper (Cu), iron (Fe) nickel (Ni), manganese (Mn), lead (Pb), and other metals known for its toxicity characteristics would have acute and/or chronic adverse effects to human health [6–8].

Several technologies for measurement of the presence of metals in GW exists such as Inductively Coupled Plasma—Mass Spectrometry (ICP—MS), Inductively Coupled Plasma—Optical Emission Spectrometry (ICP—OES), and Atomic Absorption Spectrometry (AAS). These approaches are laboratory-based and require several days before result of analysis is available. This condition become not suitable for field monitoring and measurements [9] in sub-urban, rural areas, areas where access is a challenge and when time is a primary criterion in the analysis. These conditions require in-situ measurements, and accurate detection become a critical component in monitoring the GW quality. In-situ measurements provide observations on a rapid phase as well as covering the wider areas especially those with difficulty in access. This is in contrast of laboratory—based methods with significant limitations such as expensive instrumentation making limited availability, complex sample preparation and applicability in field conditions [10]. There are on-site detection and monitoring techniques such as electrochemical analysis [10,11], cyclic voltammetry (CV) [12], anodic stripping voltammetry (ASV) [13], square wave anodic stripping voltammetry (SWASV) [14], electro-chemical impedance spectroscopy (EIS) [15], electrochemiluminescence (ECL) [16] and the use of piezoelectric biosensors [17]. However, these techniques have drawbacks such as background noise control, unable to fulfill the current requirements for selectivity [9], detection limits of CV [12], insolubility of metals and the multiple peaks of ASV [13], complicated interferences and complex matrices in SWASV [14], EIS's inability to identify different ions [15], frequent fouling of electrodes in the case of ECL [16], and only few enzymes are sensitive to heavy metals for the case of biosensors [17]. Hence, the use of portable x-ray fluorescence spectrometry (pXRF) technique in onsite metals detection and analysis is appropriate in rugged condition yet provides user of accurate and rapid analysis. This is in contrast of laboratory-based methods with significant limitations such as expensive instrumentation, complex sample preparation and applicability in field conditions [18]. Therefore, non-destructive analytical technique, such as relatively simple spectra line void of many interferences and rapid multi-element analyses [19], contributes significantly to the successful implementation of this study.

Concentration maps were frequently used tool for spatial monitoring. Spatial information in water resources are limited, and GW quality data can be obtained only through spot sampling. However, this procedure often requires extensive manpower and resources [10]. The issue in this practice and the determination of sample locations density influences the accuracy of the generated spatial maps [19]. The integration of in–situ measurements and GIS—based spatial interpolation techniques offer an improvement in the presentation and display of the status of GW quality in an area. The use of this integrated approach provides a clear and intelligent base—maps which can be utilized by researchers, policy makers, implementors for planning proposals [20] and creation of strategic programs.

Several studies on GW monitoring and assessment implemented using GIS—based approaches focused on different Southeast Asian countries such as the Philippines [1,21], Thailand [22], Malaysia [23], Singapore [24], Indonesia [25], Cambodia [26], Laos [27], and Vietnam [28]. This article illustrates the quantification and mapping of the concentrations of heavy metals such as Ba, Cu, Fe, Mn, and Zn in GW and presents the utilization of in-situ GW quality monitoring that uses a hybrid machine learning-geospatial interpolation technique and pXRF. This type of technique and analyses give prompt, accurate data and information on the current GW quality. This is to address the challenges encountered on-ground during sampling activities, the scarcity of instruments in the Philippines due to its price, and the complex samples preparation required by some laboratory-based instruments. Analyzing heavy metal concentrations in a faster, accurate and convenient method can help the researchers, authorities, water utility companies and local government units in making prompt decision, guidelines and strategic programs.

2. Materials and Methods

2.1. Description of the Study Area

The study area is Calapan City, in the province of Oriental Mindoro, Philippines. This is a third-class city and one of only two cities in the MIMAROPA region of the Philippines. It is the capital of the island province of Mindoro and located on the island's northeastern shore. It has a population of about 150,000 people (about 25,000 households) and 62 barangays (the smallest local government unit) [29]. The city lies within 13°22′ N Latitude and 121°9′ E Longitude and Mindoro is located approximately 13°11′ N Latitude and 121°53′ E Longitude south of Mainland Luzon. The island of Mindoro is popularly known of rice production. Calapan City has an area of 217.30 square kilometers. Deep and shallow wells, in addition to piped water supply, are currently the primary sources of water in the city.

2.2. *Collection/Treatment of Groundwater Samples*

The GW samples were collected from twenty-eight (28) suburban deep and shallow well sites following the USEPA SESDPROC-301-R3/SESDPROC-111-R4 [30] as shown in Figure 2. The GW samples were collected using stainless steel sampler and polyethylene (PE) bottles. The PE bottles were thoroughly pre-washed with Type 1 water. Each PE bottle was carefully labeled, sealed, and placed temporarily in coolers for metals concentration detection. This is in preparation for the detection of the presence of metals concentration in all collected GW samples.

Figure 2. Map of the Study Area and Sampling Sites.

2.3. *Physicochemical and Metal Concentrations Analysis*

Temperature, pH, electric conductivity (EC), and total dissolved solids (TDS) of the samples were determined onsite using a multi-parameter water analyzer (HANNA HI 9811-5) with HI1285-5 probe (electrode) and HI7007, HI70031, HI70032, buffer solutions for calibration [31]. The HI7007, HI70031, HI70032 solution were used for pH, EC and TDS calibration, respectively. While HI700661 solution was used for cleaning the electrode. The physicochemical values detected in groundwater were compared to the permissible limits specified in the 2017 Philippine National Standards for Drinking Water (PNSDW) [32] and the WHO Drinking Water Guidelines [33,34]. These water parameters were used in the hybrid machine learning technique.

The heavy metal concentration analysis employed the use of portable handheld Olympus Vanta X-Ray Fluorescence Spectrometry. This pXRF is a rapid onsite accurate elemental analyzer that could be used for various environmental media including water [35–39]. The pXRF was set on geochem mode and recorded metals concentration in ppm (mg/L) detected from GW samples. Target metals were Ba, Cu, Fe, Mn and Zn.

2.4. Spatial Concentration Mapping Using Machine Learning Informed Empirical Bayesian Kriging (EBK) Method

Hybrid machine learning—geostatistical interpolation (MLGI) method was employed. The EBK technique was used to produce spatial concentration maps of the GW's physicochemical properties and heavy metal concentrations. By sub-setting and replicating observed data, the EBK automates the most time-consuming aspects of constructing a viable kriging model. EBK provides a distribution of semi-variogram models and compensates for semi-variogram estimate uncertainty. The EBK is more realistic and superior to other current geostatistical modeling methods owing to its dependence on limited maximum likelihood estimation. This is in contrast to other existing kriging models that rely on weighted least squares estimation. EBK has many significant benefits, including a low need for interactive modeling, more accurate prediction of standard error and projection for small datasets as compared to other traditional kriging techniques, and exact prediction of substantially non-static data [40].

The Artificial Neural Network (ANN) approach is a subset of methods for artificial intelligence inspired by biological neurons. It is capable of quickly acquiring patterns and forecasting the result of a problem in a multi-dimensional environment. ANN models are trained using datasets [41] to show the efficacy. The training algorithm and the transfer function that was utilized in the model are two critical components of the ANN model. The Levenberg—Marquardt (LM) algorithm was chosen as the training algorithm since it is the quickest function for training a network, and the hyperbolic tangent sigmoid function was used as the transfer function because it is the recommended transfer function for rapid processes [42,43].

Particle Swarm Optimization (PSO) is a population-based stochastic optimization technique inspired by biological communities' collaborative nature. The PSO is initiated using a community of randomly generated particles as solution options. It looks for global optima via iterations in which particles with their own velocity fly around the search space following the current optimum particles, which is the best approach for finding the best solution. The PSO was integrated to the ANN to determine the weights and biases which gives the minimum error [44].

This hybrid technique integrated to the EBK method generated the spatial concentration maps of the target study area. The Neural Network—Particle Swarm Optimization (NN-PSO) approach was applied to generate the spatial concentration maps of physicochemical parameters and HM concentrations.

3. Results

Subsequent sections elaborate the results of the study and in comparison, of the WHO and PNSDW guidelines.

3.1. Physicochemical Groundwater Parameters

The recorded physical and chemical properties of groundwater of the 28 sampling points are shown in Table 1, and in comparison to WHO (2017) and PNSDW (2017) guidelines. The detailed description of the sampling locations of the study area with the physicochemical properties of GW were exhibited in Table A1 of Appendix A.

Table 1. The groundwater physical and chemical properties.

Sampling No.	Temperature (°C)	pH	EC (μS/cm)	TDS (ppm)
1	28.3	7.9	130	60
2	28.5	8.4	130	60
3	26.2	8.5	120	50
4	29.2	7.6	660	320
5	30.7	7.8	1200	590
6	27.1	8.8	130	50
7	27.2	7.9	200	90
8	32.4	7.1	900	440
9	31.6	7.5	350	160
10	32.5	7.5	970	480
11	31.4	7.4	1820	900
12	31.1	7.1	780	380
13	31.0	7.0	990	490
14	32.0	7.5	600	290
15	28.1	7.5	220	100
16	31.5	7.3	570	270
17	30.6	7.3	820	400
18	29.2	8.1	690	340
19	30.2	7.4	410	200
20	30.1	7.6	140	160
21	28.4	8.3	180	150
22	33.6	7.7	910	450
23	28.7	8.3	140	60
24	27.4	8.4	140	60
25	32.9	7.6	750	370
26	29.7	7.3	500	240
27	29.7	7.7	100	40
28	30.3	6.7	1140	560
WHO [45]	30.0	6.5–8.5	400	1000
PNSDW [46]		6.5–8.5	-	600

The GW temperatures ranged from 26.2 to 33.6 degree Celsius which could lead to increased release rates of metals concentration especially within the water temperature of 30–35 degrees Celsius [47]. Furthermore, the study of Zhu, et al. in 2010 [48] attributed the high temperature in GW to the boom of urbanization that was also observed in the City of Calapan. The recorded pH of GW ranged from 6.7 to 8.8 which is within the pH range guidelines set by the WHO and PNSDW [49]. The release rates of metals were affected by a lower water pH. Lower pH of water means acidic water and known to be aggressive, enhancing the breakdown of Fe and Mn resulting in an unpleasant taste in water [50]. This condition could have adverse effects including heavy metal poisoning and toxicity [51–53]. The majority of the water samples are slightly basic which could be attributed to the existence of carbonates and bicarbonates [54]. The TDS and EC range recorded was 40–900 ppm and 100–1820 μS/cm, respectively. The TDS in GW found to be below WHO guidelines; however, beyond PNSDW guidelines. The TDS and EC had been found to have positive correlation [55]. The elevated EC of 1820 μS/cm has been attributed to inorganic chemicals in ionized form in water [54] such as metal elements.

3.2. Heavy Metal Concentrations

Presence of heavy metals were investigated in GW samples collected from the 28 sampling sites as indicated in Figure 2. Detected concentrations were compared to the existing maximum allowable limit of the WHO and the PNSDW 2017. These limits are enumerated in Table 2. The toxicants found in the GW samples are discussed in more detail in the subsections below.

Table 2. Permissible limits of metals in groundwater.

Parameter, mg/L	WHO	USEPA (2009)	PNSDW 2017
Ba	0.7 [56]	2.00	0.70
Cu	1.30 [45]	1.30	1.00
Fe	-	0.30	1.00
Mn	0.40 [1]	0.05	0.40
Zn	-	5.00	-

[1] Guidance value.

3.2.1. Barium

All sampling locations observed Ba concentrations below permissible limits (Figure 3) of WHO, USEPA and PNSDW. The presence of Ba in GW has been attributed to the weathering of rocks such as igneous rocks, sandstone, shale, and coal [57,58].

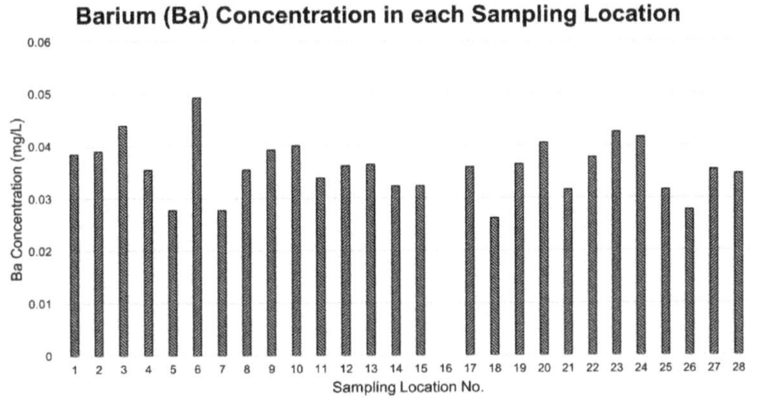

Figure 3. Concentration of Ba in GW samples.

3.2.2. Copper

The Cu concentrations (Figure 4) in GW samples from all sampling locations did not exceed the WHO guideline of 1.3 mg/L. The Cu in trace amounts in GW was associated to the kind of rock that forms the aquifer [59]. Another possible source of copper is the pipeline. Also, Cu concentrations at all sampling sites were within the acceptable limit of PNSDW (1 mg/L) and WHO guidelines.

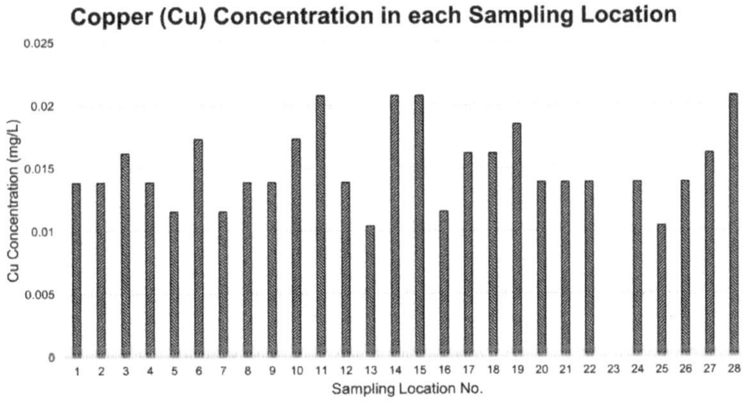

Figure 4. Concentration of Cu in GW samples.

3.2.3. Iron

Iron stains laundry and plumbing fixtures at concentrations more than 0.3 mg/L; it can also give metallic taste [60]. Hence, USEPA set an allowable concentration limit of 0.3 mg/L. The majority of the Fe in GW comes from minerals and sediments which may be in the form of particulate or dissolved [61]. The Fe concentration in each sampling location is presented in Figure 5. Sampling location 8 recorded an elevated Fe concentration. This was attributed to a longer residence time [62] which is associated to the type of subsurface (aquifer) that promotes longer residence time and creates opportunity for metals to react through chemical and physical weathering [63]. In addition, the area shown in Figure 2 illustrates the area of sampling point 8 of having lesser active wells. This condition also contributes to longer residence time of GW.

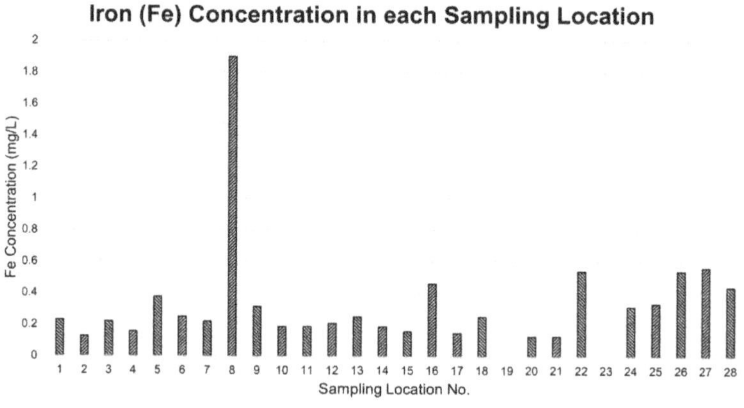

Figure 5. Concentration of Fe in GW samples.

3.2.4. Manganese

Groundwater samples collected from all sampling locations did not exceed the WHO's maximum permissible level for Mn concentration (Figure 6). The natural occurrence of Mn in GW can be influenced by several factors including TDS, GW level fluctuations, and the residence time. Agricultural operations and domestic wastewater are additional potential two sources of Mn that can adversely affect the GW quality [64,65].

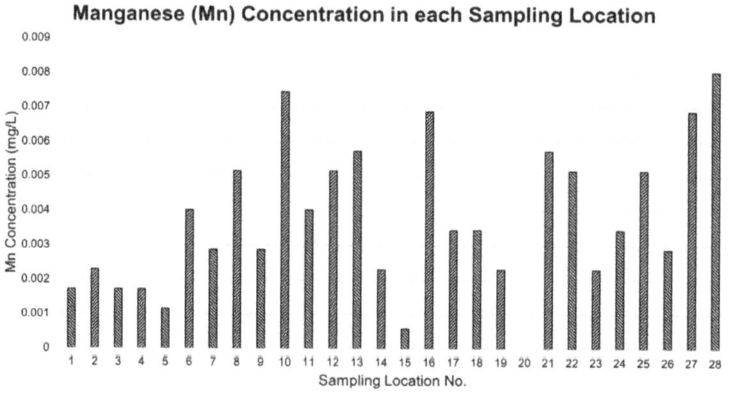

Figure 6. Concentration of Mn in GW samples.

3.2.5. Zinc

The highest concentration of Zn was recorded at Brgy. Gutad. However, this highest Zn concentration was within the WHO and PNSDW permissible limits. Several locations observed without Zn concentrations were at Brgy. Balingayan, Brgy. Maidlang, Brgy. Managpi, and Brgy. Personas. Zinc is naturally found in GW and the acidity affects the quality [64]; hence, it is important that monitoring is carried out. The acidity theory states that the higher the acidity (i.e., lower pH) of the water, the higher the Zn concentration. As observed in Table 1, GW samples from all sampling sites were slightly basic which explains the low concentration levels of Zn. The Zn concentration in each sampling location is presented in Figure 7.

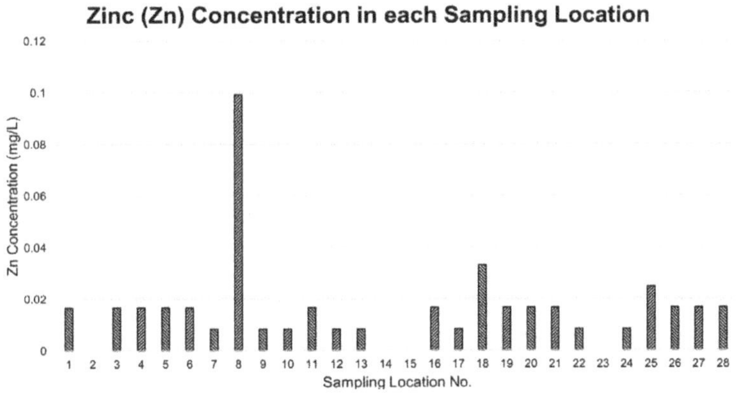

Figure 7. Concentration of Zn in GW samples.

3.3. *Correlation Analysis*

The correlations between the physicochemical characteristics were investigated using Pearson correlation analysis calculated through International Business Machine Statistical Package for Social Sciences (IBM SPSS). The r and p values were presented. The r value expresses the relationship between variables. The p value expresses the significance of the relationship. A lower *p*-value denotes statistical significance, whereas a higher *p*-value denotes the opposite. A negative correlation was found between pH and the other variables, while a positive correlation was found between the other parameters. At the 0.01 p level, all relationships were significant. The correlation matrix for the physicochemical parameters are presented in Table 3.

Table 3. Correlation matrix of the physicochemical parameters.

Parameter	Temp	pH	EC	TDS
Temp	1.000			
pH	−0.665 **	1.000		
EC	0.657 **	−0.602 **	1.000	
TDS	0.664 **	−0.603 **	0.995 **	1.000

** Correlation is significant at the 0.01 level (2-tailed).

A substantial negative correlation was observed between pH and temperature (r = −0.665), pH and EC (r = −0.602), and pH and TDS (r = −0.603) which is similar to the findings of Abou Zakhem et al. in 2017 [66] and Sunkari and Abu in 2019 [67]. On the other hand, a substantial positive correlation was observed between temperature and EC (r = 0.657), temperature and TDS (r = 0.664), and EC and TDS (r = 0.995). This correlation values agreed to the findings of Wali et al. in 2021 [68].

Similar to physicochemical parameters, Pearson correlation analysis for the relationships between Ba, Cu, Fe, Mn, and Zn was also taken. Fe was positively correlated with Mn and Zn; and Mn was positively correlated with Zn. Positive substantial correlations between these metals indicated the same origin, are mutually dependent, and have similar transport characteristics [69]. The positive p with higher r values of this study illustrates relationship between metals; however, this relationship was not significant. The presence of these metals in GW is attributed to natural weathering of rocks. The correlation matrix for metals concentrations is shown in Table 4.

Table 4. Correlation matrix for the heavy metal concentration of groundwater samples.

Metal	Ba	Cu	Fe	Mn	Zn
Ba	1				
Cu	0.055	1			
Fe	−0.136	−0.001	1		
Mn	−0.235	0.072	0.320	1	
Zn	−0.089	0.013	0.870 **	0.190	1

** Correlation is significant at the 0.01 level (2-tailed).

3.4. Spatial Concentration Mapping Using NN-PSO + EBK

The NN-PSO simulation was applied to accelerate the performance of the prediction capability of the EBK method. The simulation showed an excellent result as evident to the mean squared error (MSE) and correlation coefficient (R) values wherein the ideal value is 0 and 1, respectively. The NN-PSO simulation performed for the physicochemical parameters and heavy metal concentrations are presented in Table 5. Correlation plots of the R (validation) and R (testing) for the governing NN-PSO models of physicochemical parameters and heavy metals concentration are illustrated as Figure A1 of Appendix B.

Table 5. NN-PSO Simulation Results.

	Hidden Neurons	No. of Particles	No. of Iterations	Elapsed Time (sec)	MSE	R	
						Validation	Testing
Temp	25	3	2000	180.38559	0.01204	0.99878	0.99656
pH	29	10	2000	119.16085	0.00434	0.99078	0.99039
EC	30	1	2000	163.34882	0.00155	0.99851	0.99966
TDS	27	3	2000	337.90592	0.00032	0.99925	0.99981
Ba	29	1	2000	172.62202	2.44×10^{-6}	0.98838	0.99286
Cu	29	3	2000	366.12861	1.64×10^{-7}	0.99826	0.99636
Fe	28	3	2000	139.78693	0.00091	0.99010	0.99951
Mn	29	5	2000	115.26285	1.34×10^{-7}	0.98252	0.99725
Zn	30	1	2000	142.60618	1.08×10^{-5}	0.97945	0.99580

The relationship between the number of neurons ranging from 1 to 30 and the corresponding AIC (Akaike Information Criterion) values obtained for the physicochemical parameters (temperature, pH, EC, and TDS) as well as the heavy metal concentrations are exhibited in Figures 8 and 9, respectively. These figures represent the AIC values of all NN-PSO models for each hidden neuron that was simulated in this study. It was observed that the best models for the physicochemical parameters were determined from the 25, 29, 30, and 27 hidden neurons (HN) for temperature, pH, EC and TDS, respectively. The best models for Ba, Cu, Fe, Mn, and Zn were observed in 29, 29, 28, 29, 30 HN, respectively, for the heave metal concentrations.

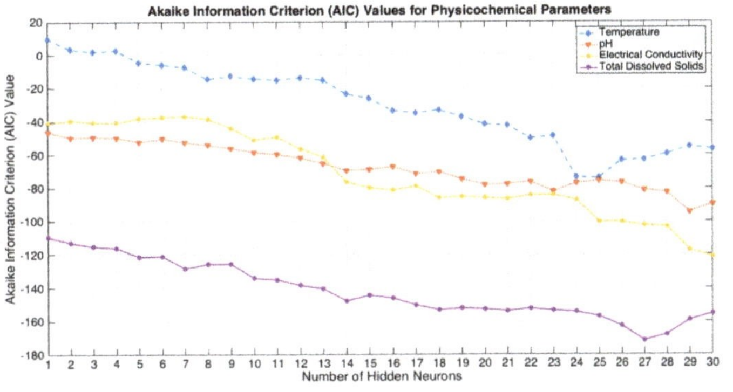

Figure 8. The AIC Values for Physicochemical Parameters.

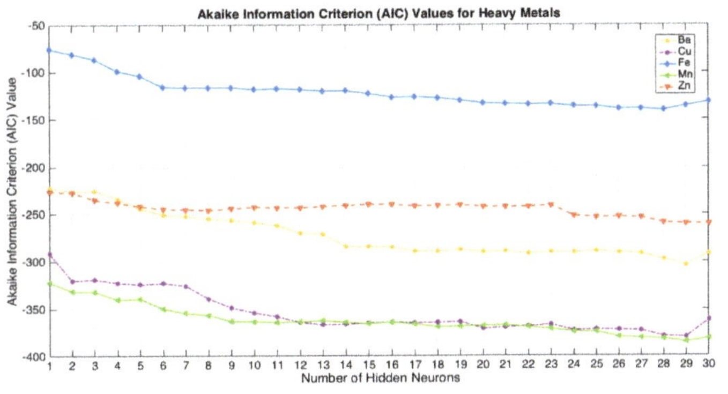

Figure 9. The AIC Values for Heavy Metals.

The spatial concentration of the physicochemical parameters of GW in Calapan City, Oriental Mindoro was mapped using NN-PSO+EBK interpolation method. The highest temperature for GW recorded in the study area was 33.6 °C which was observed in Brgy. Parang. While the least temperature was observed in Brgy. Biga with recorded temperature of 26.2 °C. The highest pH for GW was observed at Brgy. Canubing I, with pH equal to 8.8. The lowest pH was detected at Brgy. Sto. Nino with pH of 6.7. The highest EC and TDS observed in Brgy. Ibaba West with EC and TDS value of 1820 μS/cm and 900 ppm, respectively. The least observed EC and TDS concentration was 100 μS/cm and 40 ppm, respectively which was recorded in Brgy. Sta. Rita. The spatial concentration of the physicochemical parameters of GW is shown in Figure 10.

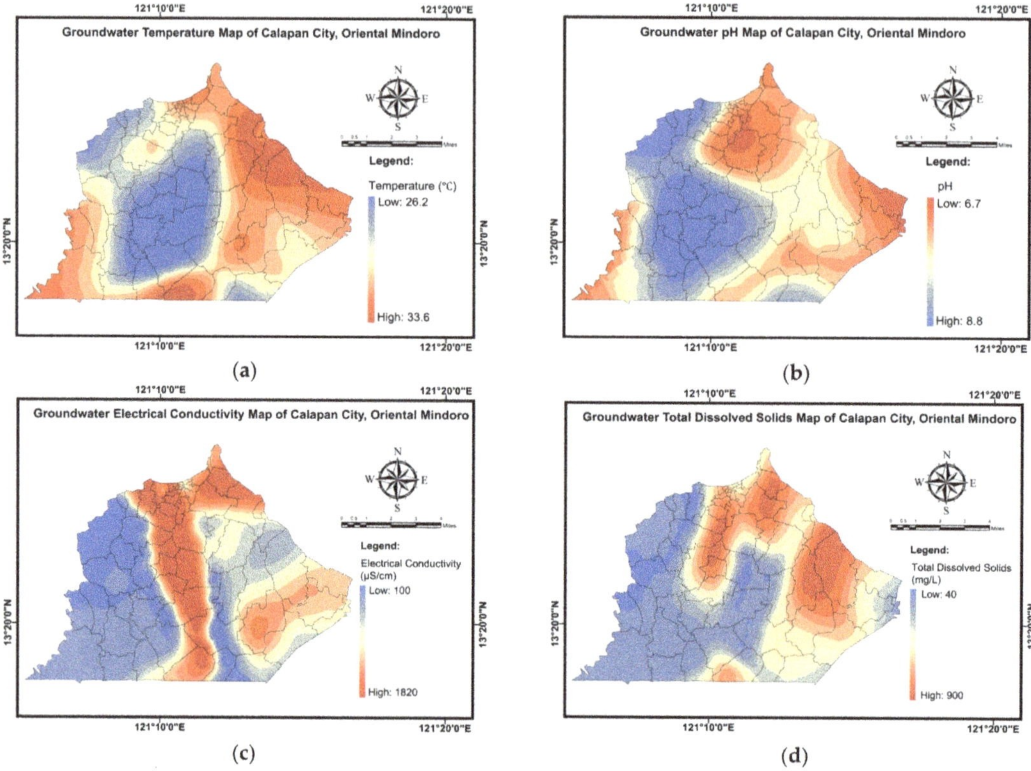

Figure 10. Physicochemical parameters map of Calapan City (**a**) Temperature, (**b**) pH, (**c**) EC, and (**d**) TDS.

The spatial concentration of the heavy metals of GW in Calapan City, Oriental Mindoro including Ba, Cu, Fe, Mn, and Zn was also mapped using the NN-PSO+EBK interpolation method. The heavy metal concentration maps generated using the NN-PSO+EBK method was presented in Figure 11.

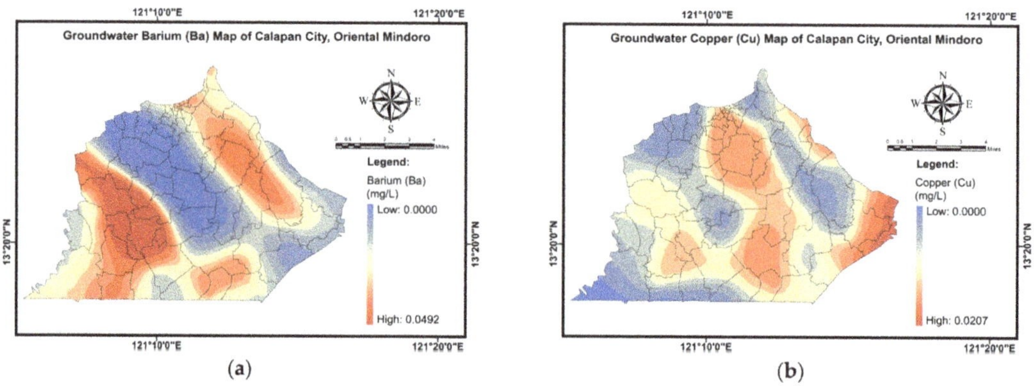

Figure 11. *Cont.*

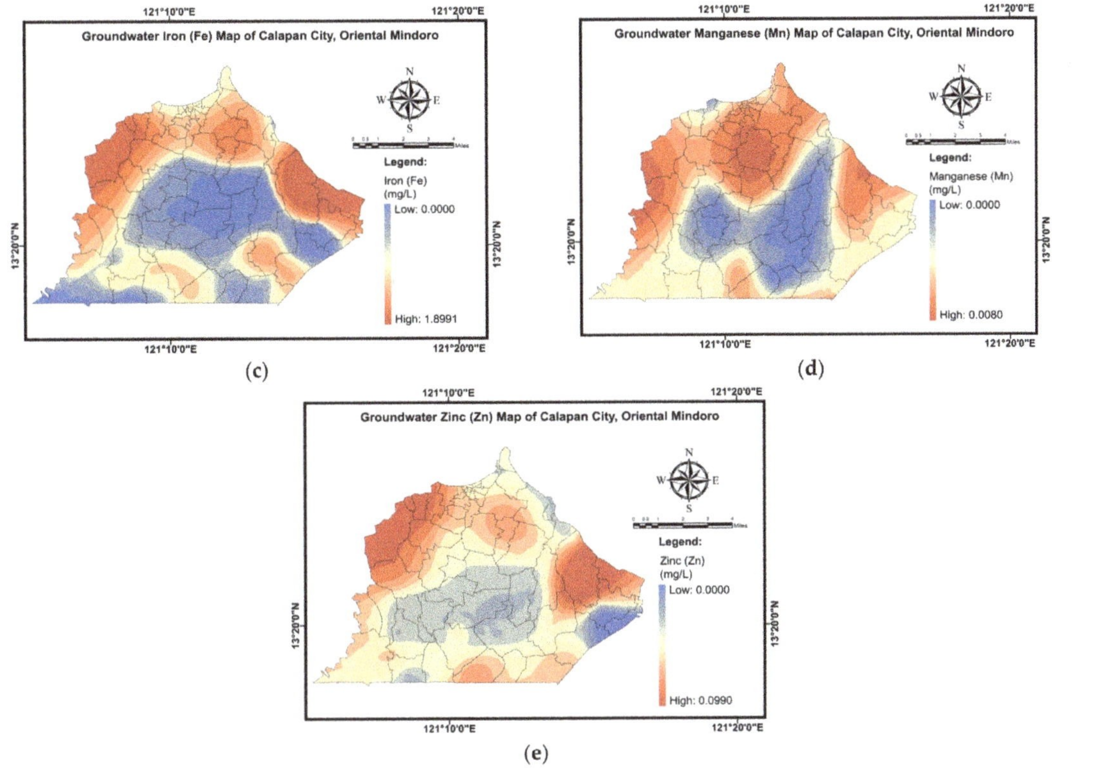

Figure 11. Heavy metal concentration map of Calapan City (**a**) Ba, (**b**) Cu, (**c**) Fe, (**d**) Mn, and (**e**) Zn.

The highest Ba concentration was measured in Brgy. Canubing I, where it was 7.9 times more than the background value for Ba measured in the research area. The average Ba concentration across all sample sites was 5.6 times more than the background value in the area of study. The highest concentrations of Cu were measured at several locations and recorded to be three times greater than the background concentration of copper. The mean concentration of Cu was found to be 2.1 times that of the background concentration. The Fe concentrations were found to be highest in Brgy. Gutad where it was found to be 6.1 times greater than the background value reported for the research region. Moreover, multiple sites were observed to exceed the WHO standards for Fe. These sites include Brgy. Camansihan, Brgy. Ibaba East, Brgy. Masipit, Brgy. Parang, Site 2 of Brgy. Personas, Brgy. San Vicente East, Brgy. Sta. Cruz, Brgy. Sta. Rita, and Brgy. Sto. Nino. Meanwhile, the mean concentration in Calapan City was just 0.5 percent more than the background level. Mn and Zn concentrations were highest in Brgy. Sto. Nino and Brgy. Gutad, respectively. However, these concentrations were still below the background concentration in the research region. The heavy metal concentration trend in the study area was observed to be Mn < Cu < Ba < Zn < Fe. Generally, these concentrations are within the WHO, USEPA and PNSDW limit except for Fe in several areas.

3.5. Cross Validation and Spot Sampling Evaluation Results

The predicted and observed values were compared to the NN-PSO+EBK method using the correctness measures to test the robustness of the predicted models. The results shown in Table 6 exhibit a robust and accurate result based on the R values close to 1. The cross-validation results suggested that all values provided more accurate spatial distribution for the study area. The cross-validation results are presented in Table 6.

Table 6. Cross Validation Results for NN-PSO+EBK simulation.

Criteria	Temp	pH	EC	TDS	Ba	Cu	Fe	Mn	Zn
R	0.989	0.990	0.934	0.977	0.994	0.976	0.985	0.975	0.984

A spot sampling analysis was performed using the data from the households in different barangays of Calapan City. A total of 21,559 households were utilized in the spot sampling analysis which is presented in Figure 12. The distribution of the households included in the spot sampling analysis per barangay is presented in Figure 13.

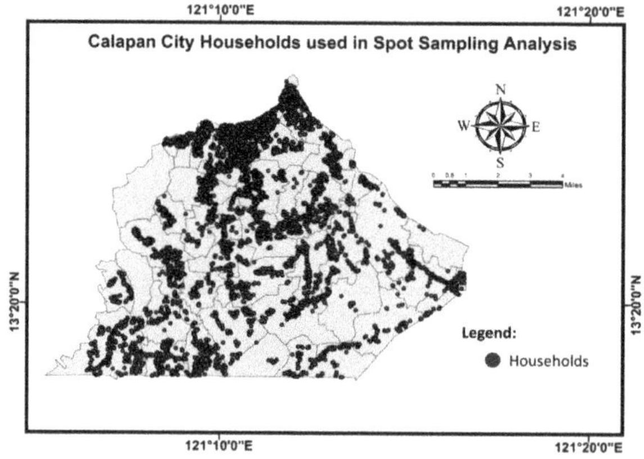

Figure 12. Concentration of Zn in GW samples.

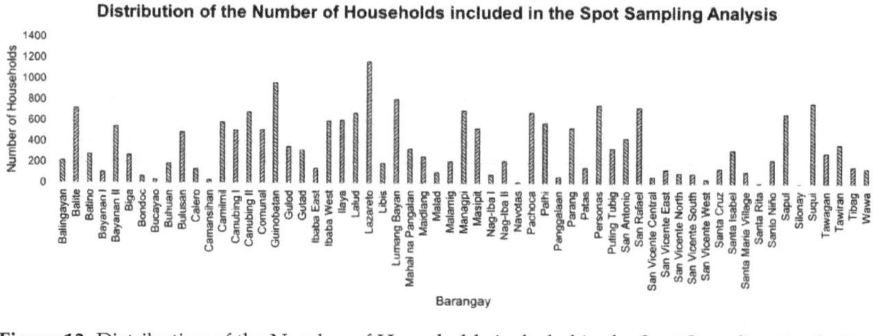

Figure 13. Distribution of the Number of Households included in the Spot Sampling Analysis.

The spot sampling results was compared to the spatial concentration maps created in Figure 11. Table 7 exhibits the spot sampling comparison results for all heavy metals detected in the GW resources in Calapan City. The results showed that the created maps provided good results based on its MSE which is approaching zero when contrasted to the spot sampling values.

Table 7. Spot Sampling Comparison Results.

Criteria	Ba	Cu	Fe	Mn	Zn
MSE	0.0001404	0.0000542	0.0006260	0.0000037	0.0004141

Considering each barangay, the MSEs for each element were also obtained as presented in Table A2 of Appendix C. Figure 14 presents the summary of the spot sampling comparison results in each barangay for all heavy metals considered in the study.

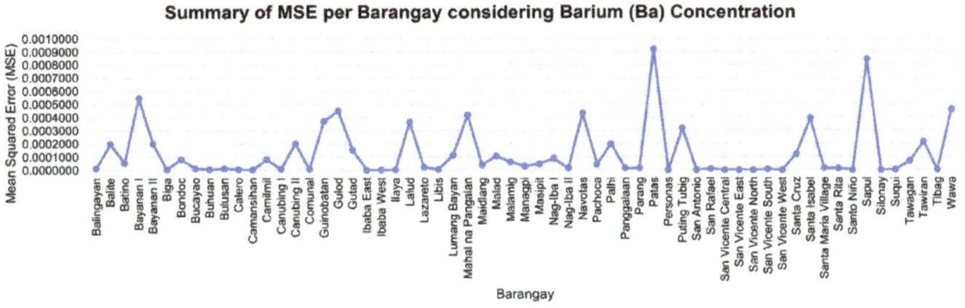

(a)

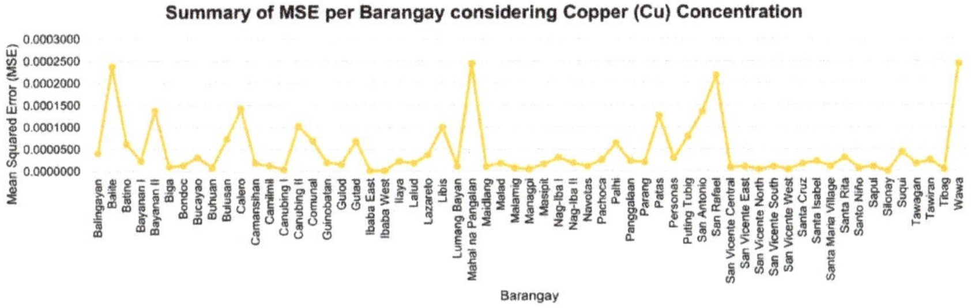

(b)

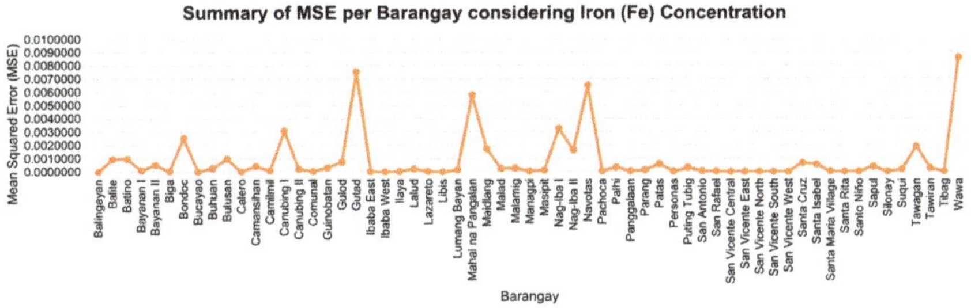

(c)

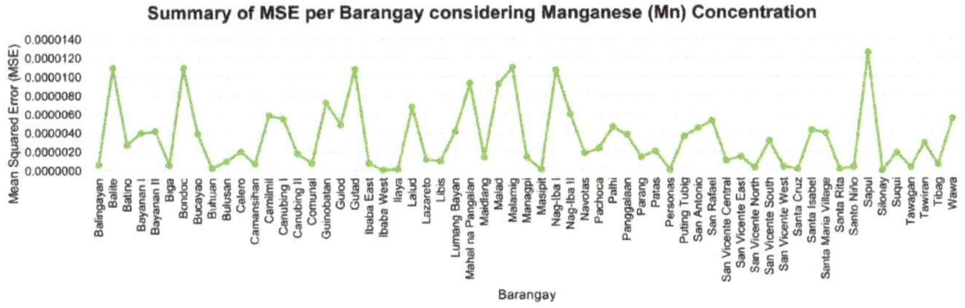

(d)

Figure 14. *Cont.*

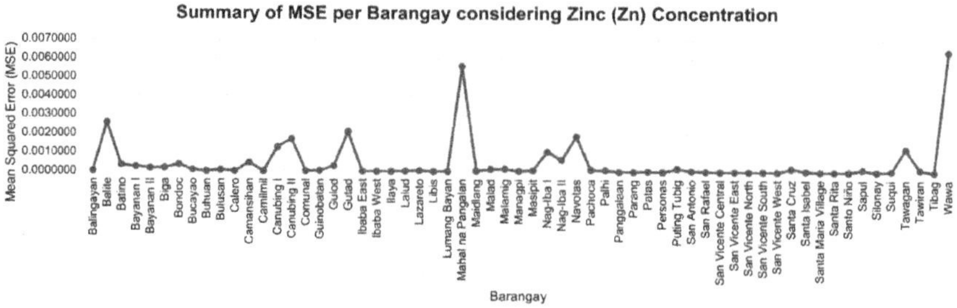

(e)

Figure 14. Summary of MSE per Barangay considering (**a**) Ba, (**b**) Cu, (**c**) Fe, (**d**) Mn, and (**e**) Zn.

4. Discussion

Oriental Mindoro, an island province, is vulnerable to GW pollution and degradation due to natural and human activities. Due to structural disadvantages and characteristics such as smaller land area and population, insufficient natural resources, geographical distribution, and other global factors beyond domestic control, a small island economy is less resilient to the threat of GW deterioration and contamination than larger and more diverse economies [70].

Water plays a critical part in achieving the United Nations' Sustainable Development Goals (SDGs). One of the problems that population has been experiencing is ensuring that everyone achieves SDG 6 (clean water and sanitation) which seeks to guarantee universal access to, and sustainable management of water. Continuous data integration and frequent monitoring remain to be critical components to achieving SDG 6 [71]. Hence, creating tools to aid in carrying out GW monitoring is significant. Tools such as the hybrid NN-PSO+EBK in making GW monitoring convenient to researchers and authorities are important.

Various heavy metals were detected at various sampling sites across Calapan City. The mean concentration of these metals in GW remained below the WHO and PNSDW acceptable levels. The recorded in-situ physicochemical characteristics were also compared to the WHO and PNSDW acceptable limits. The average GW temperature observed in the study area was 29.99 °C while the average GW pH observed was 7.69. Both figures are within the permissible range of WHO and PNSDW. One (1) sampling location exhibited pH value exceeding the maximum limit for pH of PNSDW. The average EC for the area of study was 560.36 μS/cm. This is within the permissible limit of the WHO. One (1) sampling location exhibited an EC value exceeding the WHO limit of 1500 μS/cm. The EC observation made in this location was categorized as Type II. The EC greater than 1500 μS/cm but less than 3000 μS/cm implies medium salts enrichment [72]. The average TDS is 277.15 ppm which was below the maximum allowable limit by the PNSDW and WHO. Though, TDS levels did not exceed the permissible limits set by WHO and PNSDW but were substantially lower or higher than the suggested TDS range of 600−1000 impairing palatability. Specifically, data of water samples from Ibaba West recorded TDS of 900. This number is at the high side of the limit which suggested the probability of impaired taste. On the other hand, the TDS concentrations recorded in Balingayan (60), Biga (50), Canubing (50), Comunal (90), Ibaba East Site 1 (160), Managpi (100), Personas (60), and Sta. Rita (40) were all significantly lower than the recommended range (600−1000 ppm) which may result to flat and insipid flavor.

The heavy metals concentration for GW was also observed in the study area. The Fe concentrations detected in multiple locations were above the WHO and PNSDW standards. However, the rest of the heavy metals detected were within the permissible limits. Water with elevated metals concentration has the potential to cause several public health issues. Health risks associated with elevated Fe in GW is probable. Pollutants entering the human body through drinking water have been shown in numerous studies to have detrimental health consequences for consumers. Micronutrients are essential in living organisms; however, elevated concentration adversely affects public health. Similar case with Mn which is necessary for humans; however, excessive quantities will have negative consequences. Neurological disorders, such as aberrant walking, ataxia, muscle hypotonicity, and a face devoid of lasting emotions, are frequently associated to Mn [64]. Dysfunction of liver was also reported [73]. Furthermore, excess Mn concentration has been demonstrated to produce neurotoxicity in infants receiving parenteral nourishment [74]. Excess Mn has been also linked to a lower level of IQ in children [64].

Meanwhile, asbestos-related cancer is believed to be caused by free radicals, which are produced by iron. Free radicals produced by iron can cause cancer by oxidizing DNA and causing DNA damage [75]. Additionally, elevated levels of Mn and Fe in drinking water have been associated to a decrease in birth weight in term-born infants [76]. Furthermore, since animals' intestinal mucosa is highly porous, the fast absorption into the blood has been attributed to the Ba^{2+} ions which are rapidly absorbed from the gastrointestinal system and lungs. Moreover, it has been observed that Ba poisoning mostly affects the cardiovascular system; nevertheless, renal dysfunction has been documented as well [77].

The use of in-situ and hybrid machine learning—geostatistical methods are an integral part of data integration for GW quality monitoring. The impact of GW contamination in an island province had been a threat to public health especially when GW is used as primary source of domestic, agricultural and industrial water supply. Application of NN-PSO+EBK hybrid technique enables the establishment of spatial variability map of the contaminants that contributes to the depletion of GW quality. As a result, future undesired consequences could be avoided using this monitoring technique. The NN-PSO+EBK can offer periodic and long-term data that can be utilized for permanent monitoring of GW quality and risk assessments. Also, this tool can be utilized as early warning of GW quality for detrimental effects [78] by human activities and/or natural weathering.

5. Conclusions

An in-situ approach and hybrid MLGI, i.e., NN-PSO+EBK, was applied to assess and evaluate the GW quality in Calapan City, Oriental Mindoro, Philippines. Physicochemical characteristics and metals concentrations were detected onsite at various sampling locations. Generally, the physicochemical analysis of GW samples met the WHO and PNSDW guidelines. The average values for temperature, pH, EC, and TDS were within the permissible limits though few sampling locations exceeded the permissible limits of WHO and PNSDW. The pH of all samples was within the limits set by the PNSDW. Barangays Buhuan, Camansihan, Gutad, Ibaba East (Site 2), Ibaba West, Ilaya, Lazareto, Maidlang, Masipit, Nag-iba II, Pachoca, Parang, San Vicente East, Sta. Cruz, and Sto. Nino recorded elevated EC values. This was attributed to the addition of leachable salts. Also, the recorded TDS values suggested probable impaired palatability by having values significantly below the recommended range of 600−1000 ppm. Heavy metals analysis showed that only Fe detected in multiple GW samples had concentration above the WHO and PNSDW maximum permitted levels. This condition presents health concerns to the consumers. The record on Fe concentration in Brgy. Gutad GW samples were above WHO and PNSDW limit. Other GW samples recorded target metals concentration within the WHO and PNSDW permissible limits. The spot sampling analysis results showed that the generated maps by hybrid technique such as NN-PSO+EBK were reliable in describing the heavy metal concentration in the city of Calapan based on its MSE, R and AIC values.

This study is useful as a reference to providing techniques on gathering data for GW quality monitoring to help attain SDGs 6. It is suggested to conduct a study targeting other metals and regular monitoring of its concentration using this hybrid MLGI technique. Additionally, the regular monitoring is necessary to better understanding of the possible health consequences. Furthermore, a health risk assessment based on GW quality should be conducted. Another, preliminary interventions on GW quality control is necessary.

Author Contributions: Conceptualization, D.B.S.; methodology, K.L.M.d.J. and D.B.S.; software, K.L.M.d.J. and D.B.S.; validation, D.B.S., L.C.M., K.S.E. and E.M.D.A.; formal analysis, K.L.M.d.J. and D.B.S.; investigation, L.C.M., E.M.D.A., K.S.E. and D.B.S.; resources, D.B.S.; data curation, D.B.S., L.C.M., E.M.D.A., K.S.E. and E.B.C.; writing—original draft preparation, K.L.M.d.J., L.C.M. and E.M.D.A.; writing—review and editing, E.B.C., D.B.S. and K.L.M.d.J.; visualization, K.L.M.d.J.; supervision, D.B.S.; project administration, D.B.S.; funding acquisition, D.B.S. All authors have read and agreed to the published version of the manuscript.

Funding: This research was funded by the Philippine Council for Health Research and Development of the Department of Science and Technology, Philippines.

Institutional Review Board Statement: Not applicable.

Informed Consent Statement: Not applicable.

Data Availability Statement: All data are contained in the manuscript.

Acknowledgments: This is to recognize the 'in-kind' support of Mapua University, Manila, Philippines, and the Calapan City local government units. In addition, the Philippines Mines and Geosciences Bureau by providing the hazard maps for MIMAROPA region.

Conflicts of Interest: The authors declare no conflict of interest.

Appendix A

Table A1. Observed Physicochemical Properties of Groundwater Samples.

Sampling No.	Barangay	Latitude	Longitude	Temp (°C)	pH	EC (μS/cm)	TDS (ppm)
1	Balingayan (Site 1)	13.31903° N	121.13432° E	28.3	7.9	130	60
2	Balingayan (Site 2)	13.32454° N	121.13555° E	28.5	8.4	130	60
3	Biga	13.32791° N	121.17312° E	26.2	8.5	120	50
4	Buhuan	13.31451° N	121.22395°E	29.2	7.6	660	320
5	Camansihan	13.33399° N	121.22656° E	30.7	7.8	1200	590
6	Canubing I	13.35590° N	121.14091° E	27.1	8.8	130	50
7	Comunal	13.31267° N	121.16494° E	27.2	7.9	200	90
8	Gutad	13.35518° N	121.25278° E	32.4	7.1	900	440
9	Ibaba East (Site 1)	13.41517° N	121.17836° E	31.6	7.5	350	160
10	Ibaba East (Site 2)	13.41484° N	121.17769° E	32.5	7.5	970	480
11	Ibaba West	13.41478° N	121.17676° E	31.4	7.4	1820	900
12	Ilaya	13.41181° N	121.18548° E	31.1	7.1	780	380
13	Lazareto	13.42972° N	121.19940° E	31	7	990	490
14	Maidlang	13.39711° N	121.22727° E	32	7.5	600	290
15	Managpi	13.32512° N	121.19595° E	28.1	7.5	220	100
16	Masipit	13.38917° N	121.16190° E	31.5	7.3	570	270
17	Nag-iba II	13.34643° N	121.25301° E	30.6	7.3	820	400
18	Pachoca	13.41061° N	121.16840° E	29.2	8.1	690	340
19	Palhi	13.37502° N	121.20703° E	30.2	7.4	410	200
20	Panggalaan	13.30148° N	121.19908° E	30.1	7.6	140	160
21	Panggalaan	13.30027° N	121.20041° E	28.4	8.3	180	150
22	Parang	13.40059° N	121.21769° E	33.6	7.7	910	450
23	Personas (Site 1)	13.30623° N	121.14083° E	28.7	8.3	140	60
24	Personas (Site 2)	13.30930° N	121.13945° E	27.4	8.4	140	60
25	San Vicente East	13.31045° N	121.17980° E	32.9	7.6	750	370
26	Sta. Cruz	13.31633° N	121.23461° E	29.7	7.3	500	240
27	Sta. Rita	13.35212° N	121.13091° E	29.7	7.7	100	40
28	Sto. Nino	13.40712° N	121.18545° E	30.3	6.7	1140	560

Appendix B

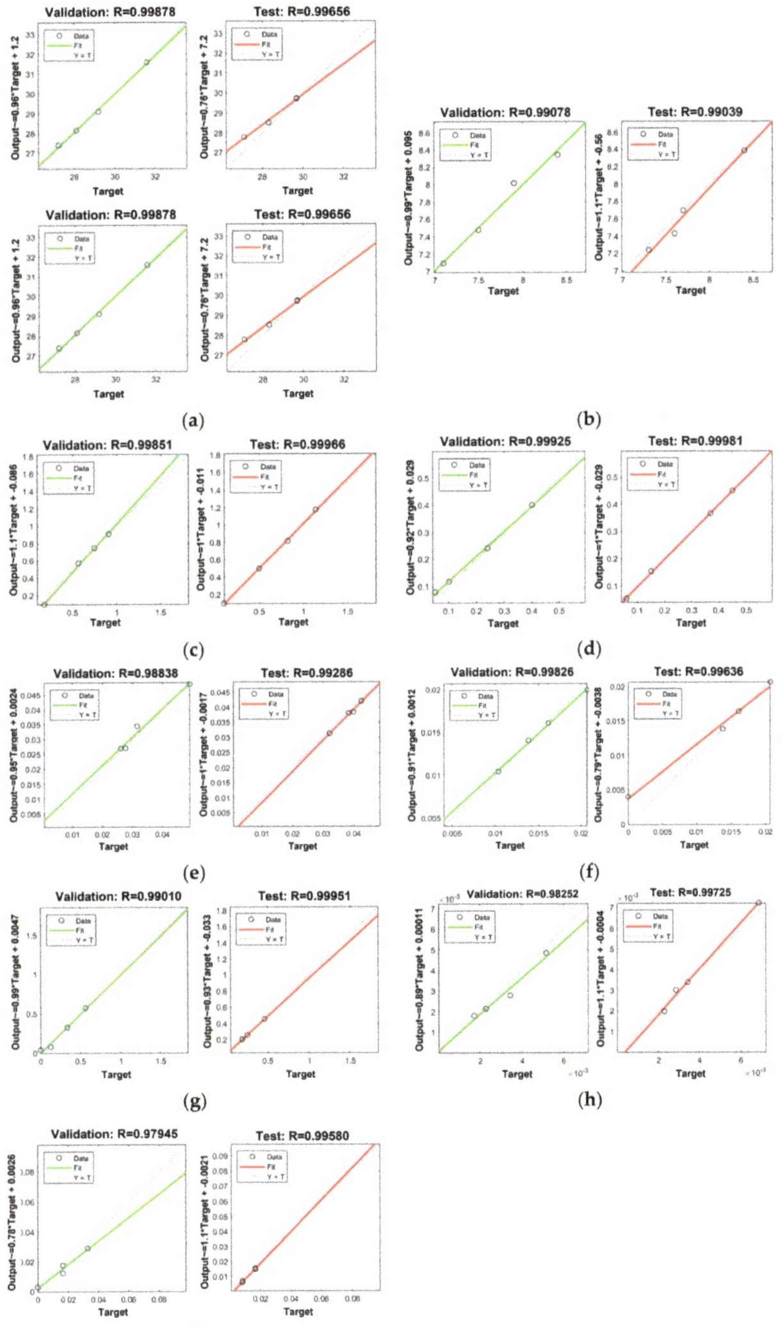

Figure A1. R Value Plots for Validation and Testing Phase of the NN-PSO: (**a**) Temperature; (**b**) pH; (**c**) EC; (**d**) TDS; (**e**) Ba; (**f**) Cu; (**g**) Fe; (**h**) Mn; (**i**) Zn.

Appendix C

Table A2. Spot Sampling Results in each Barangay of Calapan City.

Brgy. No.	Barangay	Latitude	Longitude	Elev. (m)	Average MSE				
					Ba	Cu	Fe	Mn	Zn
1	Balingayan	13.3241° N	121.1407° E	13.2	0.0000151	0.0000412	0.0000104	0.0000007	0.0000138
2	Balite	13.4131° N	121.1580° E	7.6	0.0002007	0.0002381	0.0009690	0.0000110	0.0025765
3	Batino	13.3494° N	121.2201° E	7.8	0.0000578	0.0000624	0.0009873	0.0000028	0.0003248
4	Bayanan I	13.3679° N	121.1685° E	8.8	0.0005507	0.0000240	0.0001122	0.0000040	0.0002384
5	Bayanan II	13.3560° N	121.1699° E	12.5	0.0002042	0.0001388	0.0005354	0.0000042	0.0001622
6	Biga	13.3270° N	121.1733° E	14.1	0.0000080	0.0000108	0.0000396	0.0000006	0.0001686
7	Bondoc	13.3867° N	121.2010° E	165.5	0.0000844	0.0000133	0.0025490	0.0000110	0.0003627
8	Bucayao	13.3066° N	121.1915° E	17.3	0.0000146	0.0000316	0.0000047	0.0000039	0.0000776
9	Buhuan	13.3106° N	121.1915° E	12.8	0.0000069	0.0000081	0.0002698	0.0000003	0.0000034
10	Bulusan	13.4037° N	121.2012° E	28.4	0.0000127	0.0000730	0.0009766	0.0000010	0.0000702
11	Calero	13.4159° N	121.1831° E	9.4	0.0000079	0.0001424	0.0000126	0.0000020	0.0000081
12	Camansihan	13.3428° N	121.2290° E	7.1	0.0000042	0.0000183	0.0004483	0.0000007	0.0004709
13	Camilmil	13.4061° N	121.1760° E	8.4	0.0000827	0.0000129	0.0000828	0.0000059	0.0000046
14	Canubing I	13.3554° N	121.1423° E	8.6	0.0000088	0.0000043	0.0030914	0.0000055	0.0012961
15	Canubing II	13.3261° N	121.1216° E	12.9	0.0002042	0.0001033	0.0002191	0.0000018	0.0017159
16	Comunal	13.3075° N	121.1606° E	19.5	0.0000085	0.0000694	0.0000558	0.0000008	0.0000270
17	Guinobatan	13.3829° N	121.1818° E	9.7	0.0003723	0.0000199	0.0003131	0.0000073	0.0000374
18	Gulod	13.3433° N	121.2073° E	9.0	0.0004521	0.0000160	0.0007486	0.0000049	0.0002808
19	Gutad	13.3597° N	121.2464° E	7.2	0.0001561	0.0000687	0.0075550	0.0000108	0.0021064
20	Ibaba East	13.4149° N	121.1788° E	6.6	0.0000009	0.0000016	0.0000250	0.0000008	0.0000022
21	Ibaba West	13.4146° N	121.1762° E	5.9	0.0000027	0.0000018	0.0000010	0.0000001	0.0000021
22	Ilaya	13.4129° N	121.1840° E	8.7	0.0000040	0.0000232	0.0000372	0.0000002	0.0000020
23	Lalud	13.3993° N	121.1739° E	9.1	0.0003669	0.0000186	0.0002229	0.0000068	0.0000209
24	Lazareto	13.4286° N	121.1995° E	12.9	0.0000251	0.0000371	0.0000305	0.0000012	0.0000589
25	Libis	13.4149° N	121.1847° E	9.2	0.0000074	0.0001006	0.0000018	0.0000010	0.0000036
26	Lumang Bayan	13.4009° N	121.1816° E	6.8	0.0001156	0.0000113	0.0001495	0.0000042	0.0000077
27	Mahal na Pangalan	13.4082° N	121.1502° E	9.0	0.0004217	0.0002451	0.0057978	0.0000094	0.0056021
28	Maidlang	13.3883° N	121.2339° E	8.3	0.0000427	0.0000111	0.0017492	0.0000014	0.0000401
29	Malad	13.3396° N	121.1588° E	10.8	0.0001085	0.0000189	0.0002558	0.0000093	0.0001421
30	Malamig	13.3439° N	121.1456° E	10.7	0.0000640	0.0000077	0.0002769	0.0000110	0.0001567
31	Managpi	13.3282° N	121.1997° E	18.0	0.0000321	0.0000051	0.0000465	0.0000014	0.0000330
32	Masipit	13.3869° N	121.1603° E	6.6	0.0000484	0.0000162	0.0001101	0.0000002	0.0000723
33	Nag-Iba I	13.3400° N	121.2721° E	8.5	0.0000893	0.0000313	0.0032634	0.0000108	0.0010670
34	Nag-Iba II	13.3470° N	121.2622° E	11.0	0.0000154	0.0000190	0.0016347	0.0000060	0.0006289
35	Navotas	13.3739° N	121.2486° E	7.5	0.0004349	0.0000114	0.0065142	0.0000019	0.0018669
36	Pachoca	13.4108° N	121.1677° E	6.0	0.0000441	0.0000266	0.0000465	0.0000024	0.0001095
37	Palhi	13.3750° N	121.2070° E	18.6	0.0001992	0.0000633	0.0003630	0.0000046	0.0000934
38	Panggalaan	13.3012° N	121.1990° E	17.9	0.0000164	0.0000234	0.0000402	0.0000039	0.0000126
39	Parang	13.4035° N	121.2182° E	9.5	0.0000148	0.0000209	0.0001279	0.0000015	0.0000127
40	Patas	13.3452° N	121.1222° E	12.9	0.0009213	0.0001269	0.0005874	0.0000021	0.0000333
41	Personas	13.3083° N	121.1438° E	14.5	0.0000006	0.0000300	0.0000213	0.0000001	0.0000028
42	Puting Tubig	13.3470° N	121.1887° E	7.3	0.0003178	0.0000788	0.0003261	0.0000037	0.0001840
43	San Antonio	13.4259° N	121.1956° E	14.0	0.0000013	0.0001353	0.0000519	0.0000045	0.0000587
44	San Rafael	13.4216° N	121.1911° E	6.4	0.0000099	0.0002183	0.0000171	0.0000053	0.0000280
45	San Vicente Central	13.4120° N	121.1787° E	8.6	0.0000010	0.0000086	0.0000024	0.0000011	0.0000011
46	San Vicente East	13.4098° N	121.1798° E	9.3	0.0000009	0.0000106	0.0000044	0.0000014	0.0000004
47	San Vicente North	13.4138° N	121.1785° E	6.7	0.0000003	0.0000045	0.0000005	0.0000003	0.0000001
48	San Vicente South	13.4095° N	121.1779° E	7.6	0.0000072	0.0000115	0.0000123	0.0000031	0.0000018
49	San Vicente West	13.4124° N	121.1765° E	7.8	0.0000005	0.0000040	0.0000024	0.0000005	0.0000010
50	Santa Cruz	13.3169° N	121.2370° E	12.1	0.0001231	0.0000173	0.0006697	0.0000002	0.0001946
51	Santa Isabel	13.3654° N	121.1577° E	5.3	0.0003942	0.0000233	0.0005533	0.0000043	0.0000575
52	Santa Maria Village	13.4093° N	121.1748° E	6.3	0.0000154	0.0000115	0.0000167	0.0000040	0.0000019
53	Santa Rita	13.3489° N	121.1303° E	11.1	0.0000139	0.0000317	0.0000140	0.0000002	0.0000081
54	Santo Niño	13.4066° N	121.1848° E	8.2	0.0000004	0.0000075	0.0000057	0.0000004	0.0000003
55	Sapul	13.3651° N	121.1885° E	14.3	0.0008450	0.0000108	0.0003983	0.0000126	0.0001391
56	Silonay	13.3992° N	121.2248° E	8.8	0.0000004	0.0000004	0.0000011	0.0000000	0.0000007
57	Suqui	13.4177° N	121.2040° E	11.1	0.0000057	0.0000434	0.0001749	0.0000019	0.0000464
58	Tawagan	13.3712° N	121.1448° E	8.6	0.0000693	0.0000169	0.0019052	0.0000004	0.0012349
59	Tawiran	13.3950° N	121.1680° E	7.0	0.0002160	0.0000253	0.0002604	0.0000030	0.0001365
60	Tibag	13.4123° N	121.1730° E	8.2	0.0000018	0.0000040	0.0000027	0.0000006	0.0000014
61	Wawa	13.4025° N	121.1453° E	8.8	0.0004609	0.0002441	0.0085841	0.0000055	0.0063922

References

1. Pineda-Pineda, J.J.; Martínez-Martínez, C.T.; Méndez-Bermúdez, J.A.; Muñoz-Rojas, J.; Sigarreta, J.M. Application of Bipartite Networks to the Study of Water Quality. *Sustainability* **2020**, *12*, 5143. [CrossRef]
2. Mekonnen, M.M.; Hoekstra, A.Y. Four billion people facing severe water scarcity. *Sci. Adv.* **2016**, *2*, e1500323. [CrossRef]
3. Jha, M.K.; Shekhar, A.; Jenifer, M.A. Assessing groundwater quality for drinking water supply using hybrid fuzzy-GIS-based water quality index. *Water Res.* **2020**, *179*, 115867. [CrossRef] [PubMed]
4. Department of Environment and Natural Resources—Mines and Geosciences Bureau. Geohazard Maps. Available online: http://www.region4b.mgb.gov.ph/28-geohazard-maps/98-geohazard-maps (accessed on 25 October 2021).
5. Philippine Statistics Authority. Compendium of Philippine Environment Statistics Component 2: Environmental Resources and their Use. Available online: https://psa.gov.ph/press-releases/id/163678 (accessed on 25 October 2021).
6. Kumar, A.; Cabral-Pinto, M.; Kumar, A.; Kumar, M.; Dinis, P.A. Estimation of Risk to the Eco-Environment and Human Health of Using Heavy Metals in the Uttarakhand Himalaya, India. *Appl. Sci.* **2020**, *10*, 7078. [CrossRef]
7. Kumar, A.; Jigyasu, D.K.; Kumar, A.; Subrahmanyam, G.; Mondal, R.; Shabnam, A.A.; Cabral-Pinto, M.M.S.; Malyan, S.K.; Chaturvedi, A.K.; Gupta, D.K.; et al. Nickel in terrestrial biota: Comprehensive review on contamination, toxicity, tolerance and its remediation approaches. *Chemosphere* **2021**, *275*, 129996. [CrossRef] [PubMed]
8. Kumar, A.; Cabral Pinto, M.; Candeias, C.; Dinis, P.A. Baseline maps of potentially toxic elements in the soil of Garhwal Himalayas, India: Assessment of their eco-environmental and human health risks. *Land Degrad. Dev.* **2021**, *10*, 3984. [CrossRef]
9. Hartzler, D.A.; Jain, J.C.; McIntyre, D.L. Development of a subsurface LIBS sensor for in situ groundwater quality monitoring with applications in CO 2 leak sensing in carbon sequestration. *Sci. Rep.* **2019**, *9*, 1–10.
10. Bu, X.; Koide, K.; Carder, E.J.; Welch, C.J. Rapid analysis of residual palladium in pharmaceutical development using a catalysis-based fluorometric method. *Org. Process Res. Dev.* **2013**, *17*, 108–113. [CrossRef]
11. Farghaly, O.A.; Hameed, R.A.; Abu-Nawwas, A.A.H. Analytical application using modern electrochemical techniques. *Int. J. Electrochem. Sci.* **2014**, *9*, 3287–3318.
12. Kudr, J.; Richtera, L.; Nejdl, L.; Xhaxhiu, K.; Vitek, P.; Rutkay-Nedecky, B.; Hynek, D.; Kopel, P.; Adam, V.; Kizek, R. Improved electrochemical detection of zinc ions using electrode modified with electrochemically reduced graphene oxide. *Materials* **2016**, *9*, 31. [CrossRef]
13. Estela, J.M.; Tomás, C.; Cladera, A.; Cerda, V. Potentiometric stripping analysis: A review. *Crit. Rev. Anal. Chem.* **1995**, *25*, 91–141. [CrossRef]
14. Khadro, B.; Sikora, A.; Loir, A.S.; Errachid, A.; Garrelie, F.; Donnet, C.; Jaffrezic-Renault, N. Electrochemical performances of B doped and undoped diamond-like carbon (DLC) films deposited by femtosecond pulsed laser ablation for heavy metal detection using square wave anodic stripping voltammetric (SWASV) technique. *Sens. Actuators B Chem.* **2011**, *155*, 120–125. [CrossRef]
15. Avuthu, S.G.R.; Narakathu, B.B.; Eshkeiti, A.; Emamian, S.; Bazuin, B.J.; Joyce, M.; Atashbar, M.Z. Detection of heavy metals using fully printed three electrode electrochemical sensor. In *SENSORS*; IEEE: New York, NY, USA, 2014; pp. 669–672.
16. Cheng, L.; Liu, X.; Lei, J.; Ju, H. Low-potential electrochemiluminescent sensing based on surface unpassivation of CdTe quantum dots and competition of analyte cation to stabilizer. *Anal. Chem.* **2010**, *82*, 3359–3364. [CrossRef] [PubMed]
17. Bansod, B.; Kumar, T.; Thakur, R.; Rana, S.; Singh, I. A review on various electrochemical techniques for heavy metal ions detection with different sensing platforms. *Biosens. Bioelectron.* **2017**, *94*, 443–455. [CrossRef]
18. Borgese, L.; Zacco, A.; Bontempi, E.; Pellegatta, M.; Vigna, L.; Patrini, L.; Riboldi, L.; Rubino, F.M.; Depero, L.E. Use of total reflection X-ray fluorescence (TXRF) for the evaluation of heavy metal poisoning due to the improper use of a traditional ayurvedic drug. *J. Pharm. Biomed. Anal.* **2010**, *52*, 787–790. [CrossRef]
19. De Jesus, K.; Senoro, D.B.; Dela Cruz, J.C.; Chan, E.B. A Hybrid Neural Network—Particle Swarm Optimization In-formed Spatial Interpolation Technique for Groundwater Quality Mapping in a Small Island Province of the Philippines. *Toxics* **2021**, *9*, 273. [CrossRef] [PubMed]
20. El-Rawy, M.; Fathi, H.; Abdalla, F. Integration of remote sensing data and in situ measurements to monitor the water quality of the Ismailia Canal, Nile Delta, Egypt. *Environ. Geochem. Health* **2020**, *42*, 2101–2120. [CrossRef] [PubMed]
21. Solis, K.L.B.; Macasieb, R.Q.; Parangat, R.C.; Resurreccion, A.C.; Ocon, J.D. Spatiotemporal Variation of Groundwater Arsenic in Pampanga, Philippines. *Water* **2020**, *12*, 2366. [CrossRef]
22. Tiankao, W.; Chotpantarat, S. Risk assessment of arsenic from contaminated soils to shallow groundwater in Ong Phra Sub-District, Suphan Buri Province, Thailand. *J. Hydrol. Reg. Stud.* **2018**, *19*, 80–96. [CrossRef]
23. Mogaji, K.A. Application of vulnerability modeling techniques in groundwater resources management: A comparative study. *Appl. Water Sci.* **2018**, *8*, 1–24. [CrossRef]
24. Nistor, M.M.; Rahardjo, H.; Satyanaga, A.; Hao, K.Z.; Xiaosheng, Q.; Sham, A.W.L. Investigation of groundwater table distribution using borehole piezometer data interpolation: Case study of Singapore. *Eng. Geol.* **2020**, *271*, 105590. [CrossRef]
25. Rusydi, A.F.; Onodera, S.I.; Saito, M.; Ioka, S.; Maria, R.; Ridwansyah, I.; Delinom, R.M. Vulnerability of groundwater to iron and manganese contamination in the coastal alluvial plain of a developing Indonesian city. *SN Appl. Sci.* **2021**, *3*, 1–12. [CrossRef]
26. Lado, L.R.; Polya, D.; Winkel, L.; Berg, M.; Hegan, A. Modelling arsenic hazard in Cambodia: A geostatistical approach using ancillary data. *Appl. Geochem.* **2008**, *23*, 3010–3018. [CrossRef]
27. Viossanges, M.; Pavelic, P.; Rebelo, L.M.; Lacombe, G.; Sotoukee, T. Regional mapping of groundwater resources in data-scarce regions: The case of Laos. *Hydrology* **2018**, *5*, 2. [CrossRef]

28. Giang, P.Q.; Nakhapakorn, K.; Ussawarujikulchai, A. Effectiveness of different spatial interpolators in estimating heavy metal contamination in shallow groundwater: A case study of arsenic contamination in Hanoi, Vietnam. *Environ. Nat. Resour. J.* **2011**, *9*, 31–37.
29. Agutaya, C.A.C.; Zamora, J.T. Developmental Projects in Calapan City, Philippines: Localization Perspectives. *Am. J. Educ. Res.* **2018**, *6*, 133–136.
30. United States Environmental Protection Agency (U.S. E.P.A). *Operating Procedure for In Situ Water Quality Monitoring (SESDPROC-111-R4)*. Available online: https://www.epa.gov/sites/default/files/2015-06/documents/Insitu-Water-Quality-Mon.pdf (accessed on 16 December 2021).
31. Hanna Instruments. HI9811-5 Portable pH/EC/TDS/Temperature. Available online: https://www.hannainst.com/portable-ph-ec-tds-temperature-meter-hi9811-5.html (accessed on 16 December 2021).
32. Magalona, M.L.; Peralta, M.M.; Lacsamana, M.S.; Sabularse, V.C.; Pelegrina, A.B.; De Guzman, C.C. Analysis of Inorganic Arsenic (As (III) and Total As) and Some Physicochemical Parameters in Groundwater Samples from Selected Areas in Bulacan, Batangas, and Laguna, Philippines. *KIMIKA* **2019**, *30*, 28–38. [CrossRef]
33. Groover, K.D.; Izbicki, J.A. Elemental analysis using a handheld X-ray fluorescence spectrometer (No. 2016-3043). *U.S. Geol. Surv. Fact Sheet* **2016**, 2015–3043. [CrossRef]
34. Analytical Methods Committee AMCTB No. 89. Hand-held X-ray fluorescence spectrometry. *Anal. Methods* **2019**, *11*, 2498–2501. [CrossRef]
35. Muramatsu, Y.; Izawa, R.; Nishioka, H.; Nogami, T. X-ray fluorescence analysis of dilute heavy-metals in water using a portable x-ray fluorescence spectrometer with the metal-adsorbent, tobermorite. *X-sen Bunseki No Shinpo* **2009**, *40*, 195–201.
36. Pearson, D.; Weindorf, D.C.; Chakraborty, S.; Li, B.; Koch, J.; Van Deventer, P.; de Wet, J.; Kusi, N.Y. Analysis of metal-laden water via portable X-ray fluorescence spectrometry. *J. Hydrol.* **2018**, *561*, 267–276. [CrossRef]
37. Zhou, S.; Yuan, Z.; Cheng, Q.; Zhang, Z.; Yang, J. Rapid in situ determination of heavy metal concentrations in polluted water via portable XRF: Using Cu and Pb as example. *Environ. Pollut.* **2018**, *243*, 1325–1333. [CrossRef]
38. Pearson, D.; Chakraborty, S.; Duda, B.; Li, B.; Weindorf, D.C.; Deb, S.; Brevik, E.; Ray, D.P. Water analysis via portable X-ray fluorescence spectrometry. *J. Hydrol.* **2017**, *544*, 172–179. [CrossRef]
39. Crocombe, R.A.; Leary, P.E.; Kammrath, B.W. *Portable Spectroscopy and Spectrometry, Applications*; John Wiley & Sons: Hoboken, NJ, USA, 2021; Volume 2, ISBN 978-111-963-640-3.
40. Magesh, N.S.; Chandrasekar, N.; Elango, L. Occurrence and distribution of fluoride in the groundwater of the Tamiraparani River basin, South India: A geostatistical modeling approach. *Environ. Earth Sci.* **2016**, *75*, 1–16. [CrossRef]
41. Shariati, M.; Mafipour, M.S.; Mehrabi, P.; Bahadori, A.; Zandi, Y.; Salih, M.N.; Nguyen, H.; Dou, J.; Song, X.; Poi-Ngian, S. Application of a hybrid artificial neural network-particle swarm optimization (ANN-PSO) model in behavior prediction of channel shear connectors embedded in normal and high-strength concrete. *Appl. Sci.* **2019**, *9*, 5534. [CrossRef]
42. Kayabasi, E.; Ozturk, S.; Celik, E.; Kurt, H. Determination of cutting parameters for silicon wafer with a Diamond Wire Saw using an artificial neural network. *Sol. Energy* **2017**, *149*, 285–293. [CrossRef]
43. Ozturk, S.; Kayabasi, E.; Celik, E.; Kurt, H. Determination of lapping parameters for silicon wafer using an artificial neural network. *J. Mater. Sci.: Mater. Electron.* **2018**, *29*, 260–270. [CrossRef]
44. Lin, Y.H.; Hu, Y.C. Electrical energy management based on a hybrid artificial neural network-particle swarm optimization-integrated two-stage non-intrusive load monitoring process in smart homes. *Processes* **2018**, *6*, 236. [CrossRef]
45. World Health Organization. *Guidelines for Drinking-Water Quality*, 3rd ed.; World Health Organization: Geneva, Switzerland, 2004; Volume 1, ISBN 92-4-154638-7.
46. Department of Health—Food and Drug Administration. Administrative Order No. 2017-0010—Philippine National Standards for Drinking Water of 2017. Available online: https://www.fda.gov.ph/administrative-order-no-2017-0010-philippine-national-standards-for-drinking-water-of-2017 (accessed on 25 October 2021).
47. Li, H.; Shi, A.; Li, M.; Zhang, X. Effect of pH, temperature, dissolved oxygen, and flow rate of overlying water on heavy metals release from storm sewer sediments. *J. Chem.* **2013**, *2013*, 434012. [CrossRef]
48. Zhu, K.; Blum, P.; Ferguson, G.; Balke, K.D.; Bayer, P. The geothermal potential of urban heat islands. *Environ. Res. Lett.* **2010**, *5*, 044002. [CrossRef]
49. Lomboy, M.; Riego de Dios, J.; Magtibay, B.; Quizon, R.; Molina, V.; Fadrilan-Camacho, V.; See, J.; Enoveso, A.; Barbosa, L.; Agravante, A. Updating national standards for drinking-water: A Philippine experience. *J. Water Health* **2017**, *15*, 288–295. [CrossRef] [PubMed]
50. Edwards, K.A.; Classen, G.A.; Schroten, E.H.J. The Water Resource in Tropical Africa and Its Exploitation. Available online: https://www.ilri.org/publications/water-resource-tropical-africa-and-its-exploitation (accessed on 20 November 2021).
51. Anyanwu, B.O.; Ezejiofor, A.N.; Igweze, Z.N.; Orisakwe, O.E. Heavy metal mixture exposure and effects in developing nations: An update. *Toxics* **2018**, *6*, 65. [CrossRef]
52. Fenton, T.R.; Huang, T. Systematic review of the association between dietary acid load, alkaline water and cancer. *BMJ open* **2016**, *6*, e010438. [CrossRef] [PubMed]
53. Jan, A.T.; Azam, M.; Siddiqui, K.; Ali, A.; Choi, I.; Haq, Q.M. Heavy metals and human health: Mechanistic insight into toxicity and counter defense system of antioxidants. *Int. J. Mol. Sci.* **2015**, *16*, 29592–29630. [CrossRef]

54. Patil, V.T.; Patil, P.R. Physicochemical Analysis of Selected Groundwater Samples of Amalner Town in Jalgaon District, Maharashtra, India. *E-J. Chem.* **2010**, *7*, 111–116. [CrossRef]
55. Oyem, H.H.; Oyem, I.M.; Ezeweali, D. Temperature, pH, electrical conductivity, total dissolved solids and chemical oxygen demand of groundwater in Boji-BojiAgbor/Owa area and immediate suburbs. *Res. J. Environ. Sci.* **2014**, *8*, 444. [CrossRef]
56. Ramasamy, S. Barium in Drinking Water—Background document for development of WHO Guidelines for Drinking-water Quality. Available online: https://www.who.int/water_sanitation_health/water-quality/guidelines/chemicals/barium-background-jan17.pdf (accessed on 26 October 2021).
57. Kroschwitz, J.I.; Howe-Grant, M. *Kirk-Othmer Encyclopedia of Chemical Technology*, 3rd ed.; Grayson, M., Eckroth, D., Eds.; John Wiley and Sons: New York, NY, USA, 1978; Volume 3, pp. 457–463.
58. Miner, S. *Air Pollution Aspects of Barium and Its Compounds*; Contract No. Ph-22-68-25; Litton Systems, Inc.: Bethesda, MD, USA, 1969; p. 69.
59. Pinto, V.M.; Hartmann, L.A. Flow-by-flow chemical stratigraphy and evolution of thirteen Serra Geral Group basalt flows from Vista Alegre, southernmost Brazil. *An. Acad. Bras. Cienc.* **2011**, *83*, 425–440. [CrossRef]
60. Vieira, I.F.B.; Rolim Neto, F.C.; Carvalho, M.N.; Caldas, A.M.; Costa, R.C.A.; Silva, K.S.D.; Parahyba, R.D.B.V.; Pacheco, F.A.L.; Fernendes, L.F.S.; Pissarra, T.C.T. Water Security Assessment of Groundwater Quality in an Anthropized Rural Area from the Atlantic Forest Biome in Brazil. *Water* **2020**, *12*, 623. [CrossRef]
61. Beyene, G.; Aberra, D.; Fufa, F. Evaluation of the suitability of groundwater for drinking and irrigation purposes in Jimma Zone of Oromia, Ethiopia. *Groundw. Sustain. Dev.* **2019**, *9*, 100216. [CrossRef]
62. Carasek, F.L.; Baldissera, R.; Oliveira, J.V.; Scheibe, L.F.; Dal Magro, J. Quality of the groundwater of the Serra Geral aquifer system of Santa Catarina west region, Brazil. *Groundw. Sustain. Dev.* **2020**, *10*, 100346. [CrossRef]
63. Maher, K. The dependence of chemical weathering rates on fluid residence time. *Earth Planet. Sci. Lett.* **2010**, *294*, 101–110. [CrossRef]
64. Sholehhudin, M.; Azizah, R.; Sham, A.S.S.M.; Amiruddin, Z. Analysis of Heavy Metals (Cadmium, Chromium, Lead, Manganese, and Zinc) in Well Water in East Java Province, Indonesia. *Malays. J. Med. Health Sci.* **2021**, *17*, 146–153.
65. Zhang, Z.; Xiao, C.; Adeyeye, O.; Yang, W.; Liang, X. Source and mobilization mechanism of iron, manganese and arsenic in groundwater of Shuangliao City, Northeast China. *Water* **2020**, *12*, 534. [CrossRef]
66. Abou Zakhem, B.; Al-Charideh, A.; Kattaa, B. Using principal component analysis in the investigation of groundwater hydrochemistry of Upper Jezireh Basin, Syria. *Hydrol. Sci. J.* **2017**, *62*, 2266–2279.
67. Sunkari, E.D.; Abu, M. Hydrochemistry with special reference to fluoride contamination in groundwater of the Bongo District, Upper East Region, Ghana. *Sustain. Water. Resour. Manag.* **2019**, *5*, 1803–1814. [CrossRef]
68. Wali, S.U.; Alias, N.; Harun, S.B. Reevaluating the hydrochemistry of groundwater in basement complex aquifers of Kaduna Basin, NW Nigeria using multivariate statistical analysis. *Environ. Earth Sci.* **2021**, *80*, 1–25. [CrossRef]
69. Senoro, D.B.; De Jesus, K.L.M.; Yanuaria, C.A.; Bonifacio, P.B.; Manuel, M.T.; Wang, B.N.; Kao, C.C.; Wu, T.N.; Ney, F.P.; Natal, P. Rapid site assessment in a small island of the Philippines contaminated with mine tailings using ground and areal technique: The environmental quality after twenty years. In *IOP Conference Series: Earth and Environmental Science*; IOP Publishing: Bristol, UK, 2019; Volume 351, p. 012022.
70. Wright, N. Small Island Developing States, disaster risk management, disaster risk reduction, climate change adaptation and tourism. *Background Paper for the Global Assessment Report on DRR 2013*. 2013. Available online: https://www.preventionweb.net/english/hyogo/gar/2013/en/bgdocs/Wright,%20N.,%202013.pdf (accessed on 20 November 2021).
71. Lopez-Maldonado, Y.; Batllori-Sampedro, E.; Binder, C.R.; Fath, B.D. Local groundwater balance model: Stakeholders' efforts to address groundwater monitoring and literacy. *Hydrol. Sci. J.* **2017**, *62*, 2297–2312. [CrossRef]
72. Galhardi, J.A.; Bonotto, D.M. Hydrogeochemical features of surface water and groundwater contaminated with acid mine drainage (AMD) in coal mining areas: A case study in southern Brazil. *Environ. Sci. Pollut.* **2016**, *23*, 18911–18927.
73. Williams, M.; Todd, G.D.; Roney, N.; Crawford, J.; Coles, C.; McClure, P.R.; Garey, J.D.; Zaccaria, K.; Citra, M. *Toxicological Profile for Manganese*; Agency for Toxic Substances and Disease Registry (US): Atlanta, GA, USA, 2012.
74. Erikson, K.M.; Thompson, K.; Aschner, J.; Aschner, M. Manganese neurotoxicity: A focus on the neonate. *Pharmacol. Ther.* **2007**, *113*, 369–377. [CrossRef]
75. Bhasin, G.; Kauser, H.; Athar, M. Iron augments stage-I and stage-II tumor promotion in murine skin. *Cancer Lett.* **2002**, *183*, 113–122. [CrossRef]
76. Grazuleviciene, R.; Nadisauskiene, R.; Buinauskiene, J.; Grazulevicius, T. Effects of elevated levels of manganese and iron in drinking water on birth outcomes. *Pol. J. Environ. Stud.* **2009**, *18*, 819–825.
77. Kravchenko, J.; Darrah, T.H.; Miller, R.K.; Lyerly, H.K.; Vengosh, A. A review of the health impacts of barium from natural and anthropogenic exposure. *Environ. Geochem. Health.* **2014**, *36*, 797–814. [CrossRef] [PubMed]
78. MacDonald, A.M.; Bonsor, H.C.; Ahmed, K.M.; Burgess, W.G.; Basharat, M.; Calow, R.C.; Dixit, A.; Foster, S.S.D.; Gopal, K.; Lapworth, D.J.; et al. Groundwater quality and depletion in the Indo-Gangetic Basin mapped from in situ observations. *Nat. Geosci.* **2016**, *9*, 762–766. [CrossRef]

applied sciences

MDPI

Article

Chaotic Characteristic Analysis of Vibration Response of Pumping Station Pipeline Using Improved Variational Mode Decomposition Method

Li Jiang [1,2], Zhenyue Ma [1], Jianwei Zhang [2,*], Mohd Yawar Ali Khan [3], Mengran Cheng [2] and Libin Wang [2]

1 School of Hydraulic Engineering, Dalian University of Technology, Dalian 116024, China; jiangli@ncwu.edu.cn (L.J.); dmzy@dlut.edu.cn (Z.M.)
2 School of Water Conservancy, North China University of Water Resources and Electric Power, Zhengzhou 450046, China; cmr6688@126.com (M.C.); wanglb1994@126.com (L.W.)
3 Department of Hydrogeology, King Abdulaziz University, Jeddah 21589, Saudi Arabia; makhan7@kau.edu.sa
* Correspondence: zjwcivil@126.com

Citation: Jiang, L.; Ma, Z.; Zhang, J.; Khan, M.Y.A.; Cheng, M.; Wang, L. Chaotic Characteristic Analysis of Vibration Response of Pumping Station Pipeline Using Improved Variational Mode Decomposition Method. *Appl. Sci.* **2021**, *11*, 8864. https://doi.org/10.3390/app11198864

Academic Editors: Amit Kumar, Santosh Subhash Palmate and Rituraj Shukla

Received: 18 August 2021
Accepted: 16 September 2021
Published: 23 September 2021

Abstract: The measured vibrational responses of the pumping station pipeline in the irrigation site were chosen to confirm the chaotic characteristics of the pumping station pipeline vibration and to determine the vibrational excitation that makes it chaotic. First, the chaotic properties of the pipeline vibration responses were investigated using a saturation correlation dimension and the maximum Lyapunov exponent. The vibration excitation with chaotic features was obtained using an improved variational mode decomposition (IVMD) method to examine the multi-time-scale chaotic characteristics of the pipeline vibration responses. The results show that the vibrational responses of each measuring point of the pipeline under different operating conditions have clear chaotic characteristics, where the chaotic characteristics of the axial points and bifurcated pipe points are relatively strong. The vibration of the operating conditions and measurement points affected by the unit's operation and flow state change is further complicated. The intrinsic mode function (IMF) produces a low-dimensional chaotic attractor after the IVMD disrupts the vibration response. Still, the vibration excitation of the remaining components on behalf of the units does not have chaotic properties, implying that water pulsation excitation makes the pumping station pipeline vibrations chaotic. The vibration excitation caused by the unit's operation covers the chaotic characteristics of the pipeline vibration and increases its uncertainty. The outcomes of this study provide a theoretical basis for further exploration of the vibration characteristics of pumping station pipelines, and a new method of chaos analysis is proposed.

Keywords: pumping station pipeline; chaotic characteristic; IVMD; vibration response; correlation dimension; Lyapunov exponent

1. Introduction

High-lift pumping stations and water-diversion irrigation areas have been built in many water-deficient areas due to the continuous development of electric water-lifting equipment and water-diversion irrigation technology in China. These projects have created enormous economic, ecological and social benefits. Thus, ensuring their safe and stable operation is the main task of modernising and developing water conservation in China [1]. Natural and human forces create varying degrees of pipeline vibration during long-term operation at pumping stations [2]. Long-term irregular pipe vibration will lead to the loosening of the pipelines and their auxiliary system, causing catastrophic damage in severe cases [3]. Therefore, it is of great research interest to analyse the vibration characteristics of the pumping station pipeline to avoid its adverse vibrations.

Chaos is a unique mechanical phenomenon in the vibration of strongly nonlinear structures. Most researchers believe that the vibrations of pipelines are weakly nonlinear, so they mainly focus on studying chaotic phenomena caused by the flow of water and other excitation sources such as flow. There is a scarcity of research on the chaotic processes in the pipeline itself when researching nonlinear problems. It is found that the chaotic phenomena of pipelines do not depend solely on the strength of structural nonlinearity; for some weak nonlinear or even linear structures, chaos occurs [4]. PaïDoussis studied the dynamics of a cantilever pipeline with nonlinear constraints and constant internal flow, which discovered the chaotic motion of the system [5]. Tang obtained the chaotic characteristics of the transport pipeline by increasing the nonlinear force and found that the occurrence of chaos is mainly affected by the flow velocity in the pipeline [6]. B.G. Sinir investigated the nonlinear vibrations of slightly curved pipes that transport fluid with constant velocity [7]. The periodic and chaotic movements have been observed in the transverse vibrations of slightly curved pipes transporting fluid. Zhao analysed the chaotic phenomenon in the pipeline vibration caused by the flow pulsation excitation under thermal load and then obtained the relationship between the frequency response and flow velocity [8].

Research on the chaotic characteristics of pipeline systems mainly focuses on oil-gas pipelines and the mathematical models of pipelines with specific nonlinear constraints. In contrast, the chaotic characteristics of pumping station pipeline systems are rarely studied. Most of the previous research achievements have only analysed the chaotic characteristics of the vibration system but have not further explored the vibrational excitation that caused the chaos. In this paper, the measured vibration responses of the pumping station pipeline in an irrigation area are taken as the research objective. The chaotic characteristics of the vibratory responses of the pumping station pipeline under different working conditions are analysed by using the saturation correlation dimension and the largest Lyapunov exponent. In addition, the IVMD method is used to decompose the vibration responses of the measurement points under typical working conditions. The chaotic characteristics of the IMFs are analysed to obtain the vibratory excitation that causes the chaotic characteristics of the pumping station pipeline.

2. Theoretical Aspects

2.1. Identification Method of Chaotic Characteristics

There are many methods for identifying chaotic characteristics, which are roughly divided into qualitative and quantitative analysis. Orbit observation, Poincare surface of section, and power spectral analysis are examples of qualitative approaches [9–11]. These methods are feasible and straightforward, but they are limited in determining whether the system has chaotic characteristics and cannot perform transverse comparisons under different operating conditions. Quantitative methods, such as the saturation correlation dimension method [12] and the largest Lyapunov exponent method [13], can reflect the vibration complexity and the degree of chaos under different conditions by comparing the values of the parameters. The saturation correlation dimension and the largest Lyapunov exponent are chosen as the chaotic identification indexes of pipeline vibration responses in pumping stations to improve the trustworthiness of the result.

2.1.1. Saturation Correlation Dimension

The correlation dimension characterises the compactness of a dynamic system and is used to reflect the system's complexity. When the saturation correlation dimension is fractional, the system is said to have chaotic properties. For an m-dimensional phase space, its correlation function can be defined as follows:

$$C(r) = \lim_{M\to\infty} \frac{2}{M(M-1)} \sum_{1 \le i \le j \le M} H(r - \|Y_i - Y_j\|) \quad (1)$$

where $M = N - (m-1)\tau$ is the number of phase points, $H(u)$ is the Heaviside function, N is the time series; r is the vector point in the time series; M is the embedding dimension; Tau is the time delay; Y is the reconstruction vector.

When the time series is chaotic, for the positive, the relationship between the correlation function $C(r)$ and r is

$$C(r) \propto \alpha r^{D_2} \tag{2}$$

where α is a constant, D_2 is the correlation dimension which can be obtained by the slope of the $\log_2 C(r) \sim \log_2 r$ curve, that is

$$D_2 = \lim_{r \to 0} \frac{\log_2 C(r)}{\log_2 r} \tag{3}$$

Due to the noise in the measured signal, the embedding dimension is generally controlled to rise gradually. The apparent straight line segments in the $\log_2 C(r) \sim \log_2 r$ curve are fitted using the least square method for each embedding dimension. The slope of each segment increases with the rising embedding dimension, and eventually reaches saturation, the saturation correlation dimension.

2.1.2. Largest Lyapunov Exponent

The Lyapunov exponent determines the chaotic characteristics of the system based on the diffusion of the phase trajectory. Generally, the direction represented by the positive Lyapunov exponent supports the attractors. In contrast, the contraction direction corresponding to the negative Lyapunov exponent contributes to the attractor dimension's fractional part after counteracting the expansion direction's effect. Thus, the positive Lyapunov exponent is a prominent feature of chaos. Suppose λ_1 as the largest Lyapunov exponent of a system, then the chaotic components of the system can be found if λ_1 is positive, and its value reflects the chaos degree.

Rosenstein [14] proposed the small data sets for computing λ_1. Its basic steps are as follows:

Choose suitable τ and m to reconstruct the phase space and find the nearest neighbor point $Y_{\hat{i}}$ of each Y_i in the phase space. Short separation limitation is as follows:

$$d_i(0) = \min_{\hat{i}} \|Y_i - Y_{\hat{i}}\| (|i - \hat{i}| > p) \tag{4}$$

where p is the average period of time series, i is the vector in space, $\hat{i}$ is the vector of the nearest neighbor of the second vector.

Define the distance of Y_{i+j} and $Y_{\hat{i}+j}$ as

$$d_i(j) = \|Y_{i+j} - Y_{\hat{i}+j}\| \tag{5}$$

where $j = 0, 1, 2, \cdots, \min(M - i, M - \hat{i})$.

For each j, compute the $\ln d_i(j)$ average as follows:

$$y(i) = \frac{1}{q\Delta t} \sum_{i=1}^{q} \ln d_i(j) \tag{6}$$

where q is the number of nonzero $\ln d_i(j)$. The slope of the regression line made by the least square method is λ_1.

2.2. *Improved Variational Mode Decomposition (IVMD)*

Variational mode decomposition (VMD) is a new method of multi-component adaptive signal decomposition [15]. Compared to traditional signal decomposition methods, it effectively avoids modal aliasing and over-decomposition defects and has a higher utilization value [16]. VMD comprises two processes comprising the establishment of variational constraints and iteration to find the optimal solution. The specific operation process is as follows: VMD decomposes a given signal f into K modal functions using variational constraints $m_k(t)$. The bandwidth of each IMF is limited, and each IMF is distributed around the central pulsating frequency. The variational constraint model is as follows [17]:

$$\begin{cases} \min\limits_{m_k, w_k} \left\{ \sum\limits_k \left\| \partial_t \left[\left(\sigma(t) + \frac{j}{\pi t} \right) m_k(t) \right] e^{-jw_k t} \right\|_2^2 \right\} \\ s.t. \sum\limits_k m_k = f \end{cases} \quad (7)$$

where $\{m_k\}$ represents the decomposed K IMF components, $\{m_k\} = \{m_1, m_2, \cdots, m_k\}$; $\sigma(t)$ is a pulse function; $\{w_k\}$ is the central frequency of each IMF, $\{w_k\} = \{w_1, \ldots w_k\}$.

To complete the adaptive decomposition of input signals f and to obtain the IMFs with the minimum sum of bandwidth, the following expanded Lagrange expression is introduced:

$$L(m_k, w_k, \lambda) = \alpha \sum_k \|\partial(t) \left[\left(\delta(t) + \frac{j}{\pi t} \right) m_k(t) \right] e^{-jw_k t}\|_2^2 + \|f(t) - \sum_k m_k(t)\|_2^2 + \left\langle \lambda(t), f(t) - \sum_k m_k(t) \right\rangle \quad (8)$$

where α is the penalty factor to ensure the accuracy of signal reconstruction; $\lambda(t)$ is a Lagrange multiplier used to strengthen the constraint; $\langle\rangle$ represents the inner product operation.

To solve the above variational constraint problem, the dual decomposition and alternate direction multiplication sub-algorithm are used [18]. Keep updating m_k, w_k and $\lambda(t)$ to find the saddle point of Equation (8), that is, the optimal solution of Equation (7). The modal component function m_k and the central frequency w_k are

$$m_k^{n+1}(w) = \frac{f(w) - \sum\limits_{i \neq k} m_i(w) + \frac{\lambda(\omega)}{2}}{1 + 2\alpha(w - w_k)^2} \quad (9)$$

$$w_k^{n+1} = \frac{\int_0^\infty w|m_k(w)|^2 dw}{\int_0^\infty |m_k(w)|^2 dw} \quad (10)$$

$$\lambda^{n+1} = \lambda^n + \tau \left(f(w) - \sum_k m_k^{n+1}(w) \right) \quad (11)$$

When VMD decomposes the vibration response sequence, determining the total modal number is a crucial step. The selection of modal parameters K greatly affects the accuracy of the results [19]. A parameter K is usually challenging to determine. If K is greater than the number of useful components obtained by signal decomposition, information superposition will occur; if K is smaller than it, a part of the limited bandwidth of the solid modulus cannot be decomposed. An IVMD method based on the mutual information (MI) method is proposed for K selection.

MI reflects the correlation between two random variables and allows better identification of the degree of correlation [15]. MI is as follows:

$$I(X, Y) = H(Y) - H(X|Y) \quad (12)$$

where $H(Y)$ is the entropy of Y, and $H(Y|X)$ is the conditional entropy of Y when X is known. When $I(X, Y) = 0$, X and Y are independent of each other.

The mutual information I_k of the original signal and each IMF obtained by the IVMD decomposition is calculated and normalised by Equation (13). Then the correlation between each modal component and the original signal is judged, that is, whether the original signal is completely decomposed.

$$\sigma_i = \frac{I_i}{\max(I_i)} \quad (13)$$

where σ_i is the normalized mutual information value of each IMF, i $=1, 2, \ldots$ k. Refer to reference [20], when σ_i is less than 0.02, it is considered that the IMF does not contain valid feature information. The original signal has been decomposed completely.

The specific algorithm for adaptive determination of K using MI method is as follows:

Step 1: Initialize $n = n + 1$, assign $K = 1$;
Step 2: $K = K + 1$, perform outer circulation;
Step 3: Initialize m_k^1, w_k^1, λ^1 and n, assign $n = 0$;
Step 4: Order $n = n + 1$ to execute the inner loop;
Step 5: For all $w \geq 0$, according to Equations (9) and (10), m_k and w_k are updated, respectively;
Step 6: Update λ according to Equation (11);
Step 7: For a given discriminate accuracy $e > 0$, if the iteration condition $\sum_k \frac{\|m_k^{n+1} - m_k^n\|_2^2}{\|m_k^n\|_2^2} < \mathrm{e}$ is satisfied, the process is terminated, otherwise loop step 2 to step 6;
Step 8: Circulate step 2 to 7 until the set threshold σ is greater than the normalized mutual information σ_i, that is, if $I(f - \sum m_k, f) < \sigma$, end the cycle.

The flow chart of the above calculation steps is shown in Figure 1.

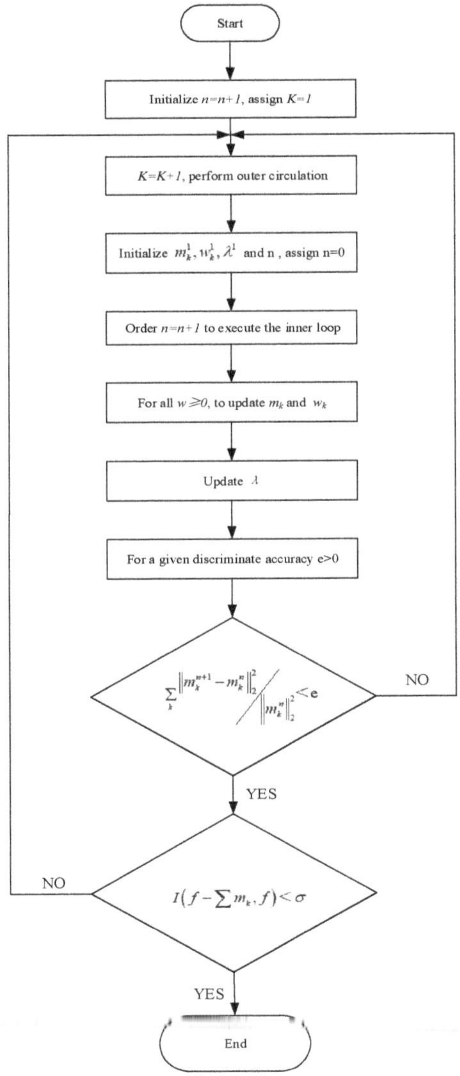

Figure 1. Flow chart for adaptive determination of K.

3. Chaotic Characteristics Analysis of Pipeline Vibration Response

The pipe material is stainless steel. Model 891–2 vibration sensors are used in the test, which are divided into four grades: small speed, medium speed, large speed and acceleration. So speed sensors are used in this test. Taking the No. 2 pressure pipeline of the Jingdian Project pumping station No. 3 as a research objective, the No. 4 and No. 5 units of a 1200S–56 horizontal centrifugal pump are connected with the branch pipe. Six measurement points are selected on the main pipe and two branches of the pipeline. Each point is equipped with vibration sensors in X, Y and Z directions. The measuring points are arranged as shown in Figure 2.

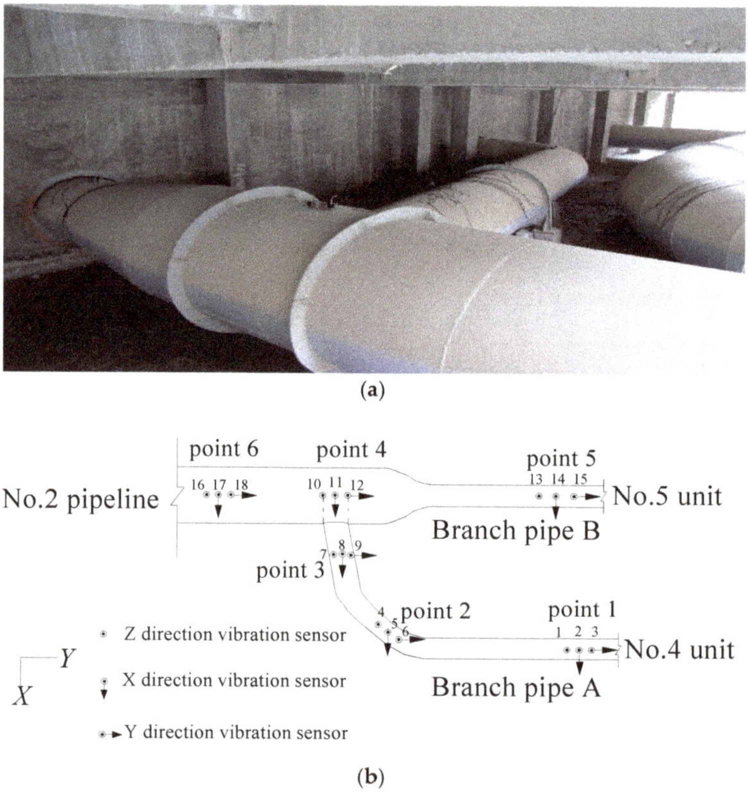

Figure 2. Layout of pipeline measuring points (**a**) Field test of pipeline, and (**b**) Measuring points layout (Note: 1~18 is the sensor number).

In the prototype test, four working conditions were selected to collect the vibration responses of the pipeline. The descriptions of each working condition, sampling time and sampling frequency are shown in Table 1.

Table 1. Four working conditions.

Cases	Description of Working Conditions	Sampling Time/s	Sampling Frequency/Hz
1	No. 4 unit stable operating	900	512
2	No. 4 unit opening	1800	512
3	No. 4 unit closing	1800	512
4	No. 4 and 5 units stable operating	900	180

The velocity-time history of points under typical conditions is shown in Figure 3. The chaotic characteristic analysis of the vibration responses under different conditions is carried out as follows:

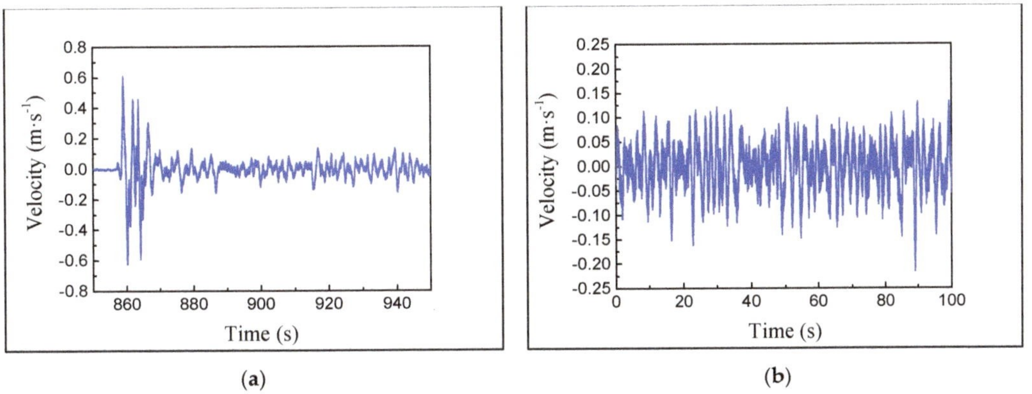

Figure 3. Velocity time history of points under typical conditions (**a**) Z-axis vibration of point 1 under condition 2, and (**b**) Z-axis vibration of point 1 under condition 4.

First, the reconstruction of the phase space of the time series is performed, that is, the calculation of the time delay τ and embedding dimension m. The CAO method essentially uses the minimum error method to determine the embedding dimension, which was proposed by Liangyue Cao in 1997. This paper calculates τ by the autocorrelation function method and chooses the CAO method to obtain m [21]. The calculation process of τ and m is illustrated by taking the Z-axis vibration of point 1 under condition 4 as an example.

In the process of calculating τ by the autocorrelation function method, when the value drops to 1–1/e of the initial value, the corresponding time delay is τ. The result of the autocorrelation function is shown in Figure 4.

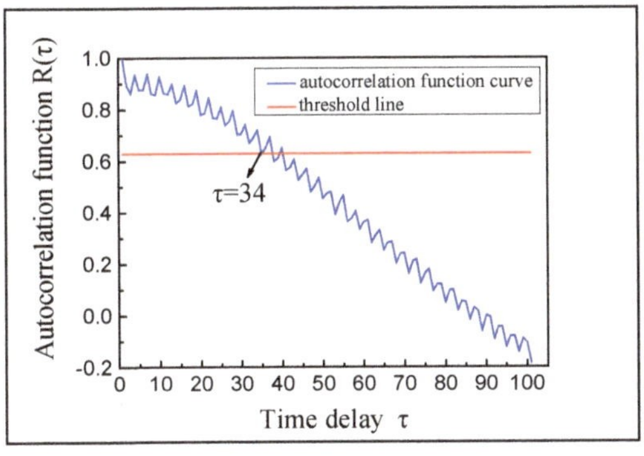

Figure 4. τ Calculation of point 1 Z-axis vibration under condition 4.

After obtaining τ, the embedding dimension is determined by the CAO method. $E_1(m)$ represents the minimum embedding dimension. $E_2(m)$ represents the characteristics of time series. When $E_1(m)$ obviously no longer changes with the increase, and the $E_2(m)$ value tends towards 1, the corresponding m is the optimal embedding dimension. From Figure 5, we can see that the optimal embedding dimension m of point 1 Z-axis vibration under condition 4 is 11.

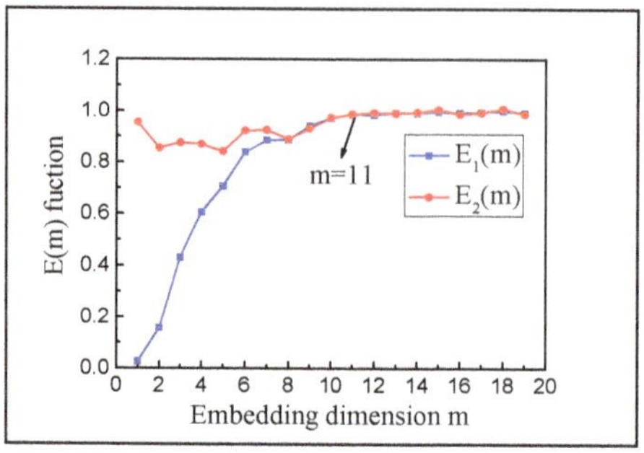

Figure 5. *m* Calculation of point 1 Z-axis vibration under condition 4.

The G-P algorithm [22] and the small data sets are chosen to calculate the saturation correlation dimension and the largest Lyapunov exponent. Two types of indexes are used to analyse the chaotic characteristics of time series.

The G-P algorithm is a chaotic eigenvalue calculation method proposed by Grassberger and Procaccia to calculate the saturation correlation dimension D_2.

The embedding dimension is selected as $m = 2, 4, 6, \cdots, 20$ and the τ has been calculated above. According to the correlation function relation in Equation (3), the $\log_2 C(r) \sim \log_2 r$ double logarithmic relation graph of different m is plotted, respectively. The slope fitted by the near line segment of the curve is the correlation dimension under the corresponding embedding dimension. As the embedding dimension increases, it is the saturation dimension D_2 when the correlation dimension reaches saturation. Figure 6 is the diagram representing the calculation of the saturation correlation dimension of specific points.

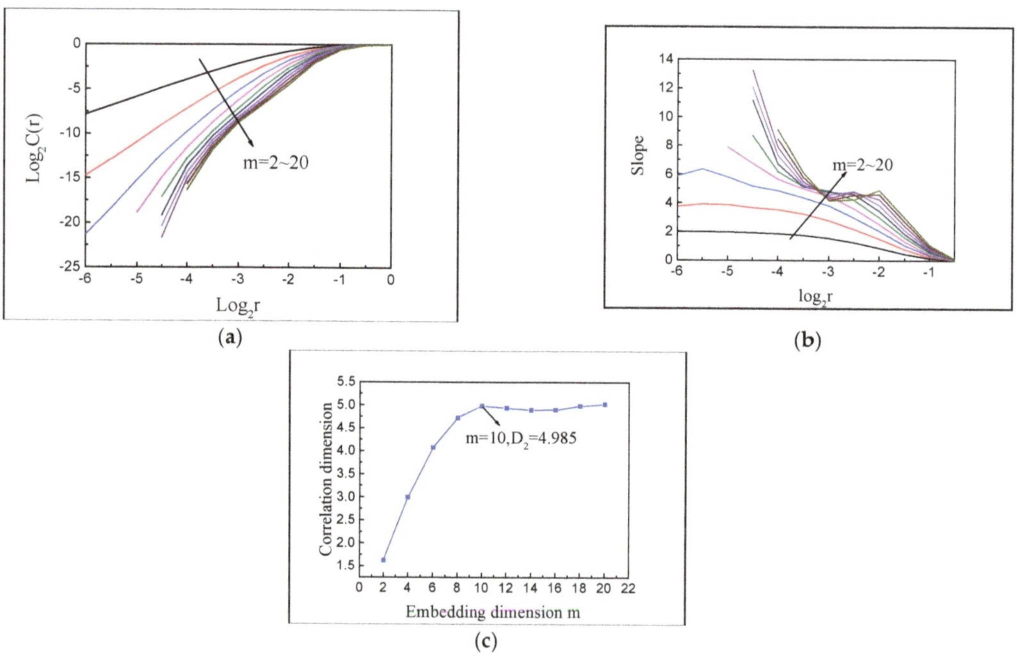

Figure 6. Calculation of point 1 z-axis vibration under condition 4 (**a**) Double logarithmic curve, (**b**) Slope of double logarithmic curve, and (**c**) Relation between D_2 and m.

To reveal the distribution law of the saturation correlation dimension, the D_2 variation curves of points in each direction under different conditions are shown in Figure 7.

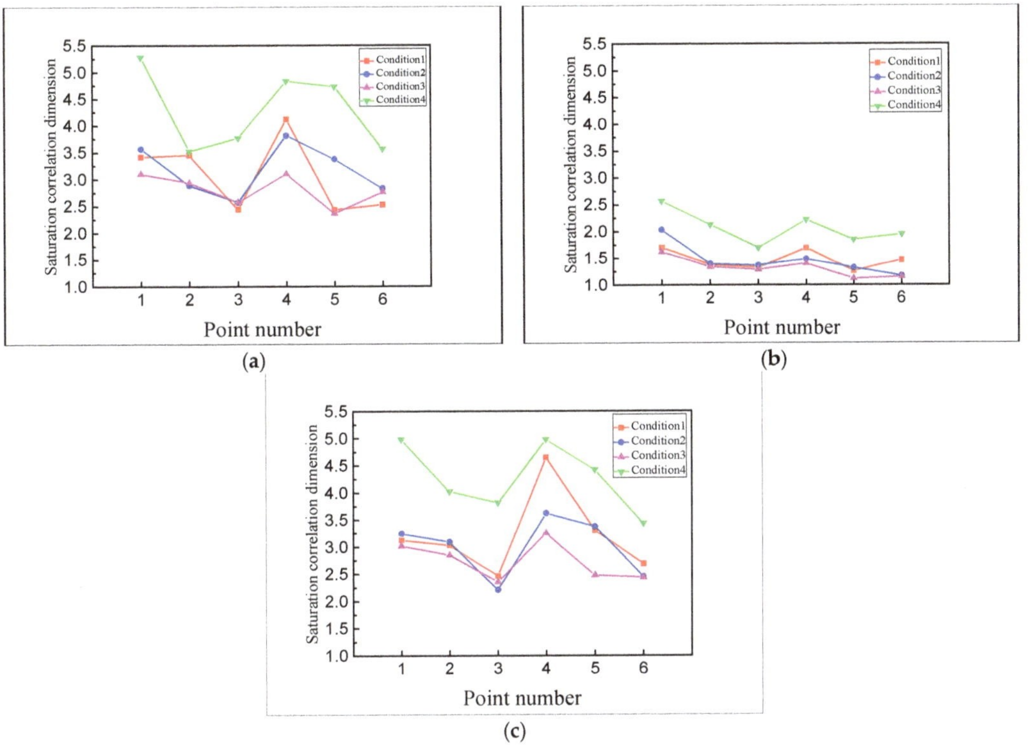

Figure 7. Correlation dimension curves of points in different directions (**a**) X-axis points, (**b**) Y-axis points, and (**c**) Z-axis points.

As can be seen from Figure 7:

(1) In general, D_2 ranges from 1.156 to 5.283, and they are fractional, indicating that the responses of the pipeline in all directions are chaotic;

(2) Compared with the other two directions, the correlation dimension of the axial measurement points (Y-axis) of the main pipe is obviously smaller than that of the other two directions. It shows that the axial vibration of the pipeline has a smaller dimension chaotic attractor and requires fewer independent control variables to describe the dynamic system. This is mainly because the direction of the centrifugal force generated by the centrifugal pump of units is not in the axial direction of the main pipeline;

(3) At the same points, the D_2 of each point in condition 4 (No. 4 and 5 units in stable operation) is greater than in other conditions, while the corresponding D_2 in condition 3 (No. 4 unit in closing) is less. It indicates that the pipeline vibration is more complicated in the stable conditions of the two units and the complexity of the pipeline vibration is relatively weak in the closed condition. Unit operation increases the uncertainty of the pipeline vibration;

(4) In the same condition, the points near the units (point 1 and 5) and the bifurcated pipe (point 4) reach a relatively larger D_2, indicating that the vibration complexity of the pumping station pipe is greatly affected by the unit's vibration and the flow pattern stability.

To verify the validity of the above analysis results, the chaotic characteristics of the pumping station pipeline are further analyzed by using the largest Lyapunov exponent λ_1. According to the time delay τ and embedding dimension m, the small data sets calculate the largest Lyapunov exponent. Figure 8 is the λ_1 calculation diagram of typical points, and the value of the separation factor $y(i)$ tends to be stable after nearly linear growth. The linear slope is adjusted by the least square method, and the value is λ_1. The λ_1 of each point in different vibration directions are shown in Figure 9.

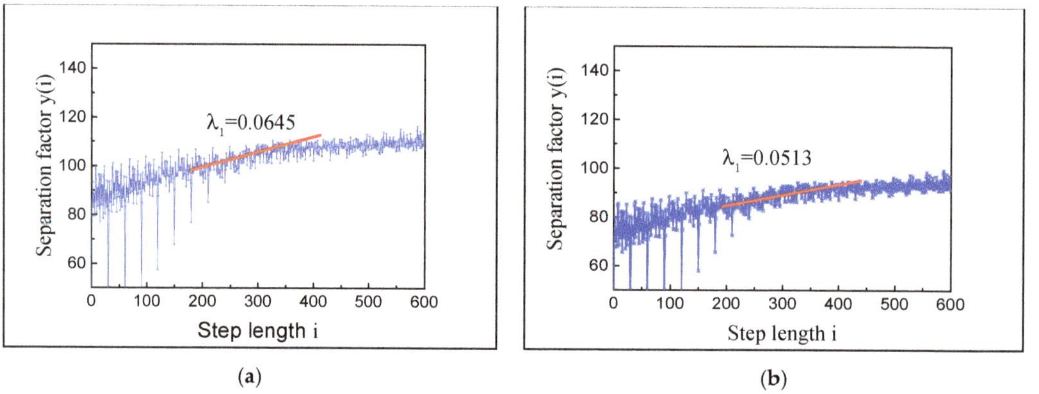

Figure 8. Calculation diagram of typical points (**a**) Point 3 Y-axis vibration under condition 1, and (**b**) Point 1 Z-axis vibration under condition 4.

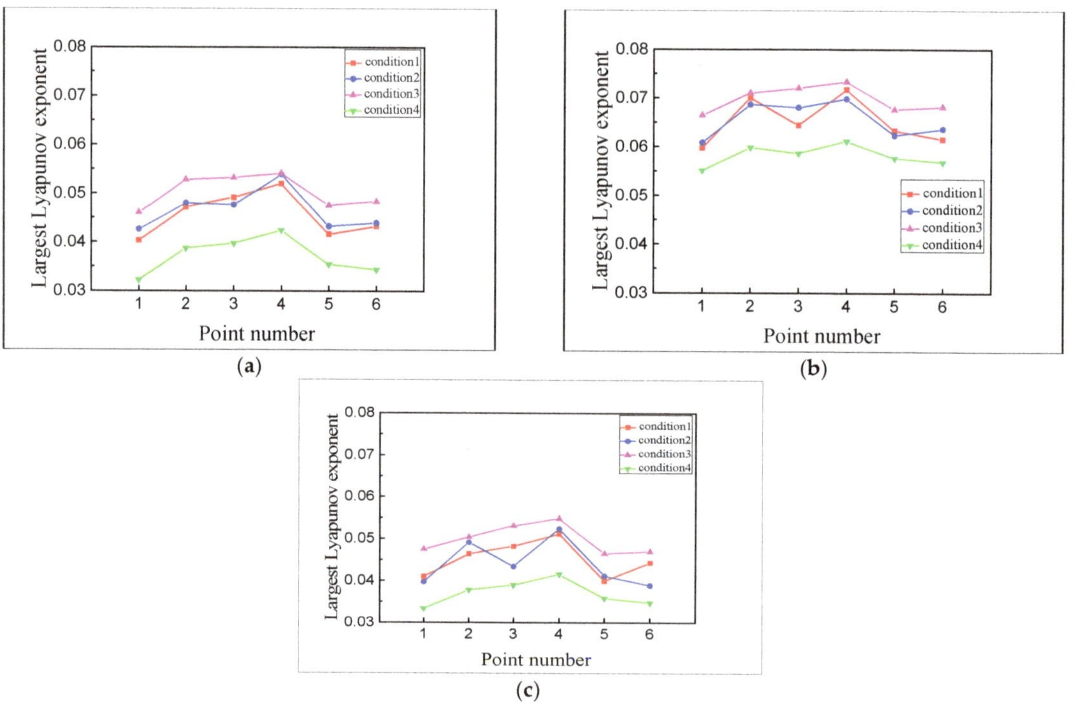

Figure 9. Largest Lyapunov exponent curves of points in different directions (**a**) X-axis points (**b**) Y-axis points, and (**c**) Z axis points.

As shown in Figure 9:

(1) The largest Lyapunov exponents λ_1 of different points are between 0.0323 and 0.0734, greater than 0. It shows that the measured vibration responses of pipelines have obvious chaotic characteristics. Also, the axial points λ_1 of the main pipeline (Y-axis) are obviously larger than the other two directions in the same condition, indicating the chaotic characteristics of the points separated from the influence of centrifugal force generated by pumping station units are more obvious;

(2) The λ_1 of measuring points under condition 4 (No. 4 and 5 units in stable operation) are lower than in other conditions, while the λ_1 of condition 3 (No. 4 unit closing) is relatively larger. The largest Lyapunov exponent λ_1 decreases with the start-up of the two units, indicating that the units' operation weakens the chaotic characteristics of the pipeline vibration;

(3) In the same condition, the λ_1 of the points near the units (point 1 and 5) are smaller. In contrast, while the λ_1 of the points at the pipeline's bifurcation (point 4) are greater than that of other measuring points, indicating that the sudden change of the flow state in the pipeline makes the vibration more chaotic, and the units' operation reduces the chaotic degree of vibration signals near the units.

The above analysis is complementary to the calculation results of the saturation correlation dimension D_2, which further confirms that the unit's operation and flow state changes greatly impact the chaotic characteristics of the pumping station pipeline.

4. The Analysis of Multi-Time-Scale Chaotic Characteristics Based on IVMD

The vibration characteristics of the pumping station pipeline are different from those of the general pipeline, which is mainly reflected in the influence of the pumping station unit on the vibration of the connecting pipeline. The vibration sources are primarily composed of low-frequency water pulsations caused by the pipeline flow and blade frequency, rotation frequency and frequency doubling produced by the unit's operation [23].

Taking the vibration response of the specific point (point 1 Z-axis vibration under condition 4) as an example, the spectrum analysis is shown in Figure 10. Concerning the author's previous article [23,24], 20, 40 and 60 Hz are the blade frequency, the rotation frequency and the frequency doubling, respectively, and 0.5 Hz is the low-frequency water pulsation. Spectrum analysis shows that the frequency band of the vibration excitation caused by water pulsation (0.5 Hz) is relatively wide. The wide-peak power spectrum is the typical characteristic of the chaotic system [25,26]. The pipeline vibration excitation produced by the unit's operation (20 Hz, 40 Hz, and 60 Hz) corresponds to the peak power spectrum and has high periodicity. Therefore, it is speculated that the chaotic characteristics of the pipeline are mainly caused by water pulsation, while the unit vibration masks the chaotic characteristics of the pump station pipeline.

The excitation components of different time scales must be effectively separated to clarify the vibration excitation with chaotic characteristics. As a new signal decomposition method, IVMD can adaptively decompose a signal into a series of IMFs with different scale characteristics. Therefore, the IVMD method is used to identify the vibration excitation that causes the chaotic characteristics of the pipeline.

The multi-time scale chaotic vibration response characteristics of the specific point (point 1 Z-axis vibration under condition 4) are analysed.

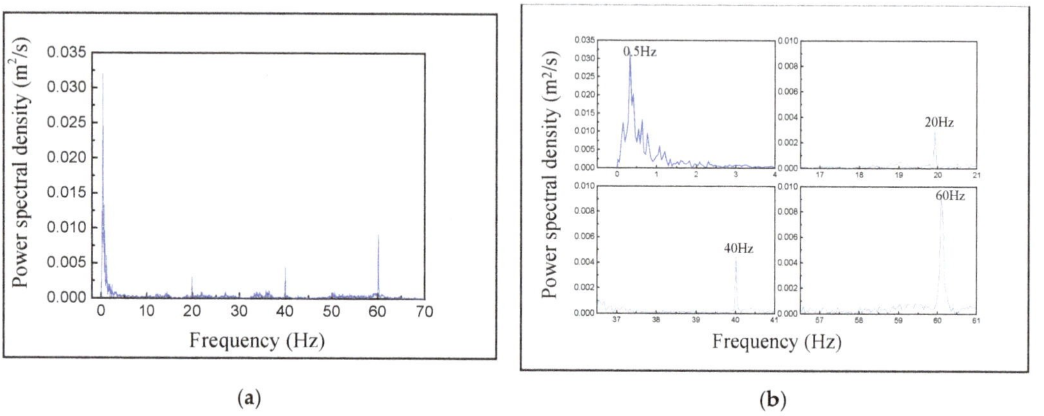

Figure 10. Spectrogram of point 1 Z-axis vibration response under condition 4 (**a**) Holistic drawing, and (**b**) Partial enlarged drawing.

The modal parameters K of IVMD are determined as 4 by the MI method. Four IMFs are obtained by the IVMD decomposition of point 1 Z-axis vibration response under condition 4. Figure 11 is the time history of decomposed IMFs.

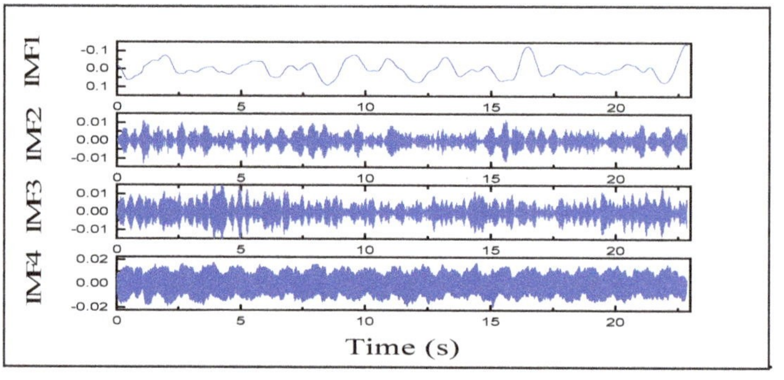

Figure 11. Time histories of IMFs decomposed by IVMD.

Mutual information value: IMF1 is 1.000, IMF2 is 0.025, IMF3 is 0.038, IMF4 is 0.0661. It can be seen that the normalized mutual information values of the IMFs are all above the threshold of 0.02, which meets the decomposition requirements. Figure 11 shows that IVMD can sequentially decompose the original vibrational response to obtain four IMFs with increasing frequency. The frequencies from IMF1 to IMF4 correspond to four major frequency bands in the original response spectrum: 0.5, 20, 40 and 60 Hz, respectively, and the decomposition effect is improved. Then the chaotic characteristics of the decomposed IMFs are analysed using the saturation correlation dimension and the largest Lyapunov exponent.

The calculation process of typical IMF chaotic eigenvalues is shown in Figure 12. From Figure 12a,b the saturation correlation dimension D_2 of IMF1 is 1.115, and the largest Lyapunov exponent λ_1 is 0.0774, indicating that IMF1 has prominent chaotic characteristics. The near-linear region of the D_2 logarithmic curve cannot be found from IMF2 to IMF4; these components have no chaotic characteristics. Due to space limitations, only the slope of the IMF2 double logarithmic curve is given in Figure 12c.

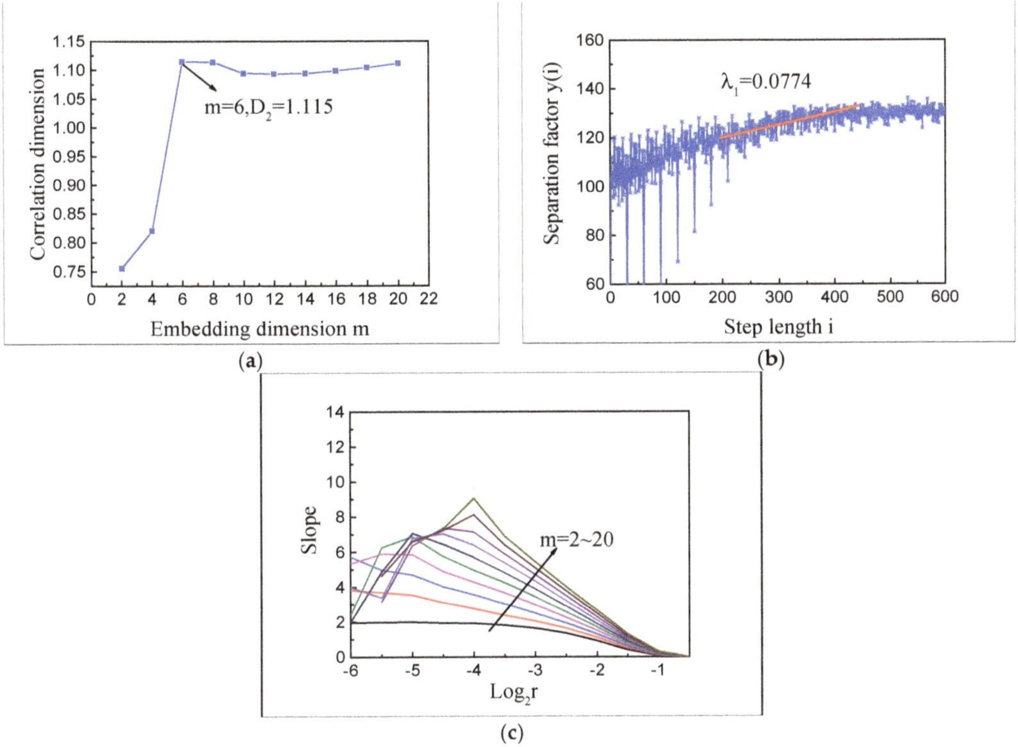

Figure 12. Calculation of chaotic eigenvalues of typical IMFs (**a**) Relation between D_2 and m of IMF1 (**b**) λ_1 Separation factor function of IMF1 (**c**) Slope of double logarithmic curve of IMF2.

By comparing the results of the chaotic eigenvalues of the IMFs with those of the vibrational response before decomposition, it can be concluded that:

(1) IMF1, which represents the water pulsation excitation, the saturated correlation dimension 1.115 is a fractal dimension, and the largest Lyapunov exponent is 0.0774 greater than zero, has prominent chaotic characteristics. IMF2 to IMF4, which represent the vibration excitation of the unit's operation, do not have any chaotic characteristics, indicating that the unit's operation cannot cause chaotic characteristics of the pumping station pipeline vibration;

(2) After eliminating the IMFs (IMF2 to IMF4) caused by the unit's operation with no chaotic characteristics, the saturation correlation dimension D_2 of the pipeline vibration response decreases from 4.985 to 1.115. At the same time the largest Lyapunov exponent increases from 0.0513 to 0.0774, that is, the complexity of the pipeline vibration decreases, and its chaotic characteristics are more evident. This shows that when the pumping station pipeline vibrates, the water pulsation excitation makes its vibration have obvious chaotic characteristics. In contrast, vibration excitation generated by the unit's operation masks the chaotic characteristics of the pumping station pipeline and increases the uncertainty of the pipeline vibration.

5. Conclusions

(1) Comparing the saturation correlation dimension D_2 among the vibration responses of the pumping station pipeline under different conditions, the D_2 of the measuring points are distributed in the range of 1.156–5.283, and all are fractions, which show that the vibration of the pumping station pipeline has chaotic characteristics. The

axial vibration of the pipeline presents a chaotic attractor with a lower dimension (1.156~2.569), and the vibration form is relatively simple. At the same time, the D_2 of conditions and points which are greatly affected by the unit's operation have a larger value (3.021~5.283), and the vibration form is more complex;

(2) The Lyapunov exponents λ_1 of measuring points under different conditions are between 0.0513 and 0.0774. With the opening of two units, the largest Lyapunov exponent λ_1 decreases accordingly, suggesting that the unit's operation weakens the chaotic characteristics of the pipeline vibration. The λ_1 of points at the bifurcation are larger than those of other points under the same condition. The chaotic characteristics of the vibration at the bifurcation are enhanced by the sudden expansion of the pipe diameter at the bifurcation and the impact of water heads at different flow velocities;

(3) After the IVMD decomposition of the vibration response of specific points under the unit's operation conditions, the chaotic characteristics of the IMFs are analysed. The results show that the saturation correlation dimension D_2 of IMF1 representing water pulsation excitation in the pipeline is 1.115, and the largest Lyapunov exponent is 0.0774. The IMF2 to IMF4 representing the blade frequency, the rotation frequency, and the frequency doubling vibration excitation generated by the unit's operation do not have chaotic characteristics. It indicates that the chaotic character of the pumping station pipeline is mainly caused by water pulsation in the pipeline, and the vibration caused by the unit masks the chaotic characteristic of the pipeline, which makes the pipeline vibration system more complex.

In this paper, the chaotic characteristics of the vibration system of the pumping station pipeline are shown by the analysis of the measured vibration responses, and the chaotic excitation is found by combination with IVMD, which provides a theoretical basis for the complete description of the vibration characteristics of the pumping station pipeline. A new way of chaotic characteristics analysis based on IVMD decomposition is also proposed.

Author Contributions: Conceptualization, L.J. and Z.M.; methodology, L.J.; software, J.Z. and L.W.; validation, L.J. and J.Z.; formal analysis, J.Z. and M.C.; investigation, L.J.; resources, Z.M.; data curation, J.Z.; writing—original draft preparation, L.J.; writing—review and editing, M.Y.A.K.; visualization, L.J.; supervision, Z.M.; funding acquisition, J.Z. All authors have read and agreed to the published version of the manuscript.

Funding: This research was funded by Program for Science & Water conservancy science and technology innovation project in GuangDong Province, grant number 2020-18.

Institutional Review Board Statement: Not applicable.

Informed Consent Statement: Not applicable.

Data Availability Statement: The data provided in this study are available from the corresponding author.

Conflicts of Interest: The authors declare no conflict of interest.

Abbreviations

IVMD	improved variational mode decomposition
IMF	intrinsic mode function

References

1. Jain, S.V.; Patel, R.N. Investigations on pump running in turbine mode: A review of the state-of-the-art. *Renew. Sustain. Energy Rev.* **2014**, *30*, 841–868. [CrossRef]
2. Lu, H.; Huang, K.; Wu, S. Vibration and Stress Analyses of Positive Displacement Pump Pipeline Systems in Oil Transportation Stations. *J. Pipeline Syst. Eng. Pract.* **2016**, *7*, 05015002. [CrossRef]
3. Pergushev, L.P.; Fattakhov, R.B.; Sakhabutdinov, R.Z. Analysis of operation of booster pumping station. *Neftyanoe Khozyaistvo* **2005**, *5*, 134–137.

4. Jiang, Q. Research on Coupled Vibration Characteristics and Vibration Control of Pipeline of Cascade Pumping Station. Master's Thesis, North China University of Water Resources and Electric Power, Zhengzhou, China, 2017.
5. PaïDoussis, M.P.; Li, G.X. Pipes Conveying Fluid: A Model Dynamical Problem. *J. Fluids Struct.* **1993**, *7*, 137–204. [CrossRef]
6. Tang, D.M.; Dowell, E.H. Chaotic oscillations of a cantilevered pipe conveying fluid. *J. Fluids Struct.* **1988**, *2*, 263–283. [CrossRef]
7. Sinir, B.G. Bifurcation and chaos of slightly curved pipes. *Math. Comput. Appl.* **2007**, *15*, 490–502. [CrossRef]
8. Zhao, D.; Liu, J.; Wu, C.Q. Stability and local bifurcation of parameter-excited vibration of pipes conveying pulsating fluid under thermal loading. *Appl. Math. Mech.* **2015**, *36*, 1017–1032. [CrossRef]
9. Liu, S.T.; Sun, F.Y.; Shen, S.L. Chaos Behaviour of Molecular Orbit. *Chin. Phys. Lett.* **2007**, *24*, 3590.
10. Hiruta, K.; Fujita, A.; Inagaki, K. 28a-B-13 Fractal Structure on Poincare Surface of Section of Chaotic Attractor. *Phys. Soc. Jpn. JPS* **1994**.
11. Yang, L.; Shan, C.L. Dual Chaos SVPWM Technology and its Power Spectral Density Analysis. *East China Electr. Power* **2013**, *41*, 0579–0583.
12. Liu, Y.; Wen, B.; Yang, J. Chaos and Fractal Dimension Research of Investment Time Series. *J. Northeast. Univ. Nat. Sci.* **2001**, *22*, 524–526.
13. Jayaraman, A.; Scheel, J.D.; Greenside, H.S. Characterization of the domain chaos convection state by the largest Lyapunov exponent. *Phys. Rev. E* **2006**, *74 Pt 1*, 016209. [CrossRef] [PubMed]
14. Rosenstein, M.T.; Collins, J.J.; Luca, C.J.D. A practical method for calculating largest Lyapunov exponents from small data sets. *Phys. D Nonlinear Phenom.* **1993**, *65*, 117–134. [CrossRef]
15. Dragomiretskiy, K.; Zosso, D. Variational mode decomposition. *Trans. Signal Process.* **2014**, *62*, 531–544. [CrossRef]
16. Zhang, N.; Zhu, Y.L.; Gao, Y.F. An On-Line Detection Method of Transformer Winding Deformation Based on Variational Mode Decomposition and Probability Density Estimation. *Power Syst. Technol.* **2016**, *40*, 297–302.
17. Liu, C.L.; Yan, X. Fault diagnosis of a wind turbine rolling bearing based on variational mode decomposition and condition identification. *J. Chin. Soc. Power Eng.* **2017**, *37*, 273–278.
18. Hestenes, M.R. Multiplier and gradient methods. *J. Optim. Theory Appl.* **1969**, *4*, 303–320. [CrossRef]
19. Tang, G.J.; Wang, X.L. Parameter optimized variational mode decomposition method with application to incipient fault diagnosis of rolling bearing. *J. Xi'an Jiaotong Univ.* **2015**, *49*, 73–81.
20. Hu, A.J. Research on the Application of Hilbert-Huang Transform in Vibration Signal Analysis of Rotating Machinery. Master's Thesis, North China Electric Power University, Baoding, China, 2008.
21. Cao, L.Y. Practical method for determining the minimum embedding dimension of a scalar time series. *Phys. D Nonlinear Phenom.* **1997**, *110*, 43–50. [CrossRef]
22. Grassberger, P.; Procaccia, I. Measuring Strangeness Strange Attractors. *Phys. D Nonlinear Phenom.* **1983**, *9*, 189–208. [CrossRef]
23. Zhang, J.W.; Jiang, Q.; Wang, T. Analysis of vibration characteristics of pipeline of trapezoid pumping station based on prototype observation. *Trans. Chin. Soc. Agric. Eng.* **2017**, *33*, 77–83.
24. Zhang, J.W.; Li, Z.Y.; Yan, P.; Li, Y.; Huang, J.L. The Method for Determining Optimal Analysis Length of Vibration Data Based on Improved Multiscale Permutation Entropy. *Shock Vib.* **2021**, *2021*, 6654089.
25. Lv, J.H.; Lu, J.A. *Chaotic Time Series Analysis and Its Application*; Wuhan University Press: Wuhan, China, 2002.
26. Wang, Z.; Zheng, L.; Junyuan, W. Application of a New Enhanced Deconvolution Method in Gearbox Fault Diagnosis. *Appl. Sci.* **2019**, *9*, 5313. [CrossRef]

applied sciences

MDPI

Article

China's Water Intensity Factor Decomposition and Water Usage Decoupling Analysis

Boyu Du [1,2,3], Xiaoqian Guo [1,2,*], Guwang Liu [1,2], Anjian Wang [1,2,*], Hongmei Duan [4] and Shaobo Guo [1,2,3]

1 MLR Key Laboratory of Metallogeny and Mineral Assessment, Institute of Mineral Resources, China Academy of Geological Science, Beijing 100037, China; dby970313@163.com (B.D.); liuguwang@mail.cgs.gov.cn (G.L.); shaobo_guo@163.com (S.G.)
2 Research Center for Strategy of Global Mineral Resources, Institute of Mineral Resources, China Academy of Geological Science, Beijing 100037, China
3 School of Earth Sciences and Resources, China University of Geosciences, Beijing 100083, China
4 Chinese Academy of International Trade and Economic Cooperation, Ministry of Commerce, Beijing 100010, China; duanhm@cugb.edu.cn
* Correspondence: xiaoqianguo88@163.com (X.G.); ajwang@126.com (A.W.); Tel.: +86-010-6899-9656 (X.G.)

Abstract: As the most populous country in the world, China has a great shortage pressure of water resources. With the acceleration of urbanization, China's water usage in different sectors will change significantly in next few years. In order to investigate the main reasons behind water usage change in China, the Logarithmic Mean Divisia Index (LMDI) model was adopted in this paper from 2000 to 2020 with provincial data. Three effects, including that of technology, industrial structure, and regional scale, were analyzed. In addition, the decoupling effect between water usage and economic growth was also considered. The results show that: (1) from 2000 to 2020, the technological effect, industrial structure effect, and regional scale effect are −376.54, −89.85 and 20.66, respectively; (2) the technical effect and industrial structure effect have the greatest impact on primary industry, followed by secondary industry; (3) the technical effect is greater than the industrial structure effect in most provinces; and (4) the decoupling state gradually changes from weak decoupling to strong decoupling. In the future, the key policy recommendations for water saving are the following: (1) technological innovation has the most efficient effect on the reduction of water usage in China, and (2) the optimization of industrial structure can be helpful in water-saving in the future.

Keywords: water intensity; LMDI model; Tapio model; technical effect; industrial structure effect; regional scale effect

Citation: Du, B.; Guo, X.; Liu, G.; Wang, A.; Duan, H.; Guo, S. China's Water Intensity Factor Decomposition and Water Usage Decoupling Analysis. *Appl. Sci.* **2022**, *12*, 7039. https://doi.org/10.3390/app12147039

Academic Editors: Rituraj Shukla, Amit Kumar and Santosh Subhash Palmate

Received: 20 June 2022
Accepted: 10 July 2022
Published: 12 July 2022

1. Introduction

China is one of the countries with the most serious water shortage pressures in the world [1–3]. Besides, the weak awareness of water saving, uneven distribution of water resources, rapid population growth, increasing water usage of residents, and climate change have all aggravated the tensions surrounding water resources in China [4,5]. With the rapid development of modern industry and the accelerating process of urbanization, the demand for water resources in different sectors will change greatly, and access to water will become an important factor restricting China's economic development [6,7].

In the last 20 years, China's water usage structure has changed significantly in line with economic development. The water usage in China has been divided into three industries. The primary industry category mainly includes agriculture, forestry, and animal husbandry and fisheries. The secondary industry category mainly refers to mining, manufacturing, and construction. The tertiary industries include everything not contained within the primary and secondary industries, including the service industry, transportation, accommodation and catering, finance, real estate, culture and sports, public administration, and social security [8,9]. The water usage in primary industry showed a downward trend, and the water usage in secondary industry increased first and then decreased [10–12]. Water usage in tertiary industry continued to rise at a rapidly increasing rate, which was 15% in 2020 [13,14]. Therefore, it is of great significance for the sustainable management of water resources to investigate the driving factors behind China's water intensity [15].

A series of publications have calculated the single resource intensity at national level, the city level, and certain industry levels. Some researchers even considered the resource intensity of overall resources in the world or in a certain country.

A series of publications regarding driving factors in different resources have been conducted by previous researchers, including structural decomposition analysis (SDA), granger causality test, and the Logarithmic Mean Divisia Index (LMDI) model. Most of the existing studies using SDA were based on the monetary I–O tables, which require a considerable amount of sector data, and the research scope is mainly in a national level [16–18]. While the Granger causality test is only a statistical estimation, not a real causality, which cannot be used as the basis for affirming or denying causality [19–21]. Therefore, the LMDI model was adopted in this paper.

The LMDI model is a factor decomposition method that does not generate residual error [22]. The application of LMDI is essentially for resources and the environment, such as carbon emissions, energy, land, and water resources [23–28]. At present, LMDI research on water resources has calculated the driving factors at national level or the city level [29–35]. The LMDI method in water resources research was mainly broken down into population scale effect, economic development effect, domestic intensity effect, production intensity, and industrial structural effect [36]. However, as the world's most populous country with serious water usage pressure, there is a shortage of research on the drivers of change in China's water intensity at industrial level with provincial data.

Based on the data of water usage, GDP, and the added value of various industries in different provinces of China from 2000 to 2020, the LMDI model was used to analyze the potential factors affecting the change of water usage within various industries. Three effects, including the technical effect, industrial structure effect, and regional scale effect will be adopted. Besides, these three effects will be applied in each province and each industry in China. The decomposition model can measure the contribution of various factors to water intensity, while it cannot directly measure the decoupling state between economic growth and water usage, and the actual decoupling situation under different policies [37]. The Tapio method is then used for decoupling analysis between water usage and GDP. Finally, the most efficient water-saving methods will also be discussed.

2. Materials and Methods

2.1. Logarithmetic Mean Divisia Index Model

In order to analyze the influence of technological progress, regional scale, industrial structure, and other factors on the water usage change of different industries in China, it is beneficial to analyze the driving factors of water usage in China by using the LMDI proposed by Ang [38]. This method has the advantages of zero value and complete decomposition, and can be completely decomposed [39–41].

Water usage index can be expressed by absolute quantity and relative quantity. The absolute quantity refers to total water usage, and the relative quantity refers to the water usage per unit of economic output, that is, water intensity. It reflects the utilization efficiency of water resources, which is influenced by economic growth, technological progress, industrial structure, regional scale, and policy factors. According to the definition of water intensity, it can be expressed as:

$$w = \sum_i \sum_j \frac{W_{ij}}{G_{ij}} \tag{1}$$

where w is the water intensity (cubic meters/10 thousand CNY); W_{ij} is the water usage (cubic meters) of the jth industry in the ith province; G_{ij} is the gross output value (10 thousand CNY) of the jth industry in the ith province.

According to LMDI analysis framework, by analyzing the influence of each effect on water intensity, we can construct Equation (2) as follows:

$$w = \sum_i \sum_j \frac{W_{ij}}{G} = \sum_i \sum_j \frac{W_{ij}}{G_{ij}} \times \frac{G_{ij}}{G_i} \times \frac{G_i}{G} = \sum_i \sum_j q_{ij} r_{ij} s_i \tag{2}$$

where w is the water intensity, W_{ij} is the total water usage of the jth industry in the ith province, G_{ij} is the gross national product of the jth industry in the ith province, G_i is the gross product of the ith province, G is the gross domestic product, q_{ij} is the water intensity of the jth industry in the ith province, r_{ij} is the proportion of the gross product of the jth industry in the gross product of the ith province, s_i is the ratio of GDP of the ith province to total GDP.

Therefore, the total effect formula of water intensity is:

$$\Delta tot = \Delta t + \Delta u + \Delta v \tag{3}$$

where Δtot is the total effect, that is, the sum of all effects, indicating the total change of water intensity; Δt refers to the technical effect, indicating the contribution of the change of resource utilization efficiency caused by technological progress to the total change of water intensity; Δu refers to the industrial structure effect, indicating the contribution of industrial structure adjustment to the total change of water intensity; Δv is the regional scale effect, which indicates the contribution of the ratio of regional economic output to GDP to the total change of water intensity.

The contribution of each effect is expressed as follows:

$$\Delta t = \sum_i \sum_j \frac{\frac{W_{ij}^T}{G^T} - \frac{W_{ij}^0}{G^0}}{\ln(\frac{W_{ij}^T}{G^T}) - \ln(\frac{W_{ij}^0}{G^0})} \ln(\frac{q_{ij}^T}{q_{ij}^0}) \tag{4}$$

$$\Delta u = \sum_i \sum_j \frac{\frac{W_{ij}^T}{G^T} - \frac{W_{ij}^0}{G^0}}{\ln(\frac{W_{ij}^T}{G^T}) - \ln(\frac{W_{ij}^0}{G^0})} \ln(\frac{r_{ij}^T}{r_{ij}^0}) \tag{5}$$

$$\Delta v = \sum_i \sum_j \frac{\frac{W_{ij}^T}{G^T} - \frac{W_{ij}^0}{G^0}}{\ln(\frac{W_{ij}^T}{G^T}) - \ln(\frac{W_{ij}^0}{G^0})} \ln(\frac{s_{ij}^T}{s_{ij}^0}) \tag{6}$$

The contribution rates of the three effects to the change of water intensity are $\Delta t/\Delta tot$, $\Delta u/\Delta tot$, and $\Delta v/\Delta tot$, respectively. When the positive and negative impacts of each effect are consistent with the total effect, it shows that this effect has a positive impact on the reduction of water intensity, and vice versa.

2.2. Decoupling Model

The decomposition model can be used to study the contribution of various factors to the change of water usage intensity, but it cannot directly measure the decoupling state between economy and water usage [22]. Therefore, the Tapio decoupling model is adopted [42–44], and the decomposition model of water usage is as follows:

$$W = \sum_i \sum_j W_{ij} = \sum_i \sum_j \frac{W_{ij}}{G_{ij}} \times \frac{G_{ij}}{G_i} \times \frac{G_i}{G} \times G \tag{7}$$

So as to decompose the changes of water usage into:

$$\Delta TOT = \sum_i \sum_j \Delta T + \Delta U + \Delta V + \Delta Q \tag{8}$$

Among them, ΔTOT is the total effect of water usage, ΔT is the technical effect, ΔU is the effect of industrial structure, ΔV is the effect of regional scale, and ΔQ is the effect of output scale. The contribution of each effect is as follows:

$$\Delta T = \sum_i \sum_j \frac{W_{ij}^T - W_{ij}^0}{\ln W_{ij}^T - \ln W_{ij}^0} \ln(\frac{W_{ij}^T / G_{ij}^T}{W_{ij}^0 / G_{ij}^0}) \tag{9}$$

$$\Delta U = \sum_i \sum_j \frac{W_{ij}^T - W_{ij}^0}{\ln W_{ij}^T - \ln W_{ij}^0} \ln(\frac{G_{ij}^T / G_i^T}{G_{ij}^0 / G_i^0}) \tag{10}$$

$$\Delta V = \sum_i \sum_j \frac{W_{ij}^T - W_{ij}^0}{\ln W_{ij}^T - \ln W_{ij}^0} \ln(\frac{G_i^T / G^T}{G_i^0 / G^0}) \tag{11}$$

$$\Delta Q = \sum_i \sum_j \frac{W_{ij}^T - W_{ij}^0}{\ln W_{ij}^T - \ln W_{ij}^0} \ln \frac{G^T}{G^0} \tag{12}$$

Decoupling elasticity index is used to discuss the decoupling relationship between economic growth and water usage. The elastic coefficient of GDP water usage is calculated as follows:

$$\omega(W, G) = \frac{\Delta W / W}{\Delta G / G} \tag{13}$$

The types of decoupling can essentially be divided into coupling, decoupling, and negative decoupling. In addition, according to the elasticity coefficient, the change of water usage and the change of GDP, the decoupling types can be subdivided into eight cases (Table 1) [45].

Table 1. Types of Tapio models.

State	Type	△W	△GDP	ω	Meaning
Connection	Decline connection (DC)	−	−	(0.8, 1.2)	Water usage is declining at the same rate as the economy.
	Expansion connection (EC)	+	+	(0.8, 1.2)	Water usage is increasing at the same rate as the economy.
Decoupling	Decline decoupling (DD)	−	−	(1.2, +∞)	Water usage is declining faster than the economic recession.
	Strong decoupling (SD)	−	+	(−∞, 0)	Economic growth is accompanied by a decline in water usage.
	Weak Decoupling (WD)	+	+	(0, 0.8)	The growth rate of water usage is slower than that of economy.
Negative decoupling	Weak negative decoupling (WND)	−	−	(0, 0.8)	The rate of water usage reduction is slower than the rate of economic recession.
	Strong negative decoupling (SND)	+	−	(−∞, 0)	Water usage is increasing but the economy is declining.
	Expansion negative decoupling (END)	+	+	(1.2, +∞)	The water usage growth rate is lower than the economic growth rate.

The decoupling elasticity index can be used to calculate the decoupling relationship between economic growth and water usage, but it cannot help to investigate the specific factors that affect the decoupling state. The LMDI model can be used to analyze the influence of various factors on water usage, but it cannot be used to analyze the decoupling effect between economic growth and water usage. Combining the LMDI model with the Tapio decoupling model, a decoupling effort index model is constructed:

$$\Delta W_{US} = \Delta TOT - \sum_i \sum_j \Delta Q = \sum_i \sum_j \Delta T + \Delta U + \Delta V \quad (14)$$

where ΔW_{US} indicates the government's efforts to save water, and refers to various measures taken by the government to reduce water usage in the process of economic development, such as improving production technology, adjusting industrial structure, and expanding regional scale.

The decoupling effort indicators are constructed as follows:

$$D_i = -\frac{\Delta W_{US}}{\sum_i \sum_j \Delta Q} \quad (15)$$

where D_i is the total decoupling effect of water usage. When $D_i > 1$, it indicates a strong decoupling effect. When $D_i < 1$, it indicates a weak decoupling effect. When $D_i < 0$, it means there is no decoupling effect.

2.3. Date

The data used in this study are the water usage and industrial added value of three major industries in each province of China from 2000 to 2020. All the data in this paper come from *the Water Resources Bulletin* issued by China's Ministry of Water Resources from 2000 to 2020 and the National Bureau of Statistics [46,47].

From 2000 to 2013, China's total annual water usage increased from 549.752 billion cubic meters to 618.394 billion cubic meters, before the water usage showed a decreasing trend. The water usage of primary industry contributes most to the total water usage and remains stable with approximately 400 billion cubic meters per year. While secondary industry is a more minor user of water in China, and it has a trend of first increasing and then decreasing. The water usage of tertiary industry continues to rise, from 57.492 billion cubic meters in 2000 to 86.310 billion cubic meters in 2020 (Figure 1).

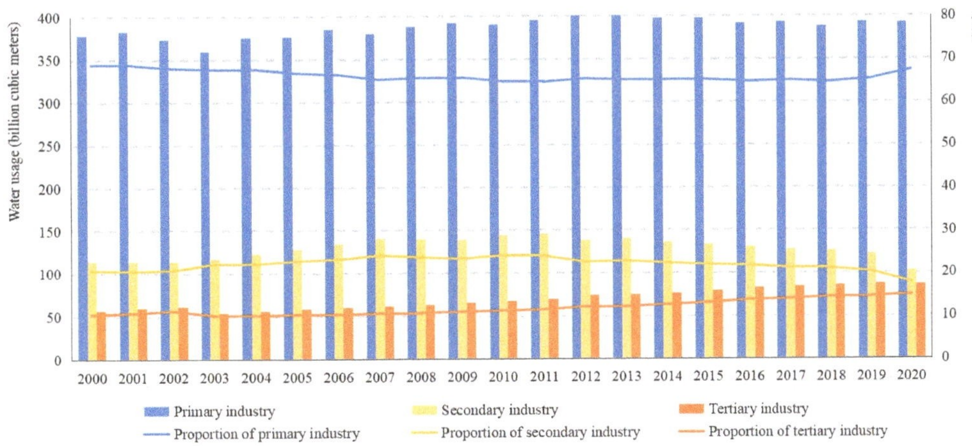

Figure 1. Water usage and proportion of various industries in China from 2000 to 2020.

The added value in each industry of the past 20 years is shown in Figure 2. China's economy maintains a high speed of development from 2000 to 2020, so the added value in each industry increases continually. The fastest growth occurs in tertiary industry, with an average annual growth rate of 0.07%, which demonstrates that China's economy has gradually shifted into tertiary industry.

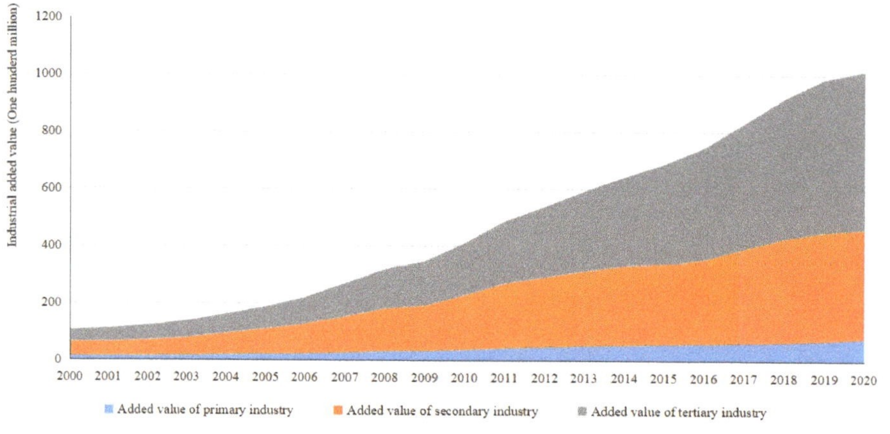

Figure 2. China's added value of various industries from 2000 to 2020.

3. Results

3.1. Water Intensity and Factor Decomposition Analysis

3.1.1. Analysis of Decomposition Effect in Each Year

According to Equations (3)–(6), three effects and their respective contribution rates from 2000 to 2020 are shown in Figure 3. The total effect of each year is negative, indicating that the water intensity is decreasing year by year, signaling water saving considerations. The total effect from 2002 to 2003 was the smallest, with value of −64.50.

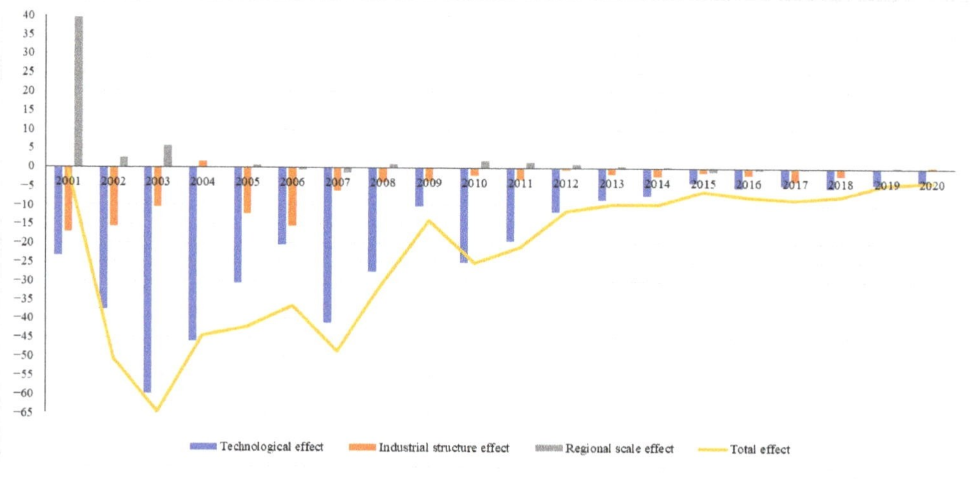

Figure 3. Effects and contribution rates in China from 2000 to 2020.

Technical effects in the last 20 years are negative and the technical effect contribution rate is the largest among three effects, indicating it has an inhibitory impact on the water intensity, while technological innovation is the most effective measure for water saving. The technical effect fluctuated greatly, with its largest value from 2002 to 2003 of −59.91 and highest contribution rate in 2001 of 2481.29%. For the industrial structure effect, it was negative except for 1.62 in 2003 and 0.38 in 2020, meaning that it restricted the water intensity in most years. As for for the regional scale effect, it fluctuated greatly from 2000 to 2001, reaching 39.49, and it was stable with values between −1 and 2 from 2003 to 2020.

The three effects in each industry were also explored in China through the LMDI in Figure 4. The technical effects are all negative for the three industries, which means that the water intensity of the three industries all declined with technological innovation. It also fluctuated greatly before 2011, with the largest absolute value of −36.24, −12.52 and −11.15, respectively, during 2002 to 2003, then it tended to be flat. Besides, it fluctuated most within primary industry, due to the largest proportion of China's primary industry in current water usage structure.

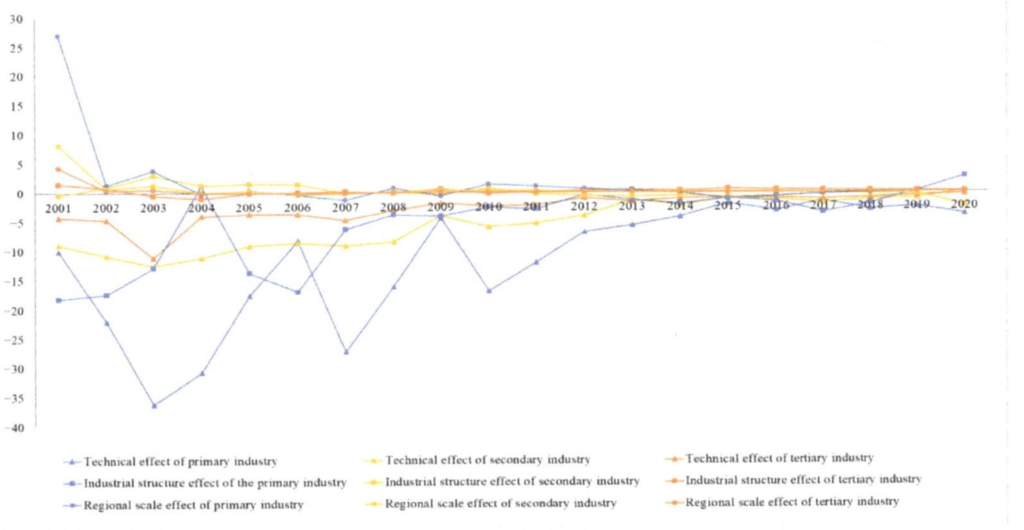

Figure 4. Decomposition analysis of China's water intensity in each industry from 2000 to 2020.

The industrial structure effect on the three major industries has different characteristics. In primary industry, it has increased from −18.21 to 2.63, shifting from a restriction effect to a promoting effect from 2018 to 2020. In secondary industry, it changed from strong promotion to weak promotion, and finally into a restriction effect, which is mainly attributed to the intensive management of industrial development with the increasing industry output. In tertiary industry, it remains essentially unchanged. Therefore, the industrial structure effect has restricted the water usage in China, indicating the industrial transformation in China has impacted on water usage reduction in the last 20 years.

3.1.2. Analysis of Decomposition Effect in Each Province

According to Equations (3)–(6), the three effects in each province are calculated in Table 2. The technical effect in each province is negative with the increasing absolute value, and it means that the technical effect in each province in China has been generally improved. Besides, due to the highest average value of −12.15 in these three effects, the technical effect is a decisive factor to promote the decline of water intensity. Among all the provinces, Xinjiang have the greatest inhibitory effect with values of −34.05, and Tianjin has the smallest inhibitory effect with values of −1.20. The industrial structure effect is also negative excepted for Anhui Province in the studied areas with values between −0.21 and

−10.35, and its absolute value is smaller than the technology effect. This indicates that the industrial structure transformation has taken effect.

Table 2. Three effects in each province in China.

Province	Technological Effect	Industrial Structure Effect	Regional Scale Effect	Total Effect
Beijing	−2.13	−1.41	0.24	−3.30
Tianjin	−1.20	−0.55	−0.05	−1.80
Hebei	−13.63	−2.69	−1.30	−17.61
Shanxi	−3.98	−0.55	0.09	−4.44
Inner Mongolia	−11.31	−3.78	1.25	−13.84
Liaoning	−8.06	−0.46	−2.69	−11.20
Jilin	−5.29	−2.75	−1.18	−9.22
Heilongjiang	−20.62	3.62	−7.05	−24.06
Shanghai	−6.20	−2.23	−0.52	−8.96
Jiangsu	−29.37	−10.35	4.59	−35.13
Zhejiang	−13.32	−4.31	0.84	−16.79
Anhui	−6.55	−9.00	2.03	−13.52
Fujian	−12.16	−3.63	1.45	−14.34
Jiangxi	−14.83	−5.38	2.71	−17.51
Shandong	−15.33	−4.35	−0.46	−20.13
Henan	−13.67	−3.87	1.15	−16.39
Hubei	−20.87	−3.82	2.68	−22.01
Hunan	−22.86	−5.70	2.65	−25.90
Guangdong	−30.22	−6.70	1.58	−35.33
Guangxi	−22.02	−3.59	1.41	−24.19
Hainan	−3.12	−0.68	0.21	−3.59
Chongqing	−4.79	−0.54	0.86	−4.46
Sichuan	−15.50	−3.49	2.25	−16.75
Guizhou	−7.78	−1.01	1.97	−6.81
Yunnan	−11.87	−1.51	1.46	−11.92
Tibet	−1.53	−1.23	0.58	−2.17
Shaanxi	−6.65	−0.99	1.33	−6.30
Gansu	−8.58	−1.20	−0.36	−10.15
Qinghai	−2.31	−0.21	0.20	−2.31
Ningxia	−6.76	−1.63	1.10	−7.29
Xinjiang	−34.05	−5.87	1.63	−38.29

The decomposition analysis of water intensity in each province from 2000 to 2020 is also obtained in Table 3 based on Equations (3)–(6). The technical effects of all industries in each province are negative, which is consistent with Table 2, indicating the restraining effect on water usage. The value of primary industry in most provinces is the smallest, with values between −31.70 and −0.06, followed by secondary industry and tertiary industry. Because the water usage of primary industry accounts for the largest proportion of the total water usage in China, the technological progress of primary industry plays a significant role. The efficiency of technological progress in secondary industry is higher than that in tertiary industry. Moreover, the industrial structure effects of primary industry are basically negative, and it has both positive and negative values in secondary industry, with an almost positive effect on tertiary industry, indicating the greater effect of industrial transition on primary industry than that in secondary or tertiary industry. In addition, the provinces with a positive industrial structure effect of secondary industry are typically underdeveloped areas, such as Tibet, Inner Mongolia, and Qinghai, which also shows that industrial transition in underdeveloped areas needs to be improved.

3.2. The Decoupling Effect of Water Usage

3.2.1. Decoupling Elasticity Index

In this paper, the elastic index of decoupling analysis between economic growth and water usage in China from 2000 to 2020 is calculated and divided into four stages (Table 4).

Table 3. Effects of various industries in various provinces.

Province	Technological Effect			Industrial Structure Effect			Regional Scale Effect		
	Primary Industry	Secondary Industry	Tertiary Industry	Primary Industry	Secondary Industry	Tertiary Industry	Primary Industry	Secondary Industry	Tertiary Industry
Beijing	−0.06	−0.80	−1.27	−1.35	−0.18	0.12	0.11	0.04	0.09
Tianjin	−0.36	−0.37	−0.47	−0.56	−0.07	0.07	−0.03	−0.01	−0.01
Hebei	−9.62	−2.00	−2.00	−2.83	−0.17	0.32	−0.99	−0.15	−0.16
Shanxi	−2.25	−1.09	−0.65	−0.56	−0.04	0.05	0.06	0.02	0.01
Inner Mongolia	−9.76	−0.73	−0.82	−3.88	0.02	0.07	1.11	0.07	0.07
Liaoning	−4.89	−1.68	−1.49	−0.45	−0.26	0.25	−1.75	−0.48	−0.45
Jilin	−3.16	−1.42	−0.71	−2.81	−0.05	0.11	−0.91	−0.16	−0.11
Heilongjiang	−14.13	−5.25	−1.24	5.12	−1.80	0.29	−5.20	−1.46	−0.38
Shanghai	−0.18	−4.81	−1.21	−0.98	−1.46	0.21	−0.08	−0.36	−0.09
Jiangsu	−13.85	−11.20	−4.31	−9.91	−1.11	0.66	2.51	1.61	0.46
Zhejiang	−6.41	−4.21	−2.70	−4.37	−0.46	0.52	0.48	0.21	0.15
Anhui	−5.89	0.91	−1.56	−5.01	−4.15	0.16	1.31	0.50	0.22
Fujian	−6.07	−4.34	−1.74	−3.88	0.12	0.13	0.90	0.36	0.19
Jiangxi	−8.36	−4.80	−1.68	−5.88	0.37	0.13	1.88	0.58	0.25
Shandong	−9.83	−3.25	−2.25	−4.43	−0.35	0.44	−0.33	−0.08	−0.06
Henan	−7.24	−3.54	−2.90	−4.23	−0.14	0.50	0.76	0.21	0.18
Hubei	−11.27	−6.91	−2.68	−3.90	−0.25	0.33	1.58	0.77	0.33
Hunan	−14.12	−5.00	−3.74	−6.12	0.11	0.32	1.84	0.47	0.34
Guangdong	−15.61	−8.39	−6.22	−6.82	−0.58	0.71	0.92	0.36	0.30
Guangxi	−15.95	−3.25	−2.82	−3.80	−0.13	0.35	1.07	0.19	0.16
Hainan	−2.32	−0.33	−0.46	−0.75	0.00	0.07	0.17	0.01	0.03
Chongqing	−1.10	−2.40	−1.29	−0.61	−0.05	0.13	0.32	0.32	0.22
Sichuan	−8.17	−4.69	−2.64	−3.84	−0.02	0.36	1.48	0.41	0.36
Guizhou	−4.08	−1.92	−1.78	−1.16	−0.06	0.21	1.18	0.43	0.37
Yunnan	−8.58	−1.57	−1.72	−1.62	−0.12	0.23	1.10	0.17	0.19
Tibet	−1.25	−0.09	−0.19	−1.26	0.02	0.01	0.52	0.02	0.05
Shaanxi	−4.39	−1.24	−1.01	−1.04	0.00	0.06	0.93	0.20	0.20
Gansu	−6.51	−1.40	−0.66	−1.17	−0.11	0.09	−0.30	−0.04	−0.03
Qinghai	−1.69	−0.38	−0.24	−0.23	0.03	−0.01	0.16	0.03	0.02
Ningxia	−6.14	−0.46	−0.16	−1.64	0.00	0.01	1.00	0.06	0.03
Xinjiang	−31.70	−0.89	−1.46	−5.96	−0.04	0.12	1.54	0.03	0.05

In these four stages, the relationships between water usage and economic growth in all provinces are decoupled, indicating the water usage is not related with the development of China's economy. There are 13 strong decoupling provinces in the first stage, with 8 in the second stage, 15 in the third stage and 23 in the fourth stage, respectively. Besides, the weak decoupling status in most provinces has gradually changed into strong decoupling status. The increasing trends of decoupling provinces in different stages is due to the gradually improvement of water efficiency with economic development. In recent years, corresponding policies in China have been issued to improve water efficiency, such as *the National Water Conservation Action Plan* and *the Water Pollution Prevention Action Plan*, and the task of water conservation has been officially put into *the 13th Five-Year Plan*, which illustrates the Chinese government's determination on the issue of water saving.

From a regional perspective, Beijing, Yunnan, and Qinghai Province have changed from strong decoupling in the first stage to weak decoupling later. The four stages of Inner Mongolia are all weak decoupling, which means that the economic development quality in water resources in these regions still need to be improved. Hebei and Ningxia Province are strongly decoupled in the four stages, which shows that the popularization of water conservation policies in these two regions is relatively effective and should be maintained. East China, such as Shanghai, Zhejiang, Jiangsu, Anhui, and Fujian; South China, such as Guangdong, Guangxi, and Hainan; and Southwest China, such as Guizhou, Sichuan, and Chongqing, have all changed from weak decoupling at first stage to strong decoupling later, meaning the areas with relatively abundant water resources are more likely to improve the local decoupling state and achieve high-quality economic development.

Table 4. Decoupling index and state of water usage and economic growth in each province.

Province	2000–2005		2005–2010		2010–2015		2015–2020	
	Decoupling Index	Decoupling Type	Decoupling Index	Decoupling Type	Decoupling Index	Decoupling Type	Decoupling Index	Decoupling Type
Beijing	1.16	SD	0.79	WD	0.71	WD	0.75	WD
Tianjin	0.81	WD	0.82	SD	0.49	WD	0.53	WD
Hebei	0.99	SD	0.92	SD	0.91	SD	0.92	SD
Shanxi	0.94	SD	0.61	WD	0.40	WD	0.86	SD
Inner Mongolia	0.95	WD	0.86	WD	0.83	WD	0.78	WD
Liaoning	0.82	SD	0.72	WD	0.81	SD	1.01	SD
Jilin	1.05	SD	0.58	WD	0.58	WD	1.14	SD
Heilongjiang	0.88	SD	0.52	WD	0.63	WD	1.20	SD
Shanghai	0.25	WD	0.14	WD	0.50	SD	0.47	SD
Jiangsu	0.54	WD	0.56	WD	0.53	WD	0.56	SD
Zhejiang	0.77	WD	0.74	SD	0.84	SD	1.06	SD
Anhui	0.51	WD	0.31	WD	0.75	SD	0.92	SD
Fujian	0.63	WD	0.58	WD	0.69	SD	1.01	SD
Jiangxi	0.97	SD	0.63	WD	0.73	WD	0.79	SD
Shandong	1.18	SD	0.82	WD	0.94	SD	0.69	WD
Henan	0.95	SD	0.62	WD	0.76	SD	0.62	WD
Hubei	0.87	SD	0.56	WD	0.63	WD	0.84	SD
Hunan	0.76	WD	0.81	SD	0.76	WD	0.94	SD
Guangdong	0.72	WD	0.66	WD	0.81	SD	0.99	SD
Guangxi	0.76	WD	0.91	SD	0.84	SD	1.19	SD
Hainan	0.93	WD	0.92	WD	0.87	WD	1.06	SD
Chongqing	0.23	WD	0.31	WD	0.87	SD	1.00	SD
Sichuan	0.78	WD	0.69	WD	0.50	WD	1.16	SD
Guizhou	0.56	WD	0.67	WD	0.95	SD	1.04	SD
Yunnan	0.88	SD	0.86	WD	0.84	WD	0.80	WD
Tibet	0.62	WD	0.90	WD	1.24	SD	0.87	WD
Shaanxi	0.92	WD	0.81	WD	0.70	WD	0.86	SD
Gansu	0.88	WD	0.90	SD	0.92	SD	1.11	SD
Qinghai	0.71	SD	0.85	SD	1.18	WD	1.16	WD
Ningxia	1.17	SD	1.06	SD	0.99	SD	0.95	SD
Xinjiang	0.88	WD	0.92	WD	0.83	WD	1.01	SD

3.2.2. Decoupling Effort Index

The decoupling effort index is used to measure the decoupling status between economic growth and water usage (Table 5).

Table 5. Decoupling effort index of China's water usage from 2000 to 2020.

Year	Technological Effect	Industrial Structure Effect	Regional Scale Effect	Output Scale Effect	Decoupling Index	Decoupling Type	ω
2000–2001	−256.68	−188.32	434.68	79.73	0.13	EC	0.88
2001–2002	−434.71	−180.49	29.23	516.51	1.13	SD	−0.13
2002–2003	−773.46	−133.11	73.95	655.57	1.27	SD	−0.26
2003–2004	−684.70	24.19	−0.26	888.14	0.74	WD	0.25

Table 5. *Cont.*

Year	Technological Effect	Industrial Structure Effect	Regional Scale Effect	Output Scale Effect	Decoupling Index	Decoupling Type	ω
2004–2005	−529.81	−213.56	11.01	817.02	0.90	WD	0.10
2005–2006	−415.81	−311.52	−14.61	903.92	0.82	WD	0.17
2006–2007	−998.16	−147.19	−36.03	1205.64	0.98	WD	0.02
2007–2008	−807.22	−111.59	29.80	980.24	0.91	WD	0.09
2008–2009	−339.02	−121.16	−5.39	520.75	0.89	WD	0.10
2009–2010	−947.96	−71.03	71.33	1004.48	0.94	WD	0.05
2010–2011	−871.84	−132.45	65.48	1023.94	0.92	WD	0.08
2011–2012	−597.65	−25.63	43.53	603.88	0.96	WD	0.04
2012–2013	−479.08	−86.17	25.23	592.19	0.91	WD	0.08
2013–2014	−456.84	−134.69	0.39	502.57	1.18	SD	−0.17
2014–2015	−256.58	−83.98	−65.77	414.73	0.98	WD	0.02
2015–2016	−377.91	−129.13	−43.03	486.99	1.13	SD	−0.12
2016–2017	−354.92	−266.43	−32.22	656.16	1.00	WD	0.00
2017–2018	−446.62	−177.31	−17.73	600.44	1.07	SD	−0.06
2018–2019	−402.14	−6.99	5.07	424.23	0.95	WD	0.04
2019–2020	−404.85	38.44	−1.81	159.92	2.30	SD	−1.26

From the perspective of contributions of these three effects, the technology effect has the greatest influence on the total decoupling effect, with the maximum absolute value of 998.16, which is bigger than the corresponding industrial structure effect with values of 66.43. This shows that technological innovation is an important measure to realize the decoupling of economic development and water usage. The influence of the regional scale effect is smallest, but it plays a driving role in most periods.

4. Conclusions and Implications

This research focused on the investigation of driving factors behind water usage intensity in China from 2000 to 2020, and the identification of decoupling status between water usage and economics. The LMDI model and Tapio model were applied jointly. The results show that:

(1) from 2000 to 2020, the technological effect, industrial structure effect, and regional scale effect are −376.54, −89.85, and 20.66, respectively. The technical effect is from −59.91 to −4.05, and the industrial structure effect is from −17.11 to 1.62, indicating these two effects constrained the increase of water usage intensity. The regional scale effect was stable with values between −1 and 2. From the perspectives of different industries, each effect has the greatest impact on primary industry, followed by secondary industry, and finally tertiary industry.

(2) From the perspective of different provinces, the development of technology and the adjustment of industrial structure have promoted the decline of water intensity. The technological effect varies in different provinces. For example, Tianjin has the value of −1.20, while Xinjiang has values of −34.05. The industrial structure effect is smaller, with the largest value of −0.21 in Qinghai and the smallest being −10.35 in Jiangsu. The technology effect is greater than the industrial structure effect, except for in Anhui Province. When the effects in each industry in different provinces were explored, the technical effect is largest in primary industry in most areas, and the industrial structure effect of primary industry is negative, with positive values in tertiary industry.

(3) The decoupling status for most provinces in China have gradually improved, from weak decoupling to strong decoupling. The technical effect is the main factor towards promoting the decoupling effect, followed by the industrial structure effect.

Therefore, two implications could be put forward. Firstly, technological innovation is the most efficient effect on the reduction of water usage intensity in China with the proliferation of water-saving facilities, and it is still the most efficient policy in China in the near future. Secondly, the optimization of industrial structure is helpful in water-saving in China, but it still needs to be strengthened.

Author Contributions: All the authors (B.D., X.G., G.L., A.W., H.D. and S.G.) have made substantial contributions to this article. Conceptualization, B.D. and X.G.; methodology, B.D. and X.G.; validation and data curation, B.D. and X.G.; writing-original draft preparation, B.D.; Guiding opinions, X.G., A.W., G.L., H.D. and S.G. All authors have read and agreed to the published version of the manuscript.

Funding: This research was funded by National Natural Science Foundation of China, grant numbers: 72088101, 71991485, 71991480. The APC was funded by grant number 72088101.

Institutional Review Board Statement: Not applicable.

Informed Consent Statement: Not applicable.

Data Availability Statement: Not applicable.

Acknowledgments: This research is supported by Institute of Mineral Resources, Chinese Academy of Geological Sciences.

Conflicts of Interest: The authors declare no conflict of interest.

References

1. Wada, Y.; Wisser, D.; Bierkens, M.F.P. Global modeling of withdrawal, allocation and consumptive use of surface water and groundwater resources. *Earth Syst. Dyn.* **2014**, *4*, 355–392. [CrossRef]
2. Wada, Y.; de Graaf, I.E.M.; van Beek, L.P.H. High-resolution modeling of human and climate impacts on global water resources. *Adv. Model. Earth Syst.* **2016**, *8*, 735–763. [CrossRef]
3. Zhang, Z.X.; Wang, W.P. Managing aquifer recharge with multi-source water to realize sustainable management of groundwater resources in Jinan, China. *Environ. Sci. Pollut. Res. Int.* **2021**, *28*, 10872–10888. [CrossRef] [PubMed]
4. Zhou, B. Analysis of the key projects of the national key R&D plan "Efficient Development and Utilization of Water Resources". *Adv. Water Sci.* **2017**, *28*, 472–478.
5. Zhou, F.; Bo, Y.; Ciais, P.; Wada, Y.; Zhou, F.; Yan, B.; Ciais, P.; Wang, A. Deceleration of China's human water use and its key drivers. *Proc. Natl. Acad. Sci. USA* **2020**, *117*, 7702–7711. [CrossRef] [PubMed]
6. Zhang, C.J.; Wu, Y.S.; Pang, Q.H.; Shi, C.F. Study on the Driving Effect of Spatial-temporal Difference of Water Consumption in the Yangtze River Economic Belt: Based on the perspective of production and life. *Resour. Environ. Yangtze Basin* **2019**, *28*, 2806–2816.
7. Wang, X.J.; Zhang, J.Y.; Gao, J.; Shahid, S.; Xia, X.H.; Geng, Z.; Tang, L. The new concept of water resources management in China: Ensuring water security in changing environment. *Environ. Dev. Sustain.* **2018**, *20*, 897–909. [CrossRef]
8. Industry Classification of National Economy in 2017. Available online: http://www.stats.gov.cn/tjsj/tjbz/hyflbz/201710/t20171012_1541679.html (accessed on 29 September 2017).
9. Notice on Amending the Provisions on the Division of Three Industries (2012). Available online: http://www.stats.gov.cn/tjsj/tjbz/201804/t20180402_1591379.html (accessed on 27 March 2018).
10. Bai, P.; Liu, C.M. Evolution and attribution analysis of water consumption structure in Beijing. *S.-N. Water Transf. Water Sci. Technol.* **2018**, *16*, 1–6.
11. Liu, B.Q.; Yao, Z.J.; Gao, Y.C. Analysis on the Changing Trend and Driving Forces of Water Consumption Structure in Beijing. *Resour. Sci.* **2003**, *25*, 38–43.
12. Huang, J.; Song, Z.W.; Chen, F. Characteristics of Water Footprint and Agricultural Water Use Structure Change in Beijing. *Acta Ecol. Sin.* **2010**, *30*, 6546–6554.
13. Guo, X.Q.; Wang, A.J.; Liu, G.W.; Du, B.Y. Expanded S-Curve Model of Relationship between Domestic Water Usage and Economic Development: A Case Study of Typical Countrie. *Appl. Sci.* **2022**, *12*, 6090. [CrossRef]
14. China's Water Resources Bulletin in 2020. Available online: http://www.mwr.gov.cn/sj/tjgb/szygb/202107/t20210709_1528208.html (accessed on 9 July 2021).
15. Meng, X.M.; Tu, L.P.; Yan, C.; Wu, L.F. Forecast of annual water consumption in 31 regions of China considering GDP and population. *Sustain. Prod. Consum.* **2021**, *27*, 713–736.
16. Wang, H.M.; Li, X.Y.; Tian, X.; Ma, L.; Wang, G.Q.; Wang, X.Z.; Wang, Z.; Wang, J.S.; Yue, Q. Socioeconomic drivers of China's resource efficiency improvement: A structural decomposotion analysis for 1997-2017. *Resour. Conserv. Recycl.* **2022**, *178*, 106028–106036. [CrossRef]
17. Meng, G.F.; Liu, H.X.; Li, J.L.; Sun, C.W. Determination of driving forces for China's energy consumption and regional disparities using a hybrid structural decompsition analysis. *Energy* **2022**, *239*, 122191–122204. [CrossRef]
18. Wachsmann, U.; Wood, R.; Lenzen, M.; Schaeffer, R. Structural decomposition of energy use in Brazil from 1970 to 1996. *Appl. Energy* **2009**, *86*, 578–587. [CrossRef]

19. Hoffmann, R.; Lee, C.C.; Ramasamy, B.; Yeung, M. FDI and Pollution: A Granger Causality Test Using Panel Data. *J. Int. Dev.* **2005**, *17*, 311–317. [CrossRef]
20. Diks, C.; Panchenko, V. A new statistic and practical guidelines for nonparametric Granger causality testing. *J. Econ. Dyn. Control* **2006**, *30*, 1647–1669. [CrossRef]
21. Barnett, L.; Seth, A.K. The MVGC multivariate Granger causality toolbox: A new approach to Granger-causal inference. *J. Neurosci. Methods* **2014**, *223*, 50–68. [CrossRef]
22. Ang, B.W. The LMDI approach to decomposition analysis: A practical guide. *Energy Policy* **2005**, *33*, 867–871. [CrossRef]
23. An, B.W. LMDI decomposition approach: A guide for implementation. *Energy Policy* **2015**, *86*, 233–238.
24. Liu, L.C.; Fan, Y.; Wu, G.; Wei, Y.M. Using LMDI method to analyze the change of China's industrial CO_2 emissions from fifinal fuel use: An empirical analysis. *Energy Policy* **2007**, *35*, 5892–5900. [CrossRef]
25. Sun, Y.D.; Hao, Q.; Cui, C.; Shan, Y.L.; Zhao, W.C.; Wang, D.P.; Zhang, Z.K.; Guan, D.B. Emission accounting and drivers in East African countries. *Appl. Energy* **2022**, *312*, 118805–118814. [CrossRef]
26. Wang, W.W.; Zhang, M.; Zhou, M. Using LMDI method to analyze transport sector CO_2 emissions in China. *Energy* **2011**, *36*, 5909–5915. [CrossRef]
27. Su, W.H.; Chen, S.B.; Balezentis, T.; Chen, J. Economy-water Nexus in Agricultural Sector: Decomposing Dynamics in Water Footprint by the LMDI. *Technol. Econ. Dev. Econ.* **2020**, *26*, 240–257. [CrossRef]
28. Li, Y.J.; Wang, S.G.; Chen, B. Driving force analysis of the consumption of water and energy in China based on LMDI method. *Sci. Direct* **2019**, *158*, 4318–4322. [CrossRef]
29. Shao, S.; Yang, L.L.; Gan, C.H.; Cao, J.H.; Geng, Y.; Guan, D.B. Using an extended LMDI model to explore techno-economic drivers of energy-related industrial CO_2 emission changes: A case study for Shanghai (China). *Renew. Sustain. Energy Rev.* **2016**, *55*, 516–536. [CrossRef]
30. Xu, S.C.; He, Z.X.; Long, R.Y. Factors that inflfluence carbon emissions due to energy consumption in China: Decomposition analysis using LMDI. *Appl. Energy* **2014**, *127*, 182–193. [CrossRef]
31. Jesus, P.M.D.O. Effect of generation capacity factors on carbon emission intensity of electricity of Latin America & the Caribbean, a temporal IDA-LMDI analysis. *Renew. Sustain. Energy Rev.* **2019**, *101*, 516–526.
32. Xu, J.H.; Fleiter, T.; Eichhammer, W.; Fan, Y. Energy consumption and CO_2 emissions in China's cement industry: A perspective from LMDI decomposition analysis. *Energy Policy* **2012**, *50*, 821–832. [CrossRef]
33. Zhao, M.; Tan, L.R.; Zhang, W.G.; Ji, M.H.; Liu, Y.; Yu, L.Z. Decomposing the inflfluencing factors of industrial carbon emissions in Shanghai using the LMDI method. *Energy* **2010**, *35*, 2505–2510. [CrossRef]
34. Wang, W.W.; Liu, X.; Zhang, M.; Song, X.F. Using a new generalized LMDI (logarithmic mean Divisia index) method to analyze China's energy consumption. *Energy* **2014**, *67*, 617–622. [CrossRef]
35. Mousavi, B.; Lopez, N.S.A.; Biona, J.B.M.; Chiu, A.S.F.; Blesl, M. Driving forces of Iran's CO_2 emissions from energy consumption: An LMDI decomposition approach. *Appl. Energy* **2017**, *206*, 804–814. [CrossRef]
36. Zhang, C.J.; Zhao, Y.; Shi, C.F.; Chiu, Y.H. Can China achieve its water use peaking in 2030? A scenario analysis based on LMDI and Monte Carlo method. *J. Clean. Prod.* **2021**, *278*, 123214–123228. [CrossRef]
37. Pan, X.F.; Guo, S.C.; Xu, H.T.; Tian, M.Y.; Pan, X.Y.; Chu, J.H. China's carbon intensity factor decomposition and carbon emission decoupling analysis. *Energy* **2022**, *239*, 122175–122191. [CrossRef]
38. Ang, B.W. Decomposition analysis for policymaking in energy. *Energy Policy* **2004**, *32*, 1131–1139. [CrossRef]
39. Ortega-Ruiz, G.; Mena-Nieto, A.; Garcia-Ramos, J.E. Is India on the right pathway to reduce CO_2 emissions decomposing an enlarged Kaya identity using the LMDI method for the period 1990–2016. *Sci. Total Environ.* **2020**, *737*, 139638–139649. [CrossRef]
40. Liu, Y.; Pan, Y.C.; Ren, X.H.; Tang, X.M. Decomposition Model and Empirical Analysis of Grain Production Factors Based on LMDI: A Case Study of Henan Province. *Acta Sci. Nat. Univ. Pekin.* **2014**, *50*, 887–894.
41. Li, Y.; Kong, X.B.; Zhang, A.L.; Zhang, X.L.; Qi, L.Y. Analysis of Influencing Factors of Provincial Grain Production Change in China Based on LMDI Model. *J. China Agric. Univ.* **2016**, *21*, 129–140.
42. Yasmeen, H.; Tan, Q.M. Assessing Pakistan's energy use, environmental degradation, and economic progress based on Tapio decoupling model. *Environ. Sci. Pollut. Res.* **2021**, *28*, 68364–68378. [CrossRef]
43. Song, Y.; Sun, J.J.; Zhang, M.; Su, B. Using the Tapio-Z decoupling model to evaluate the decoupling status of China's CO_2 emissions at provincial level and its dynamic trend. *Struct. Change Econ. Dyn.* **2020**, *52*, 120–129. [CrossRef]
44. Lai, W.W.; Hu, Q.L.; Zhou, Q. Decomposition analysis of PM2.5 emissions based on LMDI and Tapio decoupling model: Study of Hunan and Guangdong. *Environ. Sci. Pollut. Res.* **2021**, *28*, 43443–43458. [CrossRef] [PubMed]
45. Tasbasi, A. A threefold empirical analysis of the relationship between regional income inequality and water equity using Tapio decoupling model, WPAT equation, and the local dissimilarity index: Evidence from Bulgaria. *Environ. Sci. Pollut. Res.* **2021**, *28*, 4352–4365. [CrossRef] [PubMed]
46. China National Bureau of Statistics. Available online: https://data.stats.gov.cn (accessed on 25 May 2022).
47. Ministry of Water Resources of the People's Republic of China. Available online: http://www.mwr.gov.cn/ (accessed on 25 May 2022).

applied sciences

MDPI

Article

Effects of the Operational Parameters in a Coupled Process of Electrocoagulation and Advanced Oxidation in the Removal of Turbidity in Wastewater from a Curtember

Paul Alcocer-Meneses [1], Angel Britaldo Cabrera-Salazar [1], Juan Taumaturgo Medina-Collana [1], Jimmy Aurelio Rosales-Huamani [2,*], Elmar Javier Franco-Gonzales [2] and Gladis Enith Reyna-Mendoza [1]

1 Water Treatment Process Engineering Research Center, Faculty of Chemical Engineering, National University of Callao, Juan Pablo II 306 Avenue, Bellavista, Callao 07011, Peru
2 Multidisciplinary Sensing, Universal Accessibility and Machine Learning Group, Faculty of Geological, Mining and Metallurgical Engineering, National University of Engineering, Lima 15333, Peru
* Correspondence: jrosales@uni.edu.pe; Tel.: +51-1-381-5639

Citation: Alcocer-Meneses, P.; Cabrera-Salazar, A.B.; Medina-Collana, J.T.; Rosales-Huamani, J.A.; Franco-Gonzales, E.J.; Reyna-Mendoza, G.E. Effects of the Operational Parameters in a Coupled Process of Electrocoagulation and Advanced Oxidation in the Removal of Turbidity in Wastewater from a Curtember. *Appl. Sci.* **2022**, *12*, 8158. https://doi.org/10.3390/app12168158

Academic Editors: Amit Kumar, Santosh Subhash Palmate and Rituraj Shukla

Received: 28 May 2022
Accepted: 14 July 2022
Published: 15 August 2022

Abstract: The tannery industry during its process generates various polluting substances such as organic matter from the skin and chemical inputs, producing wastewater with a high concentration of turbidity. The objective of this research is to evaluate the most appropriate operational parameters of the coupled process of electrocoagulation and advanced oxidation to achieve the removal of turbidity in wastewater from a tannery in the riparian zone (tannery). This process uses a direct current source between perforated aluminum electrodes of circular geometry submerged in the effluent, which causes the dissolution of the aluminum plates. For our study, an electrocoagulation unit coupled to an ozone generator has been built at the laboratory level, where the influence of five factors (voltage, inlet flow to the reactor, initial turbidity, pH, and ozone flow) has been studied with three levels with regarding turbidity, using the Taguchi experimental methodology. The optimal conditions for the removal of turbidity were obtained at 10 volts, 7.5 pH, 360 L/h of wastewater recirculation flow rate; 2400 mg/h of ozone flow rate; and 1130 NTU of initial turbidity of the sample in 60 min of treatment reaching a removal of 99.75% of the turbidity. Under optimal conditions, the removal of chemical oxygen demand (COD) and biochemical oxygen demand (BOD) was determined, reaching a removal percentage of 33.2% of COD and 39.36% of BOD was achieved. Likewise, the degree of biodegradability of the organic load obtained increased from 0.467 to 0.553.

Keywords: electrocoagulation; tannery effluent; ozonation; optimization; turbidity removal; Taguchi

1. Introduction

The leather trade can be an economic problem for developing countries that produce a good type of product from animal leather, such as footwear, luggage, and clothing. However, its production has a terribly high environmental footprint [1]. In addition, considering the enormous quantity and low biodegradability of the chemical products present in the productive cycle of tannery work, the wastewater from said process represents a great environmental and technological inconvenience [2]. In [3], the authors mentioned that the Electrocoagulation (EC) is often considered as an alternate treatment methodology with several advantages such as easy instrumentality, simple operation and automation, a brief retention time, low sludge production, and no chemical necessities.

Other studies have stated that Electrocoagulation mixtures and alternative technologies have been designed to treat high concentration organic waste material such as the textile trade and mixed industries [4,5]. In [6], mentioned that the use of EC as the only treatment process could face serious practical limitations, especially if the wastewater is highly contaminated. Therefore, there is a need for an efficient and relatively inexpensive treatment process. Due to this, the use of a post- or pre-treatment process with the EC will improve its performance, as mentioned by several studies that have described more profitable combined treatment systems [6,7].

The authors of [8] published a review that includes EC combined with other treatment processes such as: electrocoagulation–ozone, electrocoagulation–adsorption, electrocoagulation–ultrasound, and electrocoagulation–pulses. In his work, the authors also mention about the performance of these combined systems.

According to [9] Electrocoagulation (EC) is used in chemical science water treatment techniques where anode electrodes (aluminum, Al, or iron, Fe) area unit are dissolved in place, which promote coagulation and succeeding removals of pollutants and also the concurrent reduction of turbidity from water and wastewater. EC relies on the physical–chemical method of destabilization of mixture systems below the action of a right away current [10].

The electrodes dissolve according to Equations (1) and (2) to provide coagulant metal ions (Al^{3+} or Fe^{2+}/Fe^{2-}) into the water, and these instantaneously carries rapid hydrolysis.

Anode reactions:

$$Al_{(s)} \rightarrow Al^{3+}_{(aq)} + 3e^- \quad E^0 = 1.66\ V \tag{1}$$

$$Fe_{(s)} \rightarrow Fe^{3+}_{(aq)} + 3e^- \quad E^0 = 0.04\ V \tag{2}$$

When the anode potential is sufficiently high, secondary reactions may occur, especially oxygen evolution, according to Equation (3)

$$2H_2O_{(l)} \rightarrow O_{2(g)} + 4H^+_{(aq)} + 4e^- \quad E^0 = 1.66\ V \tag{3}$$

Simultaneously with the anode reaction, water molecules H_2O break down at the cathode, producing hydrogen gas H_2 and OH^- ions, according to Equation (4).

Cathode reaction:

$$H_2O_{(l)} + 2e^- \rightarrow H_{2(g)} + OH^-_{(aq)} \quad E^0 = 1.66\ V \tag{4}$$

The electrical energy applied to the anode dissolves the aluminum into the solution which then reacts with the hydroxyl ion from the cathode to form aluminum hydroxy. The most significant advantage of electrocoagulation is avoiding any addition of chemical substances thus reducing the likelihood of secondary pollution; the dosing of coagulator depends on the cell potential (or current density) applied [11]. Other advantages are the simple equipment, so requiring less maintenance and straightforward automation of the method [12].

Standard treatments for cloudiness removal have many disadvantages, such as the use of enormous amounts of chemicals and generating large amounts of sludge that causes disposal issues and therefore the loss of water. Then in [13] mentioned that the combination of ultrasound technique with different processes such as electrocoagulation, electro-Fenton, and electrooxidation could be important to achieve effective decomposition of organic contaminants in wastewater. Independently in [14] mentioned the integrated sonoelectro-Fenton (SEF) method could be a novel methodology for the removal of paracetamol (PCT) waste material from liquid solutions through synthesized iron ore (Fe_2O_3) nanoparticles.

The novelty of our study was the design of the electrocoagulation cell with perforated plates installed vertically, improving the mixture of ozone with the residual water and the ions generated by the electrodes. In this way, reducing areas of stagnation in the electrocoagulation cell that produce passivation of the electrodes, causing a decrease in the efficiency of the process.

The objective of this study was to examine the treatment of wastewater from the tanning industry, through the electrocoagulation process, the impact of the factors electrical potential, feed flow, initial concentration of turbidity, pH, and ozone flow on the percentage reduction of turbidity and energy consumption, based on the Taguchi methodology.

2. Materials and Methods

2.1. Effluent Sample Collection

The samples were collected from the operations corresponding to the riparian zone (pre-soaking, main soaking, peeling, descaling, and purging or delivery), from the tannery located in the district of Ate Vitarte, Lima (Peru). Each sample was collected and then homogenized and allowed to stand for 3 h. These samples came from a process of transformation of sheep skins preserved with salt, with hair destruction technology.

A part of the sample was sent to a specialized laboratory, applying the corresponding monitoring protocols to know the physicochemical characteristics, as illustrated in Table 1.

Table 1. Some of the physicochemical characteristics of the wastewater of the riparian zone.

Parameters	Unit	Value
pH		9.43
Turbidity	NTU	1130
Chemical oxygen demand	mg/L	2638
Biological oxygen demand	mg/L	1232
Oils and fats	mg/L	15.1
Ammonia nitrogen	NH^{3+}-N mg/L	88.85
Sulfides	S = mg/L	21.4
Fecal coliforms	NMP/100 mL	<1.8
Aluminum	mg/L	0.29

2.2. Analytical Methods

The turbidity was measured by Ezodo model TUB-430, turbidimeter, to determine the pH, conductivity and total dissolved solids, the Multiparameter equipment (pH, EC, TDS, T °C), HANNA brand was used. To determine the voltage and current intensity, the Digital Hook Multimeter (amps, voltage, temperature, etc.) was used.

2.3. Design of Experiment

The optimization of wastewater turbidity removal using Aluminum electrodes was performed using the Taguchi Design. Five important factors such as voltage, feed flow, effluent concentration, pH, and ozone flow were used as independent variables where their combined effects were examined, while the percentage of turbidity removal was the dependent variable.

This was performed to determine the best conditions for the optimum removal of turbidity from the wastewater. The experimental design involves varying the independent variable at three different levels (−1, 0, +1). The experimental range and levels of the independent variables are presented in Table 2. In this work, a set of 27 experiments with two replicates, the mean shown in Table 3. Where the levels of the applied electrical potential were acquired from the work developed by [15] and to select the pH range the research work provided by [16] was taken.

The interactive effects of the independent (process) variables on the dependent variable (response) were examined using the analysis of variance (ANOVA) as shown in Table 4.

Table 2. Experimental range and levels of independent variables used in this study.

Factors	Unit	Notation	Levels		
			Low	Medium	High
Voltage	V	X1	4	7	10
Feed flow	L/h	X2	240	300	360
Turbidity	NTU	X3	375	580	1130
pH		X4	4.0	7.5	10.8
Ozone flow	mg/h	X5	900	1500	2400

Table 3. Presents the results of the 27 experiments carried out using the Taguchi methodology of five factors at three levels under study.

N° Tests	Factors					Response			
	X1 (V) Voltage	X2 (L/h) Feed Flow	X3 (NTU) Turbidity	X4 (pH)	X5 (mg/h) Ozone Flow	Turbidity (NTU)	% Turbidity Reduction	Faraday Aluminum Dough (g)	Energy Consumed in the Cell (KWh/m^3)
1	4	240	375	4	900	471.4	58.28	0.1176	175.17823
2	4	240	375	4	1500	487.8	56.83	0.0535	85.34855
3	4	240	375	4	2400	483.6	57.2	0.0471	76.356466
4	4	300	580	7.5	900	48.7	91.65	0.0279	45.541531
5	4	300	580	7.5	1500	55.5	90.48	0.0314	50.222793
6	4	300	580	7.5	2400	52.3	91.03	0.0397	63.811338
7	4	360	1130	10.8	900	88.3	76.45	0.0246	37.127518
8	4	360	1130	10.8	1500	91.2	75.68	0.0275	43.868429
9	4	360	1130	10.8	2400	87.1	76.77	0.0294	46.751746
10	7	240	580	10.8	900	135.2	76.49	0.2093	657.31997
11	7	240	580	10.8	1500	145.1	74.77	0.1539	499.21802
12	7	240	580	10.8	2400	139.7	75.7	0.1726	551.08302
13	7	300	1130	4	900	3.2	99.14	0.1111	365.33785
14	7	300	1130	4	1500	1.3	99.65	0.1135	362.05164
15	7	300	1130	4	2400	0	100	0.136	421.1065
16	7	360	375	7.5	900	17.5	98.45	0.1822	570.44258
17	7	360	375	7.5	1500	13.8	98.78	0.1936	604.95181
18	7	360	375	7.5	2400	15.2	98.66	0.1485	479.17558
19	10	240	1130	7.5	900	7.2	98.11	0.1402	652.35772
20	10	240	1130	7.5	1500	5.2	98.64	0.1546	748.96169
21	10	240	1130	7.5	2400	2.8	99.27	0.1492	712.034
22	10	300	375	10.8	900	207	81.63	0.2704	1178.6499
23	10	300	375	10.8	1500	195.2	82.68	0.2993	1310.3554
24	10	300	375	10.8	2400	198	82.43	0.153	726.03864
25	10	360	580	4	900	0	100	0.238	1060.6992
26	10	360	580	4	1500	1.2	99.79	0.2312	1029.1543
27	10	360	580	4	2400	1.57	99.73	0.2371	1042.1303

Table 4. Analysis of variance (ANOVA).

Source	GL	SC Sec.	Contribution	SC Ajust.	MC Ajust.	Value F	Value *p*
Model	14	5252.01	99.94%	5252.01	375.14	1337.59	0.000
Linear	5	3347.2	63.69%	3392.4	678.48	2419.14	0.000
X1	1	1669.48	31.77%	1560.97	1560.97	5565.7	0.000
X2	1	924.79	17.60%	919.82	919.82	3279.66	0.000
X3	1	514.58	9.79%	656.93	656.93	2342.3	0.000
X4	1	238.29	4.53%	254.67	254.67	908.05	0.000
X5	1	0.06	0.00%	0.06	0.06	0.23	0.643
Square	5	1903.28	36.22%	1903.28	380.66	1357.24	0.000
$(X1)^2$	1	193.78	3.69%	193.78	193.78	690.95	0.000
$(X2)^2$	1	256.89	4.89%	256.89	256.89	915.95	0.000
$(X3)^2$	1	210.76	4.01%	210.76	210.76	751.48	0.000
$(X4)^2$	1	1241.12	23.62%	1241.12	1241.12	4425.26	0.000
$(X5)^2$	1	0.72	0.01%	0.72	0.72	2.56	0.135
Error	12	3.37	0.06%	3.37	0.28		
Total	26	5255.38	100.00%				

2.4. Electrocoagulation Reactor

The EC experiments were performed by a batch process using a 7 L capacity of a cylindrical reactor, the configuration (Figure 1) of the electrochemical reactor has a cylindrical shape, aluminum electrodes were used both for the anode and for the circular cathode (Perforated plates), we work with a configuration of parallel monopolar electrodes, with a separation of 1 cm as mentioned in [17–19], and the specific area of each electrode was 0.014 cm^2. Each electrode was 10 cm (diameter) with 10 holes of 10 mm diameter each, by 0.3 cm (thickness), the number of electrodes used were four. The EC cell was configured for the vertical water flow of the feed water that was delivered by a peristaltic pump. Accessory (ACC) power supply was connected (0–15 volts). Before installation in the EC unit, each plate was weighed to allow the calculation of the mass consumed after the tests. Each experiment was continued for 60 min, which was considered enough to achieve a stable operation. Ozone was coupled to the system by means of venturi, the ozone generating equipment has a capacity of (0 to 3 g O_3/h).

All experiments were performed at room temperature (nominally 20 °C). After the seating time elapsed, the samples were removed from a depth of 2 cm using a syringe and measured using the turbidity meter. The electrodes were cleaned in a solution of low concentration hydrochloric acid (0.04 M) and another caustic soda solution (0.08 M) to remove the remains stuck on the surface of the electrodes; they were finally washed with distilled water for reuse. The arrangement of the electrodes consisted of two cathodes that were interspersed with two anodes connected by stainless steel rods to other arranged and then the samples were periodically taken every 10 min for the measurement of turbidity. The power was supplied to the electrodes with a Direct Current (DC) power supply.

An improvement over other reported works [15,20,21] is the configuration of the experimental equipment used. In this investigation, an electrocoagulation cell with perforated circular electrodes has been built. This design allows for improved mixing, longer residence time for the effluent and ozone. therefore the mechanisms used in this hybrid process are improved such as sedimentation [15,22]. A disadvantage compared to other configurations of electrocoagulation cells is the maintenance of the electrodes, which is relatively easy.

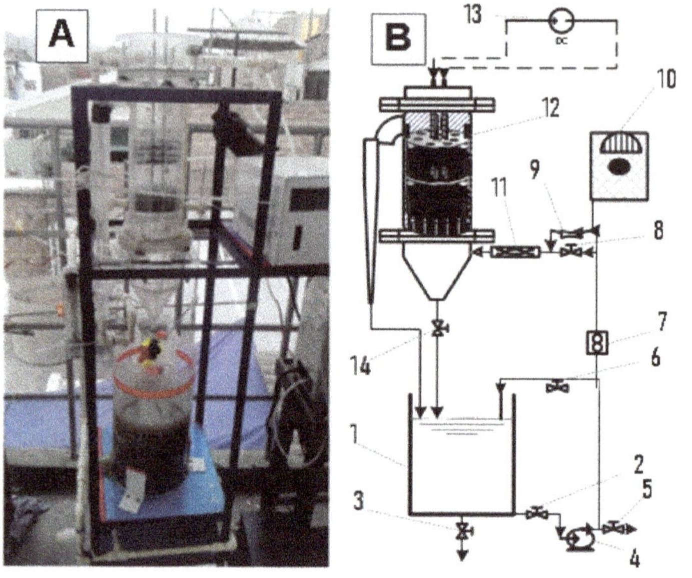

Figure 1. Electrocoagulation and ozone experimental module. (**A**) Photograph of the experimental module doing preliminary tests. The sample is fed to tank 1, followed by the sample being pumped through the flow meter, followed by the Venturi system, dynamic mixer until reaching the electrolytic reactor, once the system is filled again the sample returns to the tank. (**B**) Module diagram, where 1 is the deposit; 2, 3, 5, 6, 8, and 14 stopcocks; 4 recirculation pump; 7 flow meter; 9 Venturi; 10 ozone generator; 11 dynamic mixer; 12 electrocoagulation reactor and 13 current rectifier.

2.5. The Main Calculations of Electrocoagulation Process

The reduction rate of turbidity, expressed in percentage "T" (%), was calculated using Equation (5).

$$T(\%) = \left(\frac{T_i - T_f}{T_i}\right) \times 100\% \tag{5}$$

where T_i and T_f represent initial and final turbidity, respectively. Electrical energy consumption is a very important economical parameter in the electrocoagulation process. The electrical energy consumption was calculated using the following Equation (6) [23].

$$C.E. = \frac{U}{Vm}\int_0^{3600} I(t)dt \tag{6}$$

$C.E.$ is the energy consumption (kWh/m^3)
U is the applied voltage (V)
Vm is the treated volume of the sample (L).

The integral represents the intensity value multiplied with time in seconds.

The amount of dissolved electrode was calculated theoretically using Faraday's law [24], through the following Equation (7).

$$m = \frac{M}{nF}\int_{\text{t. inicial}}^{\text{t. final}} I(t)dt \tag{7}$$

m is the aluminum mass (g) in the electrolytic cell
I is the intensity of the current (A)
t is the electrocoagulation time (s), M is the molecular weight of the anode (g/mol)
z is The chemical equivalence, F is the faraday constant (96,500 c/mol)

(M_{Al} = 26,982 g/mol)
n is the valence of the ions of the electrode material ($n_{Al} = 3.0$).

3. Results

The results of our experiment are shown below. Complementing the results of Table 3, the standard deviation has been evaluated with respect to the mean of the percentage of turbidity removal and Energy consumption whose results are shown in Table 5. Then we show the physicochemical parameters obtained in Table 6.

Table 5. Standard deviation of percent turbidity removal and energy consumption and Energy consumption.

Response	N	Statistical Minimum	Statistical Maximum	Mean	Standard Error	Standard Deviation
Turbidity removal percentage	27	56.83	100	86.6033	2.736	14.217
Energy consumption (KW/m^3)	27	0.037	1.310	0.503	0.075	0.390

Table 6. Results of the physicochemical characterization of the treated sample.

Parameters	Unit	Value
pH		8.6
Turbidity	NTU	2.8
Chemical oxygen demand	mg/L	876
Biological oxygen demand	mg/L	485
Oils and fats	mg/L	<1.2
Ammonia nitrogen	NH^{3+}-N mg/L	32.75
Sulfides	S = mg/L	<0.002
Fecal coliforms	NMP/100 mL	<1.8
Aluminum	mg/L	44.06

3.1. Main Effect of Variables

The main effect plots for the six operating variables are given in Figure 2. The main effects of the tested variables were calculated by averaging the experiment results achieved at each level for each variable. This plot was obtained from Table 3 and is used to visualize the relation between variables and the output response.

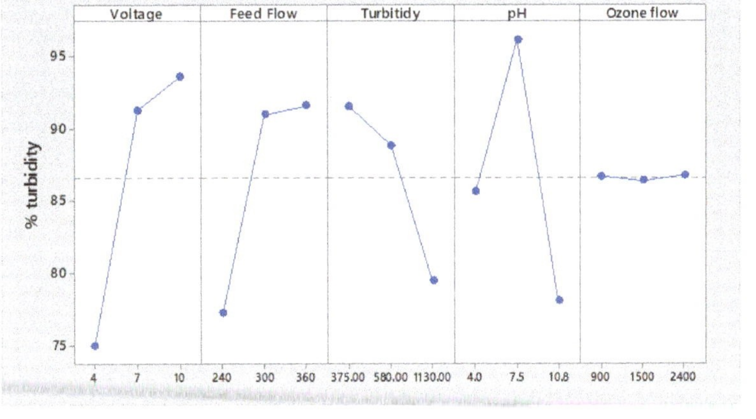

Figure 2. The effects of the operating variables on the mean turbidity removal percentage.

3.1.1. Comparison of Ozonation, Electrocoagulation, and Ozone-Assisted Electrocoagulation

From Figure 3 we observe that for the initial turbidity of 655 NTU of the sample, when the process is hybrid (electrocoagulation and ozone), a turbidity of 4.19 NTU is reached (99.36% turbidity removal). In the electrocoagulation process, a turbidity of 18.34 NTU (97.2%) is obtained and through ozone up to 196.6 NTU (69.98%) is reached, therefore it is concluded that the hybrid and electrocoagulation process reach yields above 97% for removal of turbidity. We also observed that the removal of turbidity in the three processes is achieved in the first 20 initial minutes of treatment. In the work of [25] indicated that the combined electrocoagulation/ozonation process improved both the degradation rate and the maximum removal of COD compared to the electrocoagulation and ozonation processes alone.

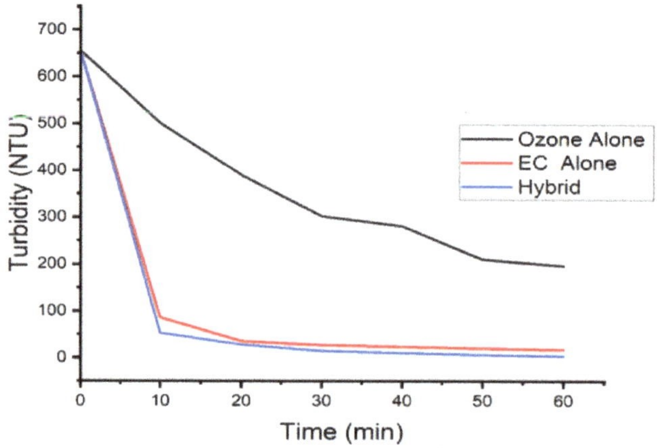

Figure 3. Turbidity reduction for separate processes such as pure ozone, electrocoagulation and coupled process of EC/O_3, operated at conditions of 10.0 volts, feed flow 360 L/h, O_3 flow 2400 mg/h, pH 6.89, and initial turbidity of 655 NTU.

3.1.2. Initial pH Effect

From Table 3, trials 1, 4, 10, 20, 24, and 27 have been plotted as they are the most representative. Then it is observed from Figure 4 that for experiments 1 and 10 the pH of the sample increases with the treatment time. For experience 27, a pH of 8.21 is reached and is attributed mainly to the increase in electrical potential (10 volts). When the initial pH is 7.5 for experiences 4 and 20, the increase is not very significant, reaching a final value of 8.54. Finally, when the sample has an initial pH of 10.8 in tests 10 and 24, a decrease is observed, reaching a value of 9.21.

The tannery industry generates effluents with a wide pH range, from pH = 3.5 to pH = 11; on the other hand, studies show that pH has a significant impact on electrocoagulation performance. The increase in pH is a consequence of the formation of Al^{3+} which precipitates due to the presence of other anions, as well as the precipitation of aluminum hydroxide; however when the pH starts at alkaline, the decrease in pH is the result of the formation of $Al(OH)_4^{-1}$ [26].

3.1.3. Effect of Initial Turbidity

According to Figure 5B, a greater reduction in turbidity is observed as the initial turbidity is less than 1130 NTU, this behavior could be explained because the amount of flocs formed is sufficient for their adsorption and thus quickly decrease turbidity. This trend is also deduced from Faraday's law, which states that Al^{3+} released to the solution for the same applied solution is constant [27].

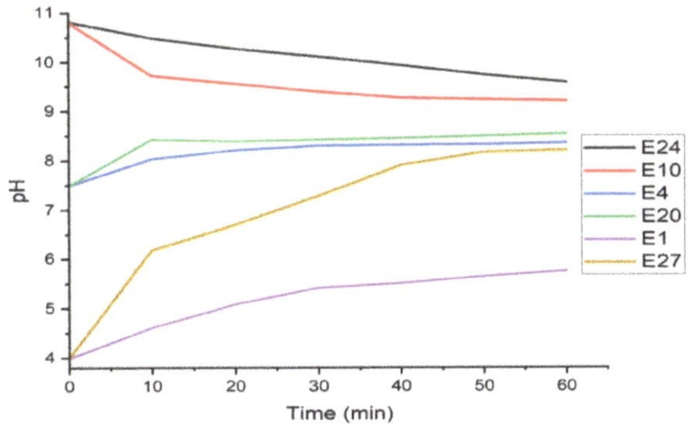

Figure 4. Variation of the pH in the time of treatment by electrocoagulation/O_3.

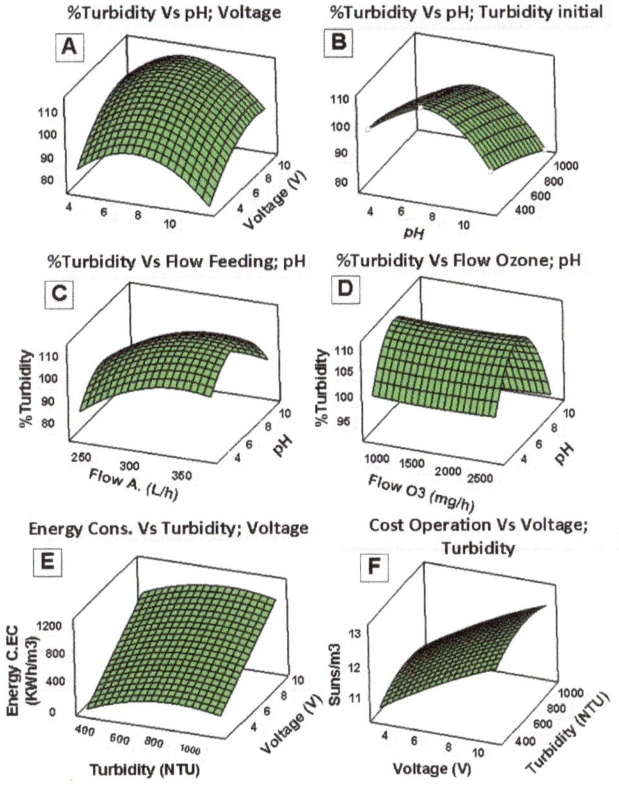

Figure 5. Representation of the effect of operational variables on % reduction in turbidity, energy consumed in the electrolytic cell, and operational cost of the module. (**A**) Variation of the pH and voltage variables against % turbidity for fixed values of 300 L/h, 752.5 NTU, and 1650 mg/h. (**B**) Variability of pH and initial turbidity versus % turbidity for fixed values of 6.5 volts, 300 L/h and 1650 mg/h of O_3 flow. (**C**) pH and feed flow variability versus % turbidity for fixed values of 6.5 volts, 752.5 NTU, and 1650 mg/h O_3 flow. (**D**) Ozone and pH flow variability versus % turbidity for fixed values of 6.5 volts, 300 L/h, and 752.5 NTU. (**E**) Variability of initial turbidity and voltage versus energy consumption in the electrocoagulation cell for fixed values of 300 L/h, 7.4 pH, and 1650 mg/h of O_3 flow. (**F**) Variability of voltage and initial turbidity against the cost of the built module for fixed values of 300 L/h, 7.4 pH, and 1650 mg/h of O_3 flow.

The proposed mechanism for the reduction of turbidity by means of the hybrid system of electrocoagulation and ozone is shown in Figure 6. This consists of destabilizing the colloidal particles and forming larger flocs, in which the contaminants are trapped and these flocs can be separated from the solution by flotation or sedimentation [28]. The dissolved air flotation mechanism is effective in reducing the organic load [29] and dissolved ozone flotation gives efficient results in the removal of suspended solids [30,31]. For soluble contaminants, aluminum-based coagulants can act as catalysts for ozone and generate hydroxyl formation [22] and also oxidize surface functional groups of colloidal contaminants that promote colloid aggregation.

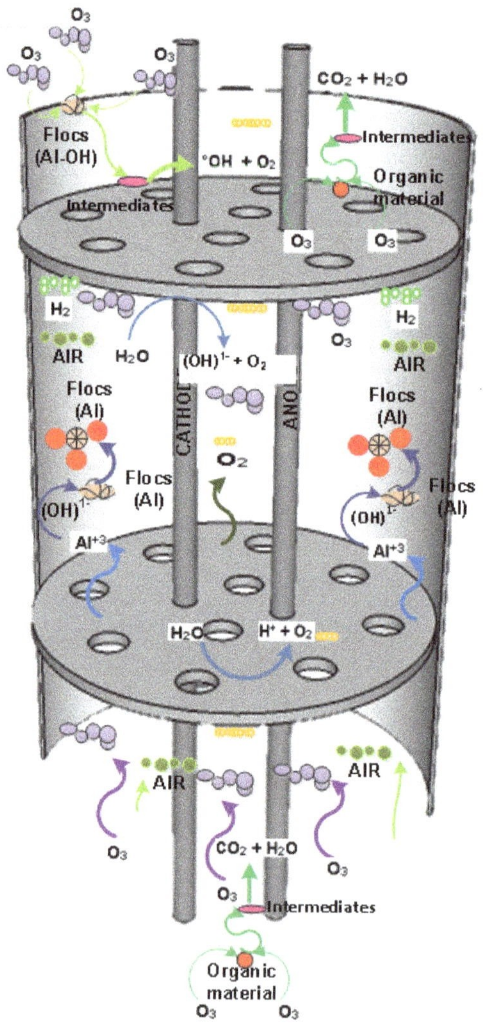

Figure 6. Mechanism of hydroxyl formation.

3.1.4. Feed Flow Effect

According to Figure 5C, as the feed flow increases (240 to 360 L/h) there is an increase in the reduction of turbidity, this could be attributed as the feed flow increases towards the reactor, there is a greater formation of bubbles, this is influenced by the principle of hydrodynamic cavitation that forms in the Venturi tube [32]. As a consequence, the flotation mechanism predominates to reduce turbidity, this formation of bubbles increases when working under acidic conditions, forming two phases (80% foam and 20% liquid) [33]. However, this generation of bubbles generates a problem in the electrodes (activation polarization), generating an increase in voltage and a decrease in electrical current, thus an increase in energy consumption [34].

3.1.5. Ozone Flow Effect

Ozone flow is one of the factors that has the least influence on reducing turbidity, as can be seen in Figure 5D. Furthermore, in Figure 2, we observe that the mass flow of ozone does not have much influence on the removal of turbidity. In [35], mentioned that for the activation of ozone and its transformation into hydroxide ion (OH^-), it is achieved through electroreduction, which in this case would help in the oxidation either directly or indirectly to the components present in the effluent (organic matter, nitrates, sulfides, etc.). To oxidize the sulfur, ozone is an alternative to the traditional ions (Fe^{2+}, O_2, etc.), as verified in the research work [36].

4. Discussion

When evaluating the five operational parameters against the reduction of turbidity according to Figure 2, it is shown that the factor with the greatest influence is the voltage, corroborating it in Table 4 of ANOVA due to its greater contribution with respect to the other parameters. By increasing the potential values from 4 to 10 volts as seen in Figure 5A, it was possible to increase the percentage of turbidity reduction reaching 56.83% and 100%, a growing effect in the elimination of turbidity. This originated effect is analogous to those reported in [37], where they worked at 6, 8.5 and 10 volts, for one hour of treatment on grey water, reaching a reduction of 68%, 73%, and 86% respectively.

On the other hand, the effect on removal is due to the increase in particle size as a function of time, studied by [38], where he reported that in a synthetic sample of kaolinite, the size formed is affected as the voltage and time are increased, allowing the generation of a higher sedimentation rate of the particles.

This ascending effect of the voltage on the turbidity can also be seen in the report presented by [39], they worked in the range of 2.9 to 11.7 mA/cm^2, for a time of 14 min on water. residues from car washes, achieving close to a 96% reduction in turbidity. On the other hand, the work presented by [15], also reported the influence of the applied potential on turbidity, where they evaluated 4 voltage levels for a period of 15 min such as: 2, 5, 10, and 15 V, achieving a reduction 83% for voltage 2 and 92% for 15 volts; therefore, as stated [40], the applied voltage is an influential and important parameter. As a main step, it ensures the production of Al^{3+} ion coagulants as a result of electrolytic oxidation of the electrode. Table 7 shows the results.

Table 7. Effect of the applied potential difference on the removal of the turbidity.

Electric Potential (V)	Turbidity (%)
2	83
5	90
10	91
15	92

From Table 3 we generate Figure 7 which shows the effect of the process parameters with respect to energy consumption in kWh/m^3. From said figure we observe that the average energy consumption in the 27 experiments was 0.5 kWh/m^3.

In addition, the factor with the greatest influence was the electrical potential applied to the electrocoagulation cell, as indicated by the diagram, the lowest energy consumption (0.069) was obtained with the electrical potential at 4 volts and the highest energy consumption (0.94) was obtained at an electrical potential of 10 volts. Likewise, it is observed that turbidity has a significant influence on energy consumption at the high level, 0.376 kWh/m^3 is consumed, whose value is below the average.

In the study carried out by [15], about the reduction of turbidity and chromium content in tannery wastewater by electrocoagulation process using aluminum electrodes at an electrical potential of 10 volts, pH of 6.1, and a time of 90 min. The authors obtained an energy consumption of 1.5 kWh/m^3, which is quite close to that obtained in our present study.

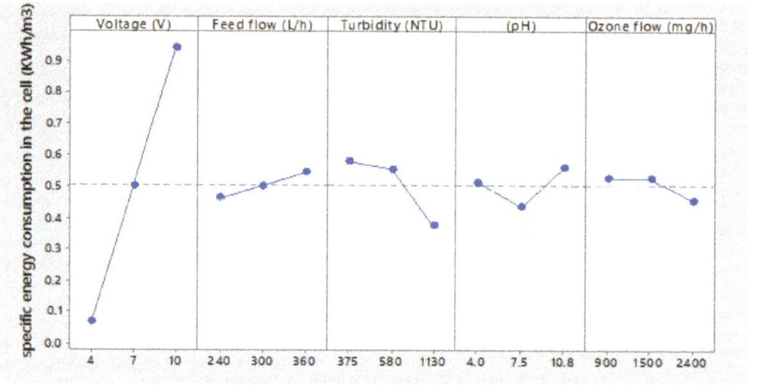

Figure 7. Effect of process parameters on specific energy consumption.

5. Conclusions

The coupled process of electrocoagulation with ozone was successfully tested in the treatment of wastewater from a tannery from the riparian zone. Parameters such as applied voltage potential, feed flow, initial turbidity concentration, pH, and ozone flux were studied on the percentage of turbidity reduction and energy consumption in the electrocoagulation cell. It was found that parameters have the greatest influence on turbidity reduction and the effects separately of each process such as ozone, electrocoagulation and ozone-assisted electrocoagulation on turbidity.

The result showed that the factor that has the greatest influence on reducing turbidity is voltage. The present study showed that the coupled electrocoagulation and ozone system reduced more turbidity than the processes alone. The optimal conditions for the removal of turbidity, Chemical Oxygen Demand (COD) and Biochemical Oxygen Demand (BOD) were obtained at 10 volts, 7.5 pH, 360 L/h of wastewater recirculation flow, 2400 mg/h of ozone flow, and 1130 NTU of initial turbidity of the sample in 60 min of treatment. Finally, under these conditions, a removal of 99.75% of turbidity, 33.2% of COD, and 39.36% of BOD was achieved. Likewise, the degree of biodegradability of the organic load obtained increased from 0.467 to 0.553.

Author Contributions: Conceptualization, P.A.-M.; Data curation, A.B.C.-S.; Investigation, J.A.R.-H.; Project administration, A.B.C.-S.; Software, J.T.M.-C.; Validation, P.A.-M.; Writing—original draft, E.J.F.-G.; Writing—review & editing, G.E.R.-M. All authors have read and agreed to the published version of the manuscript.

Funding: This research received no external funding.

Conflicts of Interest: The authors declare no conflicts of interest.

References

1. Villaseñor-Basulto, D.L.; Picos-Benítez, A.; Pacheco-Alvarez, M.; Pérez, T.; Bandala, E.R.; Peralta-Hernández, J.M. Tannery wastewater treatment using combined electrocoagulation and electro-Fenton processes. *J. Environ. Chem. Eng.* **2022**, *10*, 107290. [CrossRef]
2. Di Iaconi, C.; Lopez, A.; Ramadori, R.; Di Pinto, A.; Passino, R. Combined chemical and biological degradation of tannery wastewater by a periodic submerged filter (SBBR). *Water Res.* **2002**, *36*, 2205–2214. [CrossRef]
3. Varank, G.; Erkan, H.; Yazycy, S.; Demir, A.; Engin, G. Electrocoagulation of tannery wastewater using monopolar electrodes: process optimization by response surface methodology. *Winter* **2014**, *8*, 165–1890.
4. Chanikya, P.; Nidheesh, P.; Babu, D.S.; Gopinath, A.; Kumar, M.S. Treatment of dyeing wastewater by combined sulfate radical based electrochemical advanced oxidation and electrocoagulation processes. *Sep. Purif. Technol.* **2021**, *254*, 117570. [CrossRef]
5. Nidheesh, P.; Kumar, A.; Babu, D.S.; Scaria, J.; Kumar, M.S. Treatment of mixed industrial wastewater by electrocoagulation and indirect electrochemical oxidation. *Chemosphere* **2020**, *251*, 126437. [CrossRef]
6. Al-Qodah, Z.; Tawalbeh, M.; Al-Shannag, M.; Al-Anber, Z.; Bani-Melhem, K. Combined electrocoagulation processes as a novel approach for enhanced pollutants removal: A state-of-the-art review. *Sci. Total. Environ.* **2020**, *744*, 140806. [CrossRef]
7. GilPavas, E.; Dobrosz-Gómez, I.; Gómez-García, M.Á. Optimization and toxicity assessment of a combined electrocoagulation, H_2O_2/Fe^{2+}/UV and activated carbon adsorption for textile wastewater treatment. *Sci. Total. Environ.* **2019**, *651*, 551–560. [CrossRef]
8. Barrera-Díaz, C.E.; Roa-Morales, G.; Hernández, P.B.; Fernandez-Marchante, C.M.; Rodrigo, M.A. Enhanced electrocoagulation: New approaches to improve the electrochemical process. *J. Electrochem. Sci. Eng.* **2014**, *4*, 285–296. [CrossRef]
9. Kuokkanen, V.; Kuokkanen, M.; Hynynen, I.; Kuokkanen, T. Electrocoagulation treatment of metallurgical industry wastewater–A laboratory scale batch and pilot scale continuous study. *Hydrometallurgy* **2021**, *202*, 105596. [CrossRef]
10. Sillanpää, M.; Shestakova, M. *Electrochemical Water Treatment Methods: Fundamentals, Methods and Full Scale Applications*; Butterworth-Heinemann: Oxford, UK, 2017.
11. Mollah, M.Y.; Morkovsky, P.; Gomes, J.A.; Kesmez, M.; Parga, J.; Cocke, D.L. Fundamentals, present and future perspectives of electrocoagulation. *J. Hazard. Mater.* **2004**, *114*, 199–210. [CrossRef]
12. Mollah, M.Y.A.; Schennach, R.; Parga, J.R.; Cocke, D.L. Electrocoagulation (EC)—Science and applications. *J. Hazard. Mater.* **2001**, *84*, 29–41. [CrossRef]
13. Hassani, A.; Malhotra, M.; Karim, A.V.; Krishnan, S.; Nidheesh, P. Recent progress on ultrasound-assisted electrochemical processes: A review on mechanism, reactor strategies, and applications for wastewater treatment. *Environ. Res.* **2022**, *205*, 112463. [CrossRef] [PubMed]
14. Ghanbari, F.; Hassani, A.; Wacławek, S.; Wang, Z.; Matyszczak, G.; Lin, K.Y.A.; Dolatabadi, M. Insights into paracetamol degradation in aqueous solutions by ultrasound-assisted heterogeneous electro-Fenton process: Key operating parameters, mineralization and toxicity assessment. *Sep. Purif. Technol.* **2021**, *266*, 118533. [CrossRef]
15. Ziati, M.; Khemmari, F.; Aitbara, A.; Hazourli, S. Reduction of Turbidity and Chromium Content of Tannery Wastewater by Electrocoagulation Process: Ziati et al. *Water Environ. Res.* **2018**, *90*, 598–603. [CrossRef] [PubMed]
16. Mahmad, M.K.N.; Rozainy, M.M.R.; Abustan, I.; Baharun, N. Electrocoagulation process by using aluminium and stainless steel electrodes to treat total chromium, colour and turbidity. *Procedia Chem.* **2016**, *19*, 681–686. [CrossRef]
17. Loukanov, A.; El Allaoui, N.; Omor, A.; Elmadani, F.Z.; Bouayad, K.; Nakabayashi, S. Large-scale removal of colloidal contaminants from artisanal wastewater by bipolar electrocoagulation with aluminum sacrificial electrodes. *Results Chem.* **2020**, *2*, 100038. [CrossRef]
18. Marmanis, D.; Emmanouil, C.; Fantidis, J.; Thysiadou, A.; Marmani, K. Description of a Fe/Al Electrocoagulation Method Powered by a Photovoltaic System, for the (Pre-) Treatment of Municipal Wastewater of a Small Community in Northern Greece. *Sustainability* **2022**, *14*, 4323. [CrossRef]
19. Chen, G. Electrochemical technologies in wastewater treatment. *Sep. Purif. Technol.* **2004**, *38*, 11–41. [CrossRef]
20. Ikhlaq, A.; Javed, F.; Akram, A.; Rehman, A.; Qi, F.; Javed, M.; Mehdi, M.J.; Waheed, F.; Naveed, S.; Aziz, H.A. Synergic catalytic ozonation and electroflocculation process for the treatment of veterinary pharmaceutical wastewater in a hybrid reactor. *J. Water Process. Eng.* **2020**, *38*, 101597. [CrossRef]
21. Asaithambi, P.; Aziz, A.R.A.; Daud, W.M.A.B.W. Integrated ozone—Electrocoagulation process for the removal of pollutant from industrial effluent: Optimization through response surface methodology. *Chem. Eng. Process. Process. Intensif.* **2016**, *105*, 92–102. [CrossRef]
22. Jin, X.; Xie, X.; Liu, Y.; Wang, Y.; Wang, R.; Jin, P.; Yang, C.; Shi, X.; Wang, X.C.; Xu, H. The role of synergistic effects between ozone and coagulants (SOC) in the electro-hybrid ozonation-coagulation process. *Water Res.* **2020**, *177*, 115800. [CrossRef] [PubMed]
23. Li, G.; Yang, C.; Yao, Y.; Zeng, M. Electrocoagulation of chromium in tannery wastewater by a composite anode modified with titanium: Parametric and kinetic study. *Desalin Water Treat* **2019**, *171*, 294–301. [CrossRef]
24. Elabbas, S.; Ouazzani, N.; Mandi, L.; Berrekhis, F.; Perdicakis, M.; Pontvianne, S.; Pons, M.N.; Lapicque, F.; Leclerc, J.P. Treatment of highly concentrated tannery wastewater using electrocoagulation: Influence of the quality of aluminium used for the electrode. *J. Hazard. Mater.* **2016**, *319*, 69–77. [CrossRef]
25. Barrera-Diaz, C.; Bernal-Martinez, L.A.; Natividad, R.; Peralta-Hernández, J.M. Synergy of electrochemical/O3 process with aluminum electrodes in industrial wastewater treatment. *Ind. Eng. Chem. Res.* **2012**, *51*, 9335–9342. [CrossRef]

26. Alam, R.; Sheob, M.; Saeed, B.; Khan, S.U.; Shirinkar, M.; Frontistis, Z.; Basheer, F.; Farooqi, I.H. Use of electrocoagulation for treatment of pharmaceutical compounds in water/wastewater: A review exploring opportunities and challenges. *Water* **2021**, *13*, 2105. [CrossRef]
27. Aber, S.; Amani-Ghadim, A.; Mirzajani, V. Removal of Cr (VI) from polluted solutions by electrocoagulation: Modeling of experimental results using artificial neural network. *J. Hazard. Mater.* **2009**, *171*, 484–490. [CrossRef]
28. Hakizimana, J.N.; Gourich, B.; Chafi, M.; Stiriba, Y.; Vial, C.; Drogui, P.; Naja, J. Electrocoagulation process in water treatment: A review of electrocoagulation modeling approaches. *Desalination* **2017**, *404*, 1–21. [CrossRef]
29. Gonzalez-Galvis, J.P.; Narbaitz, R.M. Large batch bench-scale dissolved air flotation system for simulating full-scale turbidity removal. *Environ. Technol.* **2020**, *43*, 1791–1804. [CrossRef]
30. Jin, X.; Zhang, L.; Liu, M.; Hu, S.; Yao, Z.; Liang, J.; Wang, R.; Xu, L.; Shi, X.; Bai, X.; et al. Characteristics of dissolved ozone flotation for the enhanced treatment of bio-treated drilling wastewater from a gas field. *Chemosphere* **2022**, *298*, 134290. [CrossRef]
31. Yao, Z.; Jin, X.; Liang, J.; Wang, R.; Jin, P.; Li, G.; Yao, L.; Wang, Z. Application of an integrated dissolved ozone flotation process in centralised fracturing wastewater treatment plant. *Water Reuse* **2021**, *11*, 236–247. [CrossRef]
32. Wang, J.; Wang, L.; Xu, S.; Ji, B.; Long, X. Experimental investigation on the cavitation performance in a venturi reactor with special emphasis on the choking flow. *Exp. Therm. Fluid Sci.* **2019**, *106*, 215–225. [CrossRef]
33. Fukui, Y.; Yuu, S. Removal of colloidal particles in electroflotation. *AIChE J.* **1985**, *31*, 201–208. [CrossRef]
34. Skoog, D.A.; West, D.M.; Holler, F.J.; Crouch, S.R. *Fundamentals of Analytical Chemistry*; Cengage Learning: Boston, MA, USA, 2013.
35. Sivagami, K.; Sakthivel, K.; Nambi, I.M. Advanced oxidation processes for the treatment of tannery wastewater. *J. Environ. Chem. Eng.* **2018**, *6*, 3656–3663. [CrossRef]
36. Umbarila-Ortega, M.F.; Prado-Rodríguez, J.S.; Agudelo-Valencia, R.N. Remoción de sulfuro empleando ozono como agente oxidante en aguas residuales de curtiembres. *Rev. Fac. de Ing.* **2019**, *28*, 25–38. [CrossRef]
37. Vakil, K.A.; Sharma, M.K.; Bhatia, A.; Kazmi, A.A.; Sarkar, S. Characterization of greywater in an Indian middle-class household and investigation of physicochemical treatment using electrocoagulation. *Sep. Purif. Technol.* **2014**, *130*, 160–166. [CrossRef]
38. Lin, C.J.; Lo, S.L.; Kuo, C.Y.; Wu, C.H. Pilot-scale electrocoagulation with bipolar aluminum electrodes for on-site domestic greywater reuse. *J. Environ. Eng.* **2005**, *131*, 491–495. [CrossRef]
39. El-Ashtoukhy, E.; Amin, N.; Fouad, Y. Treatment of real wastewater produced from Mobil car wash station using electrocoagulation technique. *Environ. Monit. Assess.* **2015**, *187*, 628. [CrossRef]
40. Emamjomeh, M.M.; Sivakumar, M. An empirical model for defluoridation by batch monopolar electrocoagulation/flotation (ECF) process. *J. Hazard. Mater.* **2006**, *131*, 118–125. [CrossRef]

applied sciences

Article

Low-Cost RSAC and Adsorption Characteristics in the Removal of Copper Ions from Wastewater

Yan Liu, Qin Chen and Rajendra Prasad Singh *

School of Civil Engineering, Southeast University, Nanjing 210096, China; liuyian@seu.edu.cn (Y.L.); 220191095@seu.edu.cn (Q.C.)
* Correspondence: rajupsc@seu.edu.cn

Abstract: Adsorption is a typical method for treating copper-containing wastewater. Fly ash and steel slag both have a good adsorption performance, and activated clay is added in this study, too. In this study, the performance of residue and soil adsorption composite (RSAC) particles for copper ion adsorption was discussed through the substrate ratio and the influence mechanism, to achieve the win–win effect of industrial waste reuse and copper ion wastewater treatment. The results indicated that adsorption time, dosage, initial copper ion concentration, coexisting ions, and temperature showed different effects on the adsorption, respectively. Additionally, the adsorption kinetic study showed the removal of copper ions by adsorption of RSAC particles was in accordance with quasi-primary kinetic model and quasi-secondary kinetic model. The adsorption thermodynamics study shows the adsorption process of $\Delta G^0 < 0$, $\Delta H^0 > 0$ and $\Delta S^0 > 0$, indicating that the process of copper ion adsorption by RSAC particles was spontaneous, heat-absorbing, and entropy-increasing. The research demonstrates that RSAC particles have a certain adsorption capacity for copper ion.

Keywords: water pollution; adsorption; copper ions; adsorption mechanism; adsorption kinetics; thermodynamics

Citation: Liu, Y.; Chen, Q.; Singh, R.P. Low-Cost RSAC and Adsorption Characteristics in the Removal of Copper Ions from Wastewater. *Appl. Sci.* **2022**, *12*, 5612. https://doi.org/10.3390/app12115612

Academic Editors: Amit Kumar, Santosh Subhash Palmate and Rituraj Shukla

Received: 29 April 2022
Accepted: 28 May 2022
Published: 1 June 2022

1. Introduction

Water contamination through heavy metal ions is an environmental problem of great concern [1]. Adsorption is one of the most efficient methods to remove noxious heavy metal ions, especially for wastewater with large volumes and low heavy metal ion concentrations [2]. Adsorption is spontaneous and the basic principle is that the surface energy of substances could change the concentration at the phase interface. Adsorption usually relies on some adsorbent materials with a large specific surface area and a high surface energy to remove heavy metal ions [3,4]. Adsorption has two major advantages: the reaction rate is fast, and no other reagents are needed. Therefore, adsorption is regarded as an important and promising method for addressing heavy metal ions such as copper in wastewater.

The key issue of adsorption is the adsorbent. Adsorbents with good adsorption performance have such qualities as: a fast adsorption reaction rate, stable physical and chemical properties, good solid–liquid separation, an economical cost, easy recovery and regeneration, and reusability [5]. However, industrial adsorbents could not meet all of these qualities. Therefore, the core of the adsorption method focusing on treating wastewater with heavy metal ions is to find efficient adsorbents at a low cost [6,7].

Currently, common adsorbents in the water treatment domain include activated carbon, fly ash, etc. Activated carbon has a large surface area, fast filtration rate, stable structure, large adsorption capacity, a wide range of applications, and good adsorption performance. However, activated carbon has a short service life, high sludge treatment cost, and low recycling performance [8,9]. Fly ash, a waste product from power plants that use coal as the main fuel, has a loose and porous structure. It can intercept pollutant molecules and bind pollutants to the active sites on the surface by a chemical bond, resulting in excellent adsorption [10]. Whereas, Andersson et al. [11] have also considered fly ash a low-cost material for adsorption. It has been found that steel slag has a good adsorption effect on copper, nickel and zinc ions, which mainly relies on the generation of hydroxide complexes [12]. Under optimal conditions, the adsorption efficiency of modified steel slag for uranium was 98% [13]. Activated white clay is an adsorbent made from clay minerals by inorganic acidification and other means, and dried by water rinsing. It is mainly made of bentonite clay as a raw material; its appearance is presented as a milky white powder, which is a non-toxic, tasteless, odorless, strong adsorption; and it can adsorb colored substances and organic matter [14]. Bentonite is considered to be an excellent adsorbent for Cu^{2+}, and the maximum adsorption capacity was 248.9 mg/g [15].

The residue and soil adsorption composite (RSAC) particle consists of solid waste and natural minerals. The raw materials used in the preparation of RSAC granules are fly ash, steel slag, activated white clay, bonding agent, and porogenic agents. The main raw materials are fly ash, steel slag, and activated white clay, the first two of which are industrial waste substrates [16].

2. Materials and Methods

2.1. RSAC Preparation

The fly ash used in this study was obtained from a power plant in Nanjing, China, and the steel slag was from a steel mill in Nanjing, China. Table 1 presents the main physical properties of the fly ash and the steel slag, Table 2 provides the particle size distribution of the fly ash and the steel slag, and Table 3 provides the chemical composition of the fly ash and the steel slag as measured by X-ray fluorescence analysis (XRF).

Table 1. Physical properties of waste substrates.

Waste Substrates	Density (g/cm^3)	Specific Surface Area (m^2/g)	Particle Size (μm)	Color
Fly ash	2.91	580	150 or more	Grey
Steel slag	2.10	3653	210 or more	Brown

Table 2. Particle size distribution of waste substrates.

Fly ash particle size (μm)	>3350	3350~880	880~325	325~212	<212
Proportion (%)	7.45	12.76	42.93	23.55	11.87
Steel slag particle size (μm)	>1400	1400~180	180~45	<45	-
Proportion (%)	2.49	7.68	32.40	57.36	-

Table 3. Chemical composition of waste substrates (%).

Chemical Composition	Fly Ash	Steel Slag
CaO	1.31	55.0
Fe_2O_3	4.39	21.5
Al_2O_3	45.9	1.51
SiO_2	44.4	13.4
MgO	0.261	3.65
MnO	0.026	1.75
SO_3	0.666	0.512
V_2O_5	0.038	0.417
TiO_2	1.26	0.296
Na_2O	0.094	0.077
ZnO	0.021	-
CuO	0.02	-

Additionally, the physical properties and the chemical composition of the activated white clay used in this experiment are presented in Tables 4 and 5.

Table 4. Physical properties of activated white clay.

Ingredients	Density (g/cm^3)	Particle Size (Through 75 μm Sieve)	Moisture (%)	Free Acid (as H_2SO_4)	Color
Activated white clay	0.7–1.1	≥95%	12	≤0.2%	pale

Table 5. Chemical composition of activated white clay.

	SiO_2	Al_2O_3	MgO	Fe_2O_3
Content (%)	50–70	10–16	1–6	2–4

The binder used in the experiments was 525R ordinary silicate cement, which has the characteristics of a slow thickening rate, fast setting, and high strength. It can significantly improve the early strength of the composites [17]. The porogenic agent used in the experiment was a plant-based foaming agent. It is made by a saponification reaction with rosin and sodium hydroxide as the main raw materials, and it has a light yellow-brown viscous liquid appearance. The use of a plant-based foaming agent makes the adsorbent structure lose and porous, which can enhance the adsorption performance of RSPRC particles [18]. After comprehensive consideration of the adsorption and mechanical properties, the following substrate proportioning methods were set, which are shown in Table 6.

Table 6. Ratio of base material to binder for different groups (%).

Groups	Fly Ash	Steel Slag	Activated White Clay	Bonding Agent
Group I	65	15	10	10
Group II	70	10	10	10
Group III	65	10	15	10
Group IV	60	10	20	10

After the groups of adsorbents were maintained and shaped, static adsorption tests were performed. The results of the experiments are shown in Table 7 and Figure 1. The RSAC particle morphology and the residual liquid shape were observed after the completion of static adsorption experiments.

Table 7. Copper ion removal by different groups of adsorbents.

Groups	Initial Copper Ion Concentration (mg/L)	Residual Copper Ion Concentration (mg/L)	Removal Rate (%)
Group I	100	28.92	71.08
Group II	100	31.04	69.96
Group III	100	25.54	74.46
Group IV	100	18.48	81.52

The properties of the RSAC particles in each group were observed after the adsorption of copper ions. The residuals of Groups I, II, and III were clear, the RSAC particles were not abnormal, and the hardness did not change significantly. The residuals of group IV were somewhat turbid, with slight precipitation, the surface of RSAC particles showed signs of shedding, and the hardness decreased.

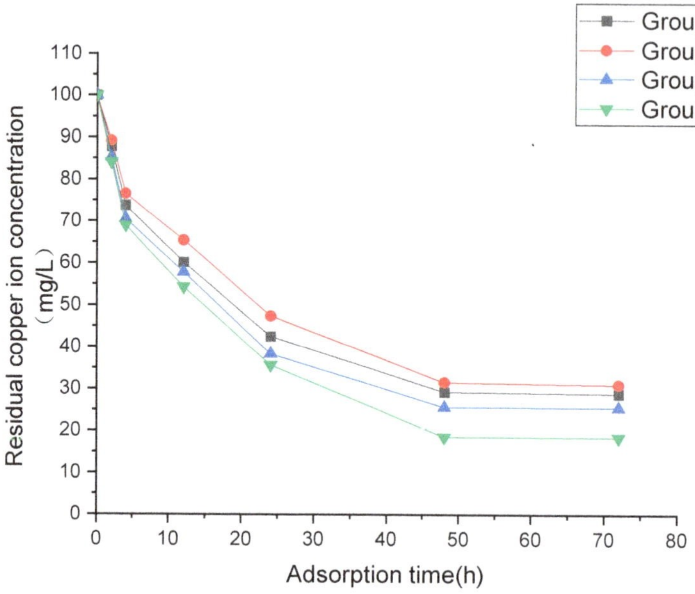

Figure 1. Comparison of adsorption effect of different groups of adsorbents.

Although Group IV had the best adsorption effect on copper ions, it was found that the residual solution had slight precipitation and the strength of the RSAC particles was reduced to break easily. On balance, Group III with the second highest removal rate was selected as the best ratio for the subsequent experiments. The process flow of the pellet is shown in Figure 2.

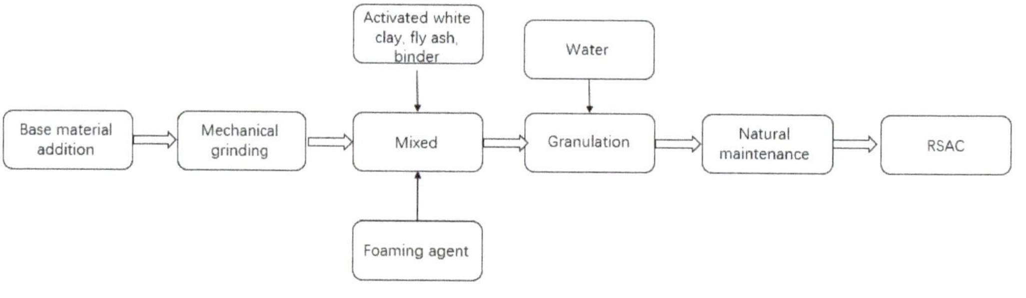

Figure 2. Experimental RSAC pellet fabrication process.

2.2. *Determination of Copper Ion Concentration*

2.2.1. Measurement Methods

The method used for the determination of copper ion concentration was the bisglyoxal oxaldihydrazone spectrophotometric method (GB/T 5750.6-2006). The measurement instrument used was a 752 UV-Vis spectrophotometer, which had a minimum detection mass concentration of 0.04 mg/L. At pH 9, copper ions (Cu^{2+}) could react with bis (cyclohexanone oxaldihydrazone) and acetaldehyde. The reaction product is a purple bis (acetaldehyde oxaldihydrazone) chelate, and the copper ion concentration is determined by the relationship between absorbance and copper ion concentration in direct proportion.

2.2.2. Reagents for Experiments

The reagents for the experiments were: copper sulfate ($CuSO_4$), acetaldehyde solution ($W(CH_3CHO) = 40\%$), ammonium chloride (NH_4Cl), ammonium hydroxide (NH_4OH), ammonium citrate ($(NH_4)_3C_6H_5O_7$), ethyl alcohol (C_2H_6O), and bis(cyclohexanone)oxaldihydrazone(BCO). All reagents were supplied by Sinopharm Chemical Reagent Co., Ltd., Nanjing, China. All solutions in these experiments were prepared with analytical grade water (R = 18 M/cm) using grade A glassware unless otherwise stated.

2.2.3. Determination Procedure

Absorb 25.0 mL water sample in 50 mL glass plug colorimetric tube; another 50 mL colorimetric tube 5, respectively, adding copper standard solution 0 mL, 0.50 mL, 1.00 mL, 1.50 mL, and 2.00 mL, diluted with deionized water to 25 mL.

Absorbing 2.0 mL ammonium citrate solution, adding each colorimetric tube, mixing evenly and adjusting pH to 9.0 with (1 + 1) ammonia. Then, 5.0 mL ammonia-ammonium chloride buffer solution was added and mixed evenly. Then, 5.0 mL BCO solution and 1.0 mL acetaldehyde were added successively. Finally, deionized water was added to the scale and mixed evenly.

Heat for 10 min in a 50 °C water bath, remove, and cool. After cooling to room temperature (standing for 20 min), under the condition of wavelength of 546 nm, the absorbance of the sample to be tested and the standard series was determined using a colorimetric dish with an optical path of 1 cm and deionized water as the reference.

The standard curve was plotted with the copper ion concentration of the standard series as the abscissa and the corresponding absorbance as the ordinate. The corresponding copper ion concentration was determined from the standard curve according to the absorbance of the water sample to be measured.

The results of the standard series measurements are presented in Table 8.

Table 8. Measurement results for the standard series.

Absorbance A	**0.019**	**0.020**	**0.374**	**0.563**	**0.751**
Copper ion concentration (mg/L)	0	2	4	6	8

The standard curve was plotted as shown in Figure 3 and the linear regression equation (Equation (1)):

$$y = 0.0914x + 0.016,\ R^2 = 0.9997 \tag{1}$$

2.3. *Copper Ion Removal Effect Examination Index*

The adsorption effect of RSAC particles on copper ions is mainly reflected in two aspects, namely the copper ion removal rate and the adsorption amount. Removal rate (η) and adsorption amount (Γ) are used in this paper to investigate the copper ion removal effect and the adsorption performance of RSAC particles, respectively.

$$\eta = \frac{C_0 - C_e}{C_0} \times 100\% \tag{2}$$

$$\Gamma = \frac{(C_0 - C_e)\,V}{m} \tag{3}$$

In Equations (2) and (3), η is the removal rate of copper ions (%), C_0 is the initial copper ion concentration of the solution (mg/L), C_e is the concentration of copper ions in solution at equilibrium (mg/L), Γ is the amount of copper adsorbed per unit mass of adsorbent (mg/g), V is the volume of the solution (L), and m is the mass of adsorbent (g).

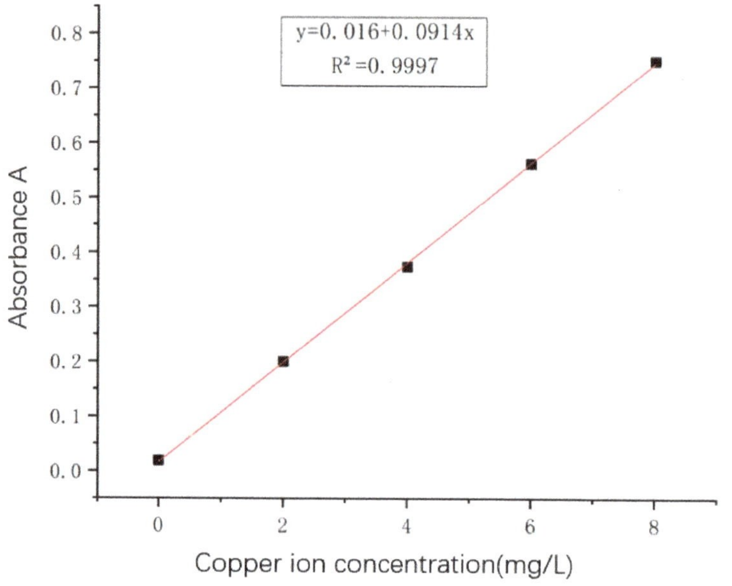

Figure 3. Copper ion standard curve.

2.4. Effect of Time on Adsorption

RSAC particles were weighted and divided into five groups: 2, 4, 6, 8, and 10 g. The groups were put into conical flasks, respectively, with 150 mL of 100 mg/L copper ion solution. Then, all the samples were mixed at room temperature (25 ± 1 °C). The concentration of residual copper ion in the supernatant of each sample was measured at a specific time and the relationship was investigated.

2.5. Effect of Dosage on Adsorption

The same samples were prepared and oscillated at room temperature (25 ± 1 °C). The concentration of residual copper ions in the supernatant of each sample was measured when the adsorption time reached 48 h.

2.6. Study of Initial Concentration on Adsorption

A series of 10, 50, 100, and 150 mg/L copper ion solutions were prepared separately. Then, a 5 g RSAC adsorbent was added into the solution series. Finally, the copper ion concentration in the supernatant of each sample was measured when the static adsorption time reached 2, 6, 12, 24, 48, and 72 h.

2.7. Study of Coexisting Metal Cations on Adsorption

A series of 150 mL of 100 mg/L copper ion solutions were prepared with the coexistence of Na^+, Mg^{2+}, Ca^{2+}, and Fe^{3+}, respectively. To each sample 5 g RSAC particles was added and then mixed at room temperature (25 ± 1 °C). The cooper ion concentration in the supernatant of each sample was measured when the adsorption time reached 48 h.

2.8. Study of Ambient Temperature on Adsorption

Three groups of 150 mL solutions with 10, 40, 80, 120, 160, and 200 mg/L cooper ion were prepared. To each sample was added 5 g RSAC particles. The three groups were mixed at 15 °C, 25 °C, and 35 °C separately. To each sample was measured a cooper ion concentration of the supernatant when reaching adsorption equilibrium.

The Freundlich and Langmuir models are often used to describe the adsorption behavior in solid–liquid systems [19]. Therefore, the adsorption isotherms of the Freundlich and the Langmuir models were plotted using nonlinear fits based on the experimental results of heavy metal ion adsorption by RSAC particles.

The expression for the Langmuir model equation is (Equation (4)):

$$q_e = \frac{q_{max} K_L C_e}{1 + K_L C_e} \tag{4}$$

In Equation (4), q_e is the adsorption equilibrium adsorption capacity (mg/g), C_e is the equilibrium concentration (mg/L), q_{max} is the maximum adsorption capacity (mg/g), and K_L is the Langmuir adsorption constant.

The expression for the Freundlich model equation is (Equation (5)):

$$q_e = K_F {C_e}^{\frac{1}{n}} \tag{5}$$

In Equation (5), q_e is the adsorption equilibrium adsorption capacity (mg/g), C_e is the equilibrium concentration (mg/L), K_F is the Freundlich adsorption constant, and n is a constant related to the adsorption capacity.

2.9. *Kinetic Study of Copper Ion Adsorption by RSAC*

Currently, quasi-primary and quasi-secondary kinetic models are often used to describe the adsorption kinetic behavior of adsorbents in solid–liquid static adsorption systems [20].

The expression for the quasi-level kinetic model equation is (Equation (6)):

$$\ln(q_e - q_t) = \ln q_e - k_1 t \tag{6}$$

In Equation (6), q_e is the amount of solute adsorbed on the adsorbent surface at adsorption equilibrium (mg/g), q_t is the amount of solute adsorbed on the adsorbent surface at the specified moment (t) during the adsorption process (mg/g), and k_1 is the adsorption rate constant (h^{-1}).

The expression for the quasi-secondary kinetic model equation is (Equation (7)):

$$\frac{t}{q_t} = \frac{1}{k_2 q_e} + \frac{t}{q_e} \tag{7}$$

In Equation (7), q_e is the amount of solute adsorbed on the adsorbent surface at adsorption equilibrium (mg/g), q_t is the amount of solute adsorbed on the adsorbent surface at the specified moment (t) during the adsorption process (mg/g), and k_2 is the adsorption rate constant (g/(mg·h)).

A 150 mL sample of a 100 mg/L copper ion was prepared and 5 g RSAC particles were added. Then, the sample was mixed at room temperature (25 ± 1 °C). The rest copper ion concentration was measured at different times to calculate the adsorbed copper ion. The experimental data of static adsorption of copper ions were fitted to the curve using quasi-primary and quasi-secondary kinetic models in turn.

2.10. *Thermodynamic Study of the Adsorption of Copper Ions by RSAC*

The thermodynamic equations are as follows (Equations (8) and (9)):

$$\Delta G^0 = \Delta H^0 - T\Delta S^0 = -RT\ln K_0 \tag{8}$$

$$\ln K_0 = \Delta S^0/R - \Delta H^0/RT \tag{9}$$

In Equations (8) and (9), T is the thermodynamic temperature (K), ΔH^0 is the enthalpy change of adsorption (kJ/mol), ΔG^0 is the free energy of adsorption (kJ/mol), ΔS^0 is the change in entropy of adsorption [J/(mol·K)], R is the molar volume constant 8.314 J/(mol·K), and K_0 is the adsorption partition coefficient, usually taken as the Langmuir constant K_L. Using the isothermal adsorption experimental data of RSAC particles at different temperatures, the adsorption equilibrium coefficient K_0 was calculated [21]. A straight line could be fitted by using the inverse of the temperature 1/T as the horizontal coordinate and $\ln K_0$ as the vertical coordinate for the graph.

3. Results and Discussion

3.1. Adsorption Experiments

3.1.1. Effect of Adsorption Time

The concentration of the residual copper ion in the supernatant of each sample as a function of adsorption time is shown in Figure 4. The residual copper ion concentration showed a similar trend for different dosage amounts. At the beginning of the adsorption reaction (0–12 h), the residual copper ion concentration decreased significantly with the increasing time; at the middle of the adsorption reaction (12–48 h), the residual copper ion concentration decreased slowly with time; at the end of the adsorption reaction (48–72 h), the adsorption equilibrium state was reached. The reason for this trend could be that the initial adsorption occurs mainly on the surface and in the pores of RSAC particles. In the initial adsorption stage, there are many active sites on the surface and in the pores of RSAC particles so copper ions could occupy the active sites rapidly and show the characteristics of a fast adsorption rate [22]. With the extension of time and the increase of adsorption capacity, the active sites become fewer [23]. It was also found that there were large functional groups on the surface of FA and MFA, such as O-H, C=C, and Si-O-Si, which played a crucial role in the process of adsorption of heavy metal ions [24]. It could be inferred that the adsorption equilibrium time of copper ions on RSAC particles is 48 h. This is consistent with the experimental results to explore the optimal ratio of RSAC particles. Therefore, the adsorption time could be set as 48 h in the subsequent static adsorption experiments.

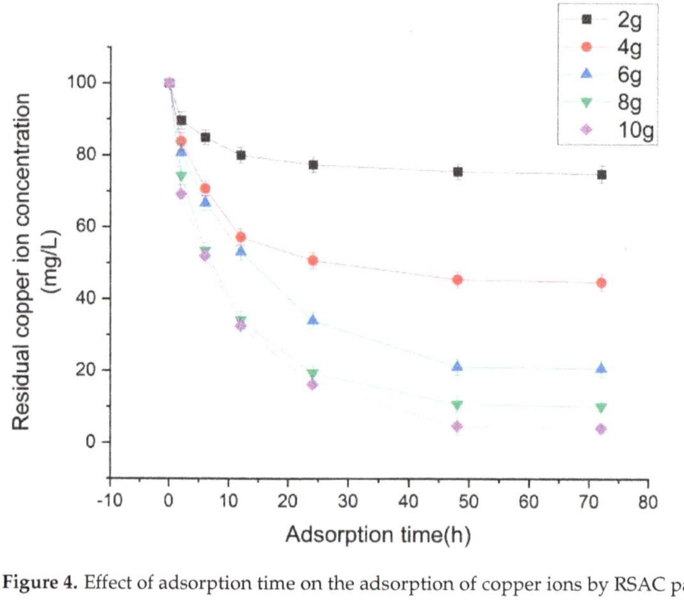

Figure 4. Effect of adsorption time on the adsorption of copper ions by RSAC particles.

3.1.2. Effect of Adsorbent Amount

Figure 5 shows the correlation between adsorbent amount and copper ion adsorption. Firstly, the copper ion removal rate increased continuously with the increase of the RSAC particle amount, but the slope decreased at the same time. Meanwhile, the adsorption amount per unit mass of RSAC particles showed a different trend: the adsorption amount per unit mass of RSAC particles increased at the beginning stage and decreased after reaching a specific point with the increase of the adsorbent amount. The two trends indicated that there was a balance point between efficiency and performance. The amount aimed best removal rate may lead to inefficient usage of RSAC particles: the adsorption capacity per mass of RSAC particles was only 1.43 mg/g at 10 g, which indicated that the adsorption performance of RSAC particles was not fully utilized. The reason may be that the copper ion removal rate increased because of the increase of RSPRC particle amount, the increase in contact area, and the increase in the number of adsorption sites [25].

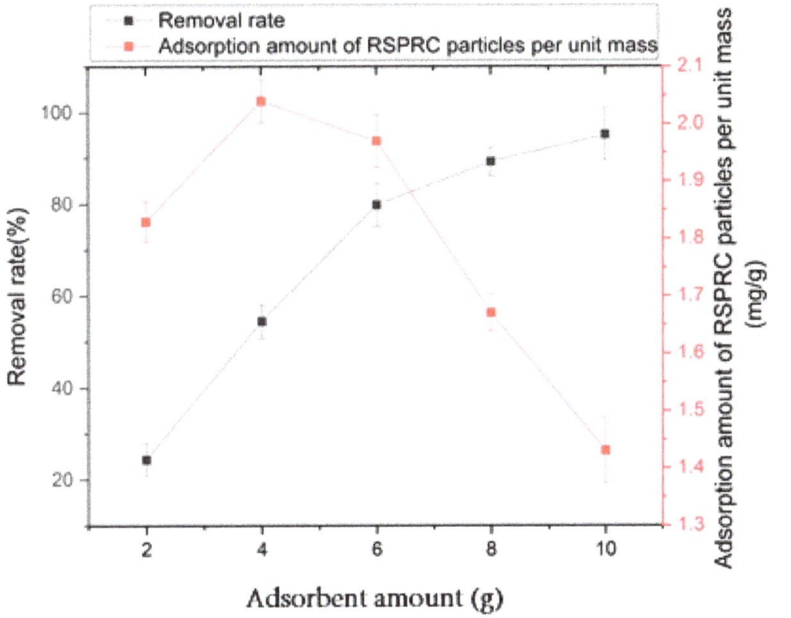

Figure 5. Effect of adsorbent amount on the adsorption of copper ions by RSAC particles.

The adsorption amount of copper ions per unit mass of adsorbent showed a different trend. This may be due to the fact that when the concentration of copper ions in the solution is constant and the adsorbent dosage is low, the adsorption sites are not significant in driving the diffusion and adsorption reaction caused by the atmosphere of copper ion concentration. With the increase in the dosage, the total adsorption sites provided to copper ions in the solid-liquid system increased and the adsorption amount per unit mass of RSPRC particles also increased [26]. Considering economic factors, the dosage of 5 g/150 mL was chosen as a balance point and used in the subsequent experiments.

3.1.3. Effect of Initial Concentration

The relationship between the initial concentration and the residual concentration of copper ions is shown in Figure 6. The removal rate of copper ions and the adsorption amount per unit mass of RSAC particles in each sample after adsorption for 72 h is presented in Table 9. As shown in Figure 6, a high initial copper ion concentration led to a corresponding steep adsorption curve and a fast adsorption rate compared with a low initial concentration sample in the pre-sorption stage. This could be attributed to high initial copper ion concentration providing a sufficient driving force for mass transport, which could make ions occupy the active site on adsorbents rapidly, facilitating the adsorption of copper ions by RSAC particles [27,28].

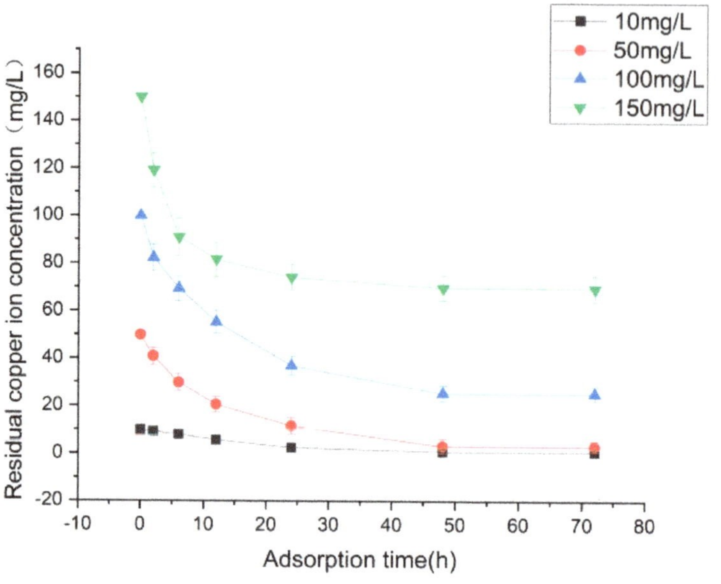

Figure 6. Effect of initial concentration on the adsorption of copper ions by RSAC particles.

Table 9. Copper ion removal for different initial concentrations.

Initial Concentration (mg/L)	Removal Rate (%)	Adsorption of RSAC Particles per Unit Mass (mg/g)
10	91.90	0.28
50	93.58	1.40
100	74.61	2.24
150	53.50	2.41

3.1.4. Effect of Coexisting Metal Cations

Figure 7 shows the relationship between the metal cation concentration and the effect on the removal of copper ions. The results revealed that the adsorption removal rate of copper ions by RSAC particles fluctuates in a small range with the increase of Na^+ concentration, which indicated that Na^+ has a weak competitive behavior against RSAC particles. The adsorption removal rate of copper ions by RSAC particles decreases with the increase of Mg^{2+} and Ca^{2+} concentration, which indicated that Mg^{2+} and Ca^{2+} may have some effect on the removal rate. The adsorption removal rate decreased significantly with the increase of Fe^{3+} concentration, from 74.65% to 62.47%, which indicated that Fe^{3+} had a significant inhibitory effect on the adsorption of copper ions.

The experimental results may be interpreted as that metal cations can replace the original cations in the RSAC particles by ion exchange into the adsorbent surface and pore channels, affecting the adsorption of copper ions on the RSAC particles by changing the adsorbent environment [29]. A higher charge number of the metal cation may result in a stronger ability to replace the original cation [30]. Ion exchange and surface adsorption may be involved in the adsorption process of copper [31].

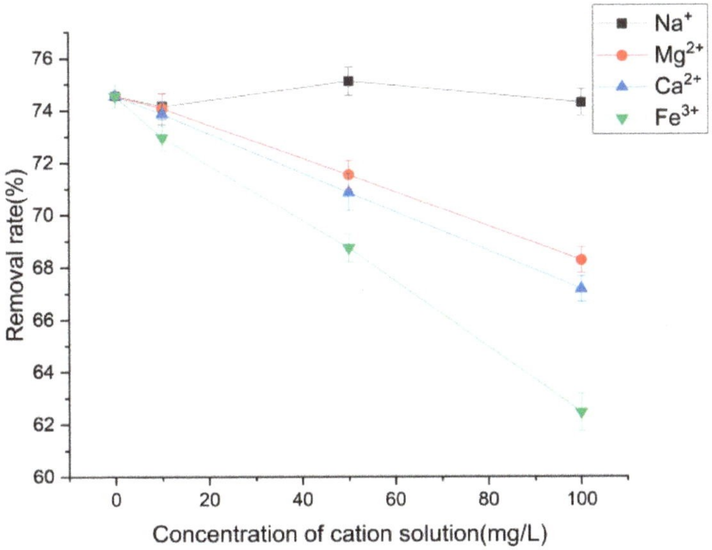

Figure 7. Effect of metal cations on the adsorption of copper ions by RSAC particles.

3.1.5. Effect of Ambient Temperature

The fitted curves at 25 °C are shown in Figure 8. The fitting parameters can be obtained from the adsorption isotherm. It can be observed from Table 10 that the fit coefficients R^2 of both Langmuir and Freundlich models are greater than 0.95 at different temperatures, which indicates that both models could well express the isothermal characteristics of the adsorption of copper ions by RSAC particles. This also implies that the adsorption isotherm characteristics of the adsorbent for copper ions could fit two or more adsorption isotherm models under certain conditions [32]. Based on the Langmuir model, the K_L and the q_{max} which increase as temperature increases indicates that the intermolecular binding and the adsorption capacity may increase as temperature increases. Based on Freunlich model, the low 1/n value indicates that the adsorption process could undertake easily. Meanwhile, the K_F which represents the absorbability increases as temperature increases [33]. The two models show that the adsorption of copper ions by RSAC particles is a heat-absorbing process. This could be explained from different aspects. Firstly, the cooper ion needs the energy to approach the RSAC particles and overcome the resistance from the liquid film of the particles to reach the internal active sites. Secondly, the physical adsorption may release heat since the intermolecular force (Van der Waals force) between adsorbates and adsorbents contributed to the main effect during adsorption, which made the molecular kinetic energy decrease by releasing thermal energy [34]. Another study also found that the adsorption process of fly ash involves physical adsorption [35]. In conclusion, chemisorption is an endothermic reaction while physical adsorption is an exothermic reaction, and the adsorbed thermal is more than the released thermal, causing the increase in temperature to promote the adsorption reaction, which is consistent with previous research conclusions [36,37].

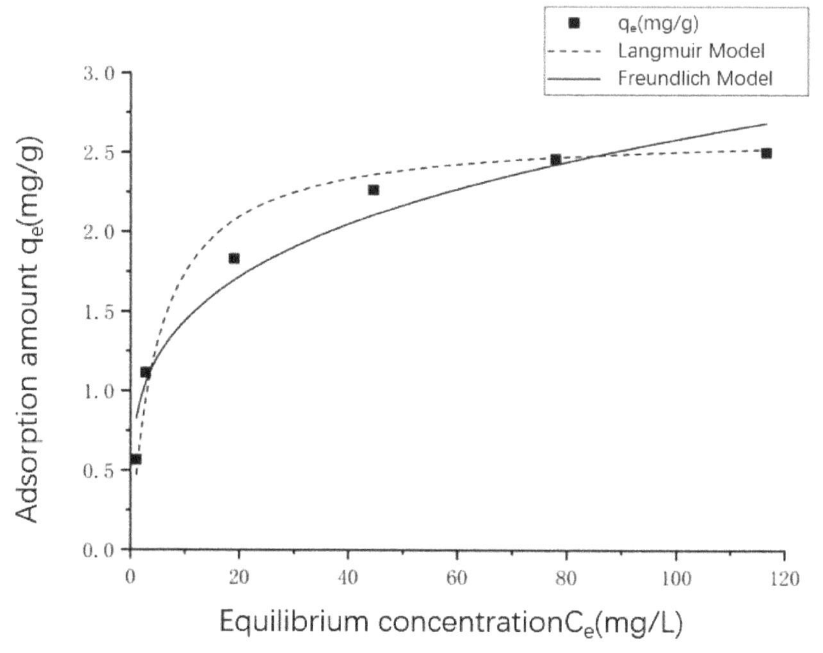

Figure 8. Adsorption isotherm of copper ions (25 °C).

Table 10. Parameters for the adsorption isotherm model fit.

Temp. (°C)	Langmuir Model			Freundlich Model		
	K_L	q_{max}	$R_L{}^2$	K_F	1/n	$R_F{}^2$
15	0.138	2.438	0.974	0.737	0.248	0.960
25	0.196	2.633	0.967	0.805	0.253	0.952
35	0.254	2.862	0.954	0.874	0.262	0.956

3.2. Kinetic Study of Copper Ion Adsorption by RSAC

The fitted curves are shown in Figure 9 and the fitted parameters of the two kinetic models are shown in Table 11. As shown in Figure 9, the adsorption amount of copper ions by RSAC particles increases rapidly with the increase of adsorption time in the early stage of the adsorption reaction. The increase of adsorption amount decreases gradually to almost 0 in the middle and the late stage of the adsorption reaction, which means the adsorption reaches the equilibrium state. In the preliminary stage of adsorption reaction, the adsorbent mainly adsorbs copper ions at the solid–liquid interface [38]. After the preliminary stage, copper ions diffuse from the adsorbent surface to the internal micropores and lattice, reach and are fixed by the internal surface active-sites, thus the adsorption rate decreases slowly [39]. Another study found that the adsorption of fly ash involves both boundary-layer diffusion and intraparticle diffusion [11]. Kai-sung Wang et al. also found fast surface adsorption was followed by a slow intra-particle diffusion adsorption of fly ash [40]. The adsorption process of copper ions by RSAC particles can reach the equilibrium state at 48 h.

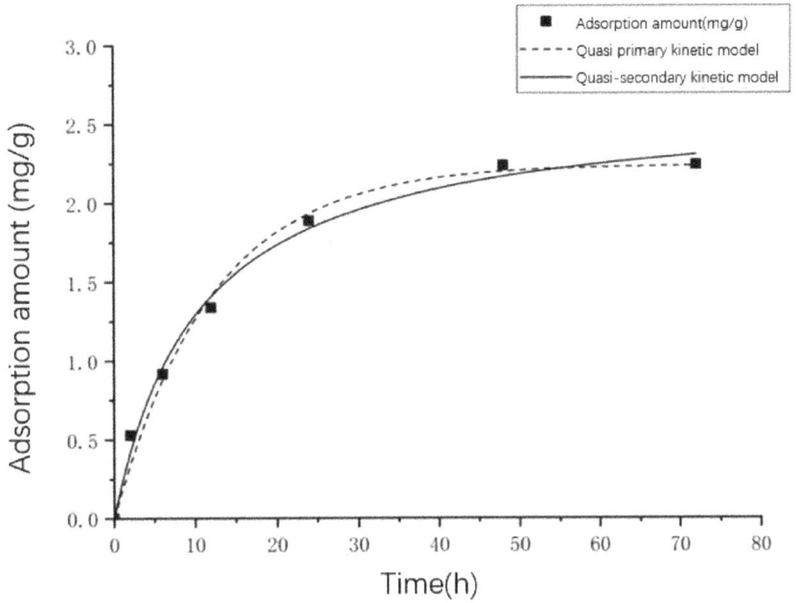

Figure 9. Adsorption kinetic curve of copper ions (25 °C).

Table 11. Sorption kinetic model fitting parameters.

Initial Concentration (mg/L)	Quasi-One Dynamical Model			Quasi-Secondary Dynamical Model (QSDM)		
	q_e	k_1	$R_1{}^2$	q_e	k_2	$R_2{}^2$
100	2.241	0.0713	0.990	2.640	0.0955	0.993

The R^2 of both models is greater than 0.95 and the difference between them is not significant (Table 11), which indicates that both kinetic models could describe the adsorption process of copper ions on RSAC particles well. It could be further inferred that the adsorption process of copper ions on RSAC particles is a mixed control: both surface diffusion and internal fine pore diffusion are important.

3.3. Thermodynamic Study of the Adsorption of Copper Ions by RSAC

As shown in Figure 10, the intercept and the slope of the straight line were calculated by ΔS^0 and ΔH^0; then, we proceeded to calculate the different temperatures of ΔG^0, and Table 12 presents the various thermodynamic parameters obtained. The adsorption free energy ΔG^0 of copper ions adsorbed by RSAC particles at different temperatures is negative, and the absolute value of ΔG^0 increases gradually with the increase of temperature (Table 12). This indicates that the adsorption of copper ions in solution by RSAC particles is a spontaneous reaction and the spontaneity increases with the increase of temperature [41]. The enthalpy change ΔH^0 during the adsorption of copper ions by RSAC particles is positive, which indicates that the adsorption process is a heat absorption reaction and therefore the increase in temperature contributes to the adsorption [42]. As presented in Table 5, the maximum adsorption capacity of copper ions increases with the increasing temperature, which also confirms that the adsorption of copper ions by RSAC particles is a heat absorption reaction.

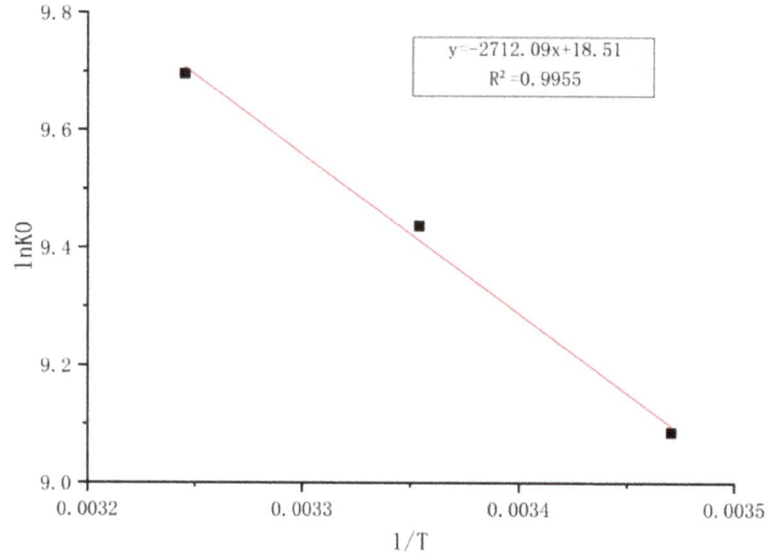

Figure 10. Relationship between $\ln K_0 - 1/T$ for the process of copper ion adsorption by RSAC particles.

Table 12. Thermodynamic parameters of copper ion adsorption on RSAC particles.

Temp. (K)	q_{max} (mmol/g)	$K_L \times 10^{-3}$ (L/mol)	ΔG^0 (kJ/mol)	ΔH^0 (kJ/mol)	ΔS^0 (J/mol-K)	R^2
288.15	0.0381	8.832	−21.79			
298.15	0.0411	12.544	−23.33	22.55	153.89	0.9955
308.15	0.0447	16.256	−24.87			

4. Conclusions

This study aimed to prepare and to apply RSAC to remove copper ions from wastewater and to discuss the influence mechanism and the microstructure for adsorption by RSAC particles. One of the more significant findings to emerge from this study is that the adsorption of copper ions in a solution by RSAC particles is a spontaneous, heat absorption reaction. The mechanism of copper ion removal by RSAC particles includes an ion exchange reaction and chemical precipitation in addition to physical adsorption. The second major finding is that metal cations can replace the original cations in the RSAC particle ion exchange, change the environment of the adsorbent, and affect the adsorption of copper ions on RSAC particles. This study has also indicated that both Langmuir and Freundlich models can well describe the isothermal characteristics of the adsorption of copper ions by RSAC particles. Both the quasi-primary kinetic model and the quasi-secondary kinetic model can describe the adsorption process of copper ions on RSAC particles well. It can be further inferred that the adsorption process of copper ions on RSAC particles is a mixed control, and both surface diffusion and internal fine pore diffusion are important.

Author Contributions: Conceptualization, Y.L.; methodology, Y.L.; software, Y.L. and Q.C.; validation, Y.L. and Q.C.; formal analysis, Q.C.; investigation, Q.C.; resources, Q.C.; data curation, Q.C.; writing—original draft preparation, Q.C. and R.P.S.; writing—review and editing, R.P.S. and Q.C.; visualization, Y.L. and Q.C.; supervision, Y.L. and R.P.S.; project administration, Y.L. All authors have read and agreed to the published version of the manuscript.

Funding: This research received no external funding.

Institutional Review Board Statement: Not applicable.

Informed Consent Statement: Not applicable.

Data Availability Statement: Not applicable.

Acknowledgments: The authors would like to show their gratitude to all those who helped during the experimental period and with the writing of this manuscript. Q.C. would like to gratefully acknowledge the help of Rajendra Prasad Singh, School of Civil Engineering, Southeast University, and appreciate his guidance, patience, encouragement, and professional instructions.

Conflicts of Interest: The authors declare no conflict of interest.

References

1. Maryam, H.; Mohammad, H. Application of three-dimensional porous aerogels as adsorbent for removal of heavy metal ions from water/wastewater: A review study. *Adv. Colloid Interface Sci.* **2020**, *284*, 102247.
2. Burakov, A.E.; Galunin, E.V.; Burakova, I.V.; Kucherova, E.A.; Agarwal, S.; Tkachev, A.G.; Gupta, V.K. Adsorption of heavy metals on conventional and nanostructured materials for wastewater treatment purposes: A review. *Ecotoxicol. Environ. Saf.* **2018**, *148*, 702–712. [CrossRef] [PubMed]
3. Chai, W.S.; Cheun, J.Y.; Kumar, P.S.; Mubashir, M.; Zaheed, M.; Banat, F.; Ho, S.; Pau, L.S. A review on conventional and novel materials towards heavy metal adsorption in wastewater treatment application. *J. Clean. Prod.* **2021**, *296*, 126589. [CrossRef]
4. Kumar, A.; Jigyasu, D.K.; Subrahmanyam, G.; Mondal, R.; Shabnam, A.A.; Cabral-Pinto, M.M.; Malyan, S.K.; Chaturvedi, A.K.; Gupta, D.K.; Fagodiya, R.K.; et al. Nickel in terrestrial biota: Comprehensive review on contamination, toxisity, tolerance and its remediation approaches. *Chemosphere* **2021**, *275*, 129996. [CrossRef]
5. Rezaei Kahkha, M.R.; Kaykhaii, M.; Rezaei Kahkha, B.; Khosravi, H.; Tohidlou, E. Simultaneous removal of heavy metals from wastewater using modified sodium montmorillonite nanoclay. *Anal. Sci.* **2020**, *36*, 1039–1043. [CrossRef]
6. Tamjidi, S.; Ameri, A. A review of the application of sea material shells as low cost and effective bio-adsorbent for removal of heavy metals from wastewater. *Environ. Sci. Pollut. Res.* **2020**, *27*, 31105–31119. [CrossRef]
7. Kumar, A.; Gupta, H. Activated carbon from sawdust for naphthalene removal from contaminated water. *Environ. Technol. Innov.* **2021**, *20*, 101080. [CrossRef]
8. Aniagor, C.O.; Elshkankery, M.; Fletcher, A.J.; Morsy, O.M.; Abdel-Halim, E.S.; Hashem, A. Equilibrium and kinetic modelling of aqueous cadmium ion and activated carbon adsorption system. *Water Conserv. Sci. Eng.* **2021**, *6*, 3. [CrossRef]
9. Singh, S.; Kumar, A.; Gupta, H. Activated banana peel carbon: A potential adsorbent for Rhodamine B decontamination from aqueous system. *Appl. Water Sci.* **2020**, *10*, 185. [CrossRef]
10. Gadore, V.; Ahmaruzzaman, M. Tailored fly ash materials: A recent progress of their properties and applications for remediation of organic and inorganic contaminants from water. *J. Water Process Eng.* **2021**, *41*, 101910. [CrossRef]
11. Ansdersson, K.I.; Eriksson, M.; Norgren, M.; Kerstin, I. Andersson. Removal of lignin from wastewater generated by mechanical pulping using activated charcoal and fly ash: Adsorption kinetics. *Ind. Eng. Chem. Res.* **2011**, *50*, 7722–7732. [CrossRef]
12. Dimitrova, S.V. Metal sorption on blast-furnace slag. *Water Res.* **1996**, *30*, 228–232. [CrossRef]
13. Chen, Q.; Wang, H.Q.; Hu, E.M.; Lei, Z.W.; Liu, W.J.; Jiang, X.Y.; Hou, W.; Wang, Q.L. Efficient adsorption of uranium (VI) from aqueous solution by a novel modified steel slag adsorbent. *J. Radioanal. Nucl. Chem.* **2020**, *323*, 73–81. [CrossRef]
14. Meng, B.; Guo, Q.; Men, X.; Ren, S.; Jin, W.; Shen, B. Modified bentonite by polyhedral oligomeric silsesquioxane and quaternary ammonium salt and adsorption characteristics for dye. *J. Saudi Chem. Soc.* **2020**, *24*, 334–344. [CrossRef]
15. Chang, Y.S.; Au, P.I.; Mubarak, N.M.; Khalid, M.; Jagadish, P.; Walvekar, R.; Abdullah, E.C. Adsorption of Cu(II) and Ni(II) ions from wastewater onto bentonite and bentonite/GO composite. *Environ. Sci. Pollut. Res.* **2020**, *27*, 33270–33296. [CrossRef]
16. Pranoto; Martini, T.; Sunardi, B.M.W.; Anandita, P. Optimization of the use of fly ash and natural zeolite activated as composite for an adsorbent of lead heavy metal (Pb). *IOP Conf. Ser. Mater. Sci. Eng.* **2019**, *578*, 012022. [CrossRef]
17. Samarakoon, M.H.; Ranjith, P.G.; Rathnaweera, T.D.; Perera, M.S.A. Recent advances in alkaline cement binders: A review. *J. Clean. Prod.* **2019**, *227*, 70–87. [CrossRef]
18. Zhu, W.; Song, H.; Jia, C.; Yao, S. Fabrication and adsorption property of novel adsorbent for potential application to wastewater with Fe (III) ions. *J. Cent. South Univ.* **2014**, *21*, 2832–2836. [CrossRef]
19. Olusegun, S.J.; de Sousa Lima, L.F.; Mohallem, N.D.S. Enhancement of adsorption capacity of clay through spray drying and surface modification process for wastewater treatment. *Chem. Eng. J.* **2018**, *334*, 1719–1728. [CrossRef]
20. Wang, J.; Guo, X. Adsorption kinetic models: Physical meanings, applications, and solving methods. *J. Hazard. Mater.* **2020**, *390*, 122156. [CrossRef]
21. Zhao, G.; Li, J.; Wang, X. Kinetic and thermodynamic study of 1-naphthol adsorption from aqueous solution to sulfonated graphene nanosheets. *Chem. Eng. J.* **2011**, *173*, 185–190. [CrossRef]
22. Pizarro, J.; Castillo, X.; Jara, S.; Ortiz, C.; Navarro, P.; Cid, H.; Rioseco, H.; Barros, D.; Belzile, N. Adsorption of Cu^{2+} on coal fly ash modified with functionalized mesoporous silica. *Fuel* **2015**, *156*, 96–102. [CrossRef]

23. Appiah-Hagan, E.; Chen, Y.; Yu, X.; Arteca, G.A.; Pizarro, J.; Mercier, L.; Wei, O.; Belzile, N. Simple and energy-saving modifications of coal fly ash to remove simultaneously six toxic metal cations from mine effluents. *J. Environ. Chem. Eng.* **2018**, *6*, 5498–5509. [CrossRef]
24. Jiang, X.; Fan, W.; Li, C.; Wang, Y.; Bai, J.; Yang, H.; Liu, X. Removal of Cr (VI) from wastewater by a two-step method of oxalic acid reduction-modified fly ash adsorption. *RSC Adv.* **2019**, *9*, 33949–33956. [CrossRef] [PubMed]
25. Li, W.; Zhao, F.; Ma, H.; Li, J.; Han, Q.; Chen, Y. Research on the adsorption treatment of the malachite green simulated wastewater on fallen acer monoes leaves. *IOP Conf. Ser. Earth Environ. Sci.* **2018**, *186*, 12010. [CrossRef]
26. Taylor, M.L.; Morris, G.E.; Self, P.G.; Smart St., R.C. Kinetics of Adsorption of High Molecular Weight Anionic Polyacrylamide onto Kaolinite: The Flocculation Process. *J. Colloid Interface Sci.* **2002**, *250*, 28–36. [CrossRef]
27. Jia, P.; Tan, H.; Liu, K.; Gao, W. Adsorption behavior of methylene blue by bone char. *Int. J. Mod. Phys. B* **2017**, *31*, 1744099. [CrossRef]
28. Li, B.; Ma, C. Study on the mechanism of SO_2 Removal by activated carbon. *Energy Procedia* **2018**, *153*, 471–487. [CrossRef]
29. Efome, J.E.; Rana, D.; Matsuura, T.; Lan, C.Q. Effects of operating parameters and coexisting ions on the efficiency of heavy metal ions removal by nano-fibrous metal-organic framework membrane filtration process. *Sci. Total Environ.* **2019**, *674*, 355–362. [CrossRef]
30. Dong, L.; Liang, J.; Li, Y.; Hunang, S.; Wei, Y.; Bai, X.; Jin, Z.; Zhang, M.; Qu, J. Effect of coexisting ions on Cr (VI) adsorption onto surfactant modified *Auricularia auricula* spent substrate in aqueous solution. *Ecotoxicol. Environ. Saf.* **2018**, *166*, 390–400. [CrossRef]
31. Zheng, W.; Li, X.M.; Yang, Q.; Zeng, G.M.; Shen, X.X.; Zhang, Y.; Liu, J.J. Adsorption of Cd(II) and Cu(II) from aqueous solution by carbonate hydroxylapatite derived from eggshell waste. *J. Hazard. Mater.* **2007**, *147*, 534–539. [CrossRef] [PubMed]
32. Harja, M.; Buema, G.; Lupu, N.; Chiriac, H.; Herea, D.; Ciobanu, G. Fly Ash Coated with Magnetic Materials: Improved Adsorbent for Cu (II) Removal from Wastewater. *Materials* **2020**, *14*, 63. [CrossRef] [PubMed]
33. Zia, Q.; Tabassum, M.; Lu, Z.; Khawar, M.T.; Song, J.; Gong, H.; Meng, J.; Li, Z.; Li, J. Porous poly(L–lactic acid)/chitosan nanofibres for copper ion adsorption. *Carbohydr. Polym.* **2020**, *227*, 115343. [CrossRef] [PubMed]
34. Ahmadifar, Z.; Koohi, A.D. Characterization, preparation, and uses of nanomagnetic Fe_3O_4 impregnated onto fish scale as more efficient adsorbent for Cu^{2+} ion adsorption. *Environ. Sci. Pollut. Res.* **2018**, *25*, 19687–19700. [CrossRef] [PubMed]
35. Wang, N.; Zhao, Q.; Xu, H.; Niu, W.; Ma, L.; Lan, D.; Hao, L. Adsorptive treatment of coking wastewater using raw coal fly ash: Adsorption kinetic, thermodynamics and regeneration by Fenton process. *Chemosphere* **2018**, *210*, 624–632. [CrossRef]
36. Li, Y.; Yue, Q.Y.; Gao, B.Y. Adsorption kinetics and desorption of Cu (II) and Zn (II) from aqueous solution onto humic acid. *J. Hazard. Mater.* **2010**, *178*, 455–461. [CrossRef]
37. Hsu, T.C. Experimental assessment of adsorption of Cu^{2+} and Ni^{2+} from aqueous solution by oyster shell powder. *J. Hazard. Mater.* **2009**, *171*, 995–1000. [CrossRef]
38. Jung, K.W.; Lee, S.Y.; Choi, J.; Lee, Y.J. A facile one-pot hydrothermal synthesis of hydroxyapatite/biochar nanocomposites: Adsorption behavior and mechanisms for the removal of copper (II) from aqueous media. *Chem. Eng. J.* **2019**, *369*, 529–541. [CrossRef]
39. Huang, Q.; Liu, M.; Zhao, J.; Chen, J.; Zeng, G.; Huang, H.; Tian, J.; Wen, Y.; Zhang, X.; Wei, Y. Facile preparation of polyethylenimine-tannins coated SiO_2 hybrid materials for Cu^{2+} removal. *Appl. Surf. Sci.* **2018**, *427*, 535–543. [CrossRef]
40. Wang, K.; Wei, M.; Peng, T.; Li, H.; Chao, S.; Hsu, T.; Lee, H.; Chang, S. Treatment and toxicity evaluation of methylene blue using electrochemical oxidation, fly ash adsorption and combined electrochemical oxidation-fly ash adsorption. *J. Environ. Manag.* **2010**, *91*, 1778–1784. [CrossRef]
41. Ben-Ali, S.; Jaouali, I.; Souissi-Najar, S.; Ouederni, A. Characterization and adsorption capacity of raw pomegranate peel biosorbent for copper removal. *J. Clean. Prod.* **2017**, *142*, 3809–3821. [CrossRef]
42. Guiza, S. Biosorption of heavy metal from aqueous solution using cellulosic waste orange peel. *Ecol. Eng.* **2017**, *99*, 134–140. [CrossRef]

applied sciences

MDPI

Article

Removal of Copper, Nickel, and Zinc Ions from an Aqueous Solution through Electrochemical and Nanofiltration Membrane Processes

Jagdeesh Kumar [1,*], Himanshu Joshi [1] and Sandeep K. Malyan [2,*]

[1] Department of Hydrology, Indian Institute of Technology Roorkee, Roorkee 247667, India; himanshu.joshi@hy.iitr.ac.in

[2] Research Management and Outreach Division, National Institute of Hydrology, Jalvigyan Bhawan, Roorkee 247667, India

* Correspondence: jkumar@hy.iitr.ac.in (J.K.); sandeepkmalyan@gmail.com (S.K.M.)

Abstract: Heavy metal contamination in water is a major health concern, directly related to rapid growth in industrialization, urbanization, and modernization in agriculture. Keeping this in view, the present study has attempted to develop models for the process optimization of nanofiltration (NF) membrane and electrocoagulation (EC) processes for the removal of copper, nickel, and zinc from an aqueous solution, employing the response surface methodology (RSM). The variable factors were feed concentration, temperature, pH, and pressure for the NF membrane process; and time, solution pH, feed concentration, and current for the EC process, respectively. The central composite design (CCD), the most commonly used fractional factorial design, was employed to plan the experiments. RSM models were statistically analyzed using analysis of variance (ANOVA). For the NF membrane, the rejection of Zn, Ni, and Cu was observed as 98.64%, 90.54%, and 99.79% respectively; while the removal of these through the EC process was observed as 99.81%, 99.99%, and 99.98%, respectively. The above findings and a comparison with the conventional precipitation and adsorption processes apparently indicate an advantage in employing the NF and EC processes. Further, between the two, the EC process emerged as more efficient than the NF process for the removal of the studied metals.

Keywords: nanofiltration; electrocoagulation; nickel; zinc; copper; heavy metals; water pollution

Citation: Kumar, J.; Joshi, H.; Malyan, S.K. Removal of Copper, Nickel, and Zinc Ions from an Aqueous Solution through Electrochemical and Nanofiltration Membrane Processes. *Appl. Sci.* **2022**, *12*, 280. https://doi.org/10.3390/app12010280

Academic Editor: Bart Van der Bruggen

Received: 15 October 2021
Accepted: 22 December 2021
Published: 28 December 2021

1. Introduction

Heavy metals are inorganic elements naturally found throughout the earth's crust [1]. Their concentration above permissible limits is considered pollution. "Heavy metals" refers to a group of elements with a density greater than 4 g cm^{-3}, including metals and metalloids [2]. Industrial discharges, agricultural runoff, storm water, mining activity, and direct inclusion of sewage/wastewater contribute to the heavy metal pollution load in fresh water, leading to various health and environmental problems. Among the commonly reported heavy metals, copper (Cu) is used widely in electroplating, batteries, pesticides, galvanized pipes, and alloys [3–7]. Regular consumption of copper-contaminated drinking water may cause stomach upsets, abdominal cramp and diarrhea. Nickel (Ni) is another metal found widely in water and wastewater. The electroplating industry, rechargeable batteries, and galvanized pipes are its main sources. High levels of nickel contamination cause serious lung and kidney problems as well as skin dermatitis and pulmonary fibrosis. In drinking water, the maximum allowable limit for nickel is 0.1 ppm [8]. Zinc (Zn) is used in many types of industry, such as metal production, galvanization, food preservation, agri-food and biological engineering, pharmaceuticals, electronics, mining and metallurgy, with major contributions coming from electroplating and mining effluents [9,10]. Zinc is not considered highly toxic but its presence in drinking water if exceeding 15 mg/L is reported to cause nausea, vomiting and diarrhea [11]. These heavy metals are ingested directly either by drinking contaminated water or indirectly through the food chain, and subsequently affect human health [12–14]. Drinking of contaminated water has been reported to lead to around 70–80% of the total diseases in developing countries [15,16], where the impact of increased pollution is particularly problematic because the population at large does not have sufficient resources to effectively treat the contaminated water or access to safe drinking water systems at their homes. According to a WHO (2017) estimate, around 844 million people do not have access to a basic drinking water source and 230 million people spend more than 30 min/day in collecting water from an improved water source, which may include piped water, boreholes, protected wells and springs, rainwater and packaged/delivered water [17]. According to the United Nations, an estimated 80% of all industrial and municipal wastewater in the developing world is released into freshwater bodies without any prior treatment [18]. Heavy metal removal can be achieved through different physical, chemical and biological methods such as fungal remediation [19], microbial remediation [12,20], phytoremediation [21,22], adsorption [23,24], flotation, coagulation–flocculation [25], chemical precipitation or ion exchange [26]; selection between these may be based on the nature and quantum of the pollution load and merits/demerits of decontamination processes along with other factors. It is noted that removal of heavy metals from water/wastewater is still an evolving research area, and there is wide scope for case-specific evaluation, optimization and integration of new and/or available technologies. In this regard, it has been noted that removal of heavy metals from aqueous solutions, especially metal-laden water or wastewater displaying high and heterogeneous concentrations is one of the major challenges. For this, nanofiltration (NF) and electrocoagulation (EC) processes have been reportedly more reliable than bioremediation in terms of the shorter time taken in providing near complete removal, ease of setup, and predictability. The primary emphasis of the present study is to explore the efficiency of removal of Cu, Ni, and Zn by NF and EC processes from their synthetic aqueous solutions in low to high concentrations. The selection of these metals for study is based on the findings of a comprehensive literature review, indicating that these comprise the major constituents in electroplating effluents or the recipient waters of these effluents.

2. Materials and Methods

2.1. Chemicals, Membranes and Electrodes

The experiments were conducted for the technical evaluation of the NF and EC processes employing a range of concentrations of metals in aqueous solution based on the available secondary data on electroplating effluent quality in the literature and also in the study area [27,28]. All chemicals used in this research were of analytical grade, and synthetic composite metal solutions were prepared by dissolving the appropriate mass of each metal in high purity Milli-Q water (18.2 MΩ cm). Copper (III) sulphate pentahydrate $CuSO_4{\cdot}5H_2O$, nickel (II) sulphate hexahydrate $NiSO_4{\cdot}6H_2O$, zinc sulphate $ZnSO_4{\cdot}7H_2O$, sodium hydroxide (NaOH), nitric acid (69–72%), sodium chloride (NaCl) and calcium carbonate extra pure ($CaCO_3$) were all obtained from Merck Specialties Private Ltd. Quicklime (CaO), nanofiltration membrane (Permionics, Flat Sheet Membrane HFN-300 AR) and stainless steel (SS-304) electrodes were procured from the market as commercially available. The stated membrane was specifically employed so as to effectively work in both acid and basic medium. Stainless steel electrodes were used, as they are less susceptible to corrosion, and have reportedly shown a better performance in earlier studies. The grade of steel used is reported to not contain Zn, and have very low amount of carbon.

2.2. Experimental Setup and Procedure

2.2.1. Experimental Setup for Nanofiltration

The NF unit was a cross-flow lab-scale system (Nilshan Nishotech Pvt. Ltd., Navi Mumbai, India). It consisted of a high-pressure pump, feed vessel, flat membrane sheet housing cell, and a temperature control unit (Figure 1a). The membrane housing cell contained a rectangular channel. The active surface area of the membrane was 0.0155 m^2. Lab experiments were conducted by filtering the multicomponent solution with NF membrane. The permeate and the concentrate streams were recirculated back into the feed tank continuously during experiments. After the completion of every single experiment, the system was appropriately cleaned by Milli-Q water. The samples were collected after each experiment.

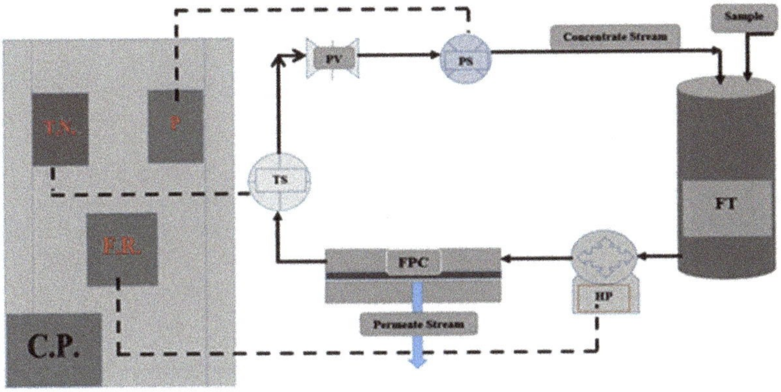

(a)

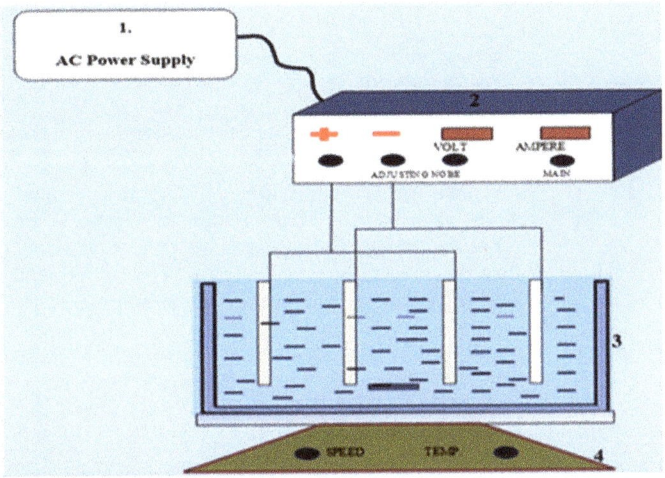

(b)

Figure 1. (**a**) Flat Plate Membrane System. (Control Panel, TN = Temperature Node, P = Pressure Node, FR = Flow Rate Controller node, TS = Temperature Sensor, PV = Pressure Valve, PS = Pressure Sensor, FPC = Flat Plate Membrane Cell, HP = High-Pressure Pump, FT = Feed Tank.). (**b**) Laboratory scale experimental setup of electrocoagulation unit. (1-AC, power scheme; 2, direct current supply; 3, treatment vessel, consists of anode and cathode in mono-polar mode, magnetic-bead; 4, magnetic stirrer).

2.2.2. Experimental Set for Electrocoagulation

The electrocoagulation (EC) experimental setup consisted of a DC power supply unit for constant DC output. The experimental reactor (11.0 cm × 11.0 cm × 15.0 cm) was made of plexiglass and four mono-polar stainless-steel plates (9.0 cm × 9.0 cm × 0.1 cm) submerged in the solution as the electrodes (Figure 1b). Plate spacing was 1 cm. The magnetic stirrer was used to provide proper stirring to maintain the uniformity of the solution throughout the reactor. Initially, at the start of each EC experimental run, 1.8 L of the synthetic solution was put into the electrolytic reaction cell after mixing it with one gm/L electrolyte (NaCl). The pH of the solution was measured and maintained by adding drops of 0.1 N NaOH and H_2SO_4 solution using a pH meter. The current was controlled through the power supply regulator. Samples were collected at the end of the electrolysis process.

Samples collected after the experiments employing different treatment processes were digested through the microwave digestion unit (Anton Par) and filtered by 0.42 μ filter papers. They were further analyzed by inductively coupled plasma mass spectroscopy (ICP-MS of Agilent). The removal efficiency was determined by calculating the difference in the concentrations measured by ICP-MS before and after each experiment.

2.3. Preparation of Working Solutions

Metal solutions were prepared by dissolving the appropriate mass of each metal in high purity Milli-Q water (18.2 MΩ cm), as mentioned earlier. All metal salts were added sequentially, after the previous metal salt had completely dissolved. Thereafter, the mixed-metal solutions with different concentrations (ppm) were prepared for each batch experiment.

2.4. Calculation of Removal Percentage

The removal efficiency of any metal can be calculated using the following equation:

$$R(\%) = \frac{C_{i}-C_{f}}{C_{i}} \times 100 \quad (1)$$

where C_i and C_f (mg/L) denote the concentration of the metal before and after the treatment process, respectively.

2.5. Experimental Design and Optimization through Response Surface Methodology

For mathematical modelling of the process, an empirical approach [29–31] employing response surface methodology (RSM) was adopted [32–34]. RSM reportedly reduces systematic errors with an estimation of investigational error and also reduces the number of experiments [30], requires lesser computer simulations, is more accessible and more efficient than the other methods based on limited components or computational complexity [31].

RSM based on the central composite design (CCD) was used to examine the efficacy of the NF membrane and EC processes. CCD helped in arriving at the operational conditions highlighting the highest removal efficiency scenarios. In the NF membrane process, the solution pH, pressure, concentration, and temperature were the key factors widely reported to contribute to the removal of metal ions [35–38] and thus design expert software was used for the experimental design with a varying range of these factors (Table 1).

Table 1. Factor and range for design experiments of NF membrane and EC.

Code	Factors	Coded Level of N.F.				
		$-\alpha$	−1	0	+1	$+\alpha$
X_1	pH	2.0	4.5	7.0	9.5	12.0
X_2	Pressure (bar)	2.5	10	17.5	25	32.5
X_3	Concentration (ppm)	2.5	10	17.5	25	32.5
X_4	Temperature (°C)	5	15	25	35	45
Code	Factors	Coded level of E.C.				
		$-\alpha$	−1	0	+1	$+\alpha$
X_1	pH	3.0	4.5	6.0	7.5	9.0
X_2	Time (min)	20	40	60	80	100
X_3	Concentration(ppm)	2.5	10	17.5	25	32.5
X_4	Current (Å)	0.5	1.0	1.5	2.0	2.5

In the EC process, solution pH, time, concentration, and current were the key factors [25] widely reported to contribute to the removal of metal ions [39–42] and thus design expert software was used for experimental design with a varying range of these factors (Table 1). The initial and final conductivity values were 2.78 and 2.35 mS/cm for the final optimum condition. The complete design matrix of the NF membrane and EC processes obtained after the application of CCD is presented in Supplementary Table S1 and Supplementary Table S2 respectively, which suggest thirty sets of runs and six centrally coded level runs for each treatment process.

3. Results and Discussion

In this study, experiments were performed for different combinations of factors for both processes, as described in the following sections.

3.1. Experimental Performance of NF

The details of the coded variables (X_1, X_2, X_3 and X_4), and their response values are presented in supplementary (Supplementary Table S1).

3.1.1. Statistical Analysis and Modelling by RSM

The NF membrane process responses were studied for the permeate flux and metal rejections (Zn, Cu, and Ni). The findings of the experimental studies were analyzed statistically through analysis of variance (ANOVA). Table 2 shows the ANOVA models.

Table 2. ANOVA analysis for the NF membrane.

Source	Sum of Squares	Degree of Freedom	Mean Square	F-Value	p-Value	R^2	R^2_{adj}	Responses
Model	5816.87	14	415.49	70.29	<0.0001	0.985	0.971	Flux
Model	17,449.52	14	1246.39	12.41	<0.0001	0.9205	0.8464	Zn Removal%
Model	10,360.61	14	740.04	5.07	0.0017	0.8255	0.6626	Ni Removal%
Model	2541.01	14	181.50	19.12	<0.0001	0.9469	0.8974	Cu Removal%

For the above models, Fisher's test statistic (*F*-Value) clarifies the scattering of actual data around the fitted models, while the *p*-value indicates the significance of the model terms. The F value of responses suggested that the respective models were significant regarding the residual error. The *p*-value of a model lower than 0.05 indicates a significant model, and higher than 0.10 means an insignificant model. The *p*-value of all responses were lower than 0.0001, suggesting that the models are highly significant. The coefficient of regression (R^2) described the system behaviour and the adequacy of the model in the range of independent variables. The high R^2 and adjusted R^2 in Table 2 also reveal that the models are highly significant.

The quadratic regression model equations for NF membrane permeate flux (Y_1), Zn removal (Y_2), Ni removal (Y_3), and Cu removal (Y_4) in terms of coded factors are presented below as Equations (2)–(5), respectively.

$$Y_1 = +26.75 + 1.84X_1 + 12.96X_2 + 0.51X_3 + 7.08X_4 - 1.06X_1^2 + 0.51X_2^2 + 1.36X_3^2 + 3.04X_4^2 + 1.06X_1X_2 - 0.53X_1X_3 + 0.82X_1X_4 + 0.33X_2X_3 + 2.5X_2X_4 + 0.77X_3X_4 \quad (2)$$

$$Y_2 = +89.32 + 24.78X_1 - 1.62X_2 - 4.67X_3 - 2.82X_4 - 6.26X_1^2 - 4.22X_2^2 - 4.71X_3^2 - 3.91X_4^2 + 0.52X_1X_2 + 1.96X_1X_3 + 0.67X_1X_4 - 0.34X_2X_3 - 1.25X_2X_4 + 0.23X_3X_4 \quad (3)$$

$$Y_3 = +84.38 + 19.27X_1 - 2.23X_2 - 3.82X_3 - 2.17X_4 - 0.43X_1^2 - 3.58X_2^2 - 3.46X_3^2 - 3.75X_4^2 + 0.61X_1X_2 + 0.59X_1X_3 + 0.31X_1X_4 - 0.52X_2X_3 + 0.076X_2X_4 + 0.26X_3X_4 \quad (4)$$

$$Y_4 = +92.65 + 8.88X_1 - 0.85X_2 - 2.02X_3 - 1.04X_4 - 3.76X_1^2 + 0.66X_2^2 + 0.47X_3^2 - 0.46X_4^2 + 0.44X_1X_2 + 1.50X_1X_3 + 0.31X_1X_4 - 0.33X_2X_3 + 0.070X_2X_4 + 0.27X_3X_4 \quad (5)$$

3.1.2. Response Surface Plots

The response surface plot for the permeate flux of NF is presented in Figure 2. The observations show an upsurge in the permeate flux simultaneously with the increase in trans-membrane pressure. It has been well established that permeate flux depends on pressure and gets amplified almost linearly with increasing pressure [32–36]. The maximum permeate flux of 59.34 $L/m^2{\cdot}h$ is obtained at the feed concentration of 25 ppm, pH 9.5, pressure 25 bar, and temperature 35 °C, as shown in Figure 2. It is typically theorized that an increased temperature accelerates the permeate flux due to one or more reasons such as a decline in solvent viscosity, a rise in solvent diffusion, intensification in the solvent diffusion coefficient, or a surge in polymer chain mobility [24]. Membrane-solvent interactions can be expected to differ with a change in solvent properties, like dielectric constant, molecular size, dipole movements, and Hildebrand solubility parameter. The rise in temperature also affects structural properties such as pore radius and membrane thickness, which have shown a much more noticeable impact on membrane performance in comparison to solvent and solute motilities [37–39]. Experiments have demonstrated a linear increase in the slope of flux with a rise in temperature, as reported by others [40,41]. Figure 2a shows a direct increase of permeate flux with an increase in trans-membrane pressure. Figure 2b demonstrates a significant increase in the permeate flux with an increase in temperature. Water permeation by micropores is an activated process that is absolutely dissimilar from viscous flow. It should be taken into account that the water molecule is one of the smallest molecules, having the same range of kinetic diameter (0.29 nm), as helium (0.24 nm) and hydrogen (0.27 nm). A portion of water molecules gets adequate thermal energy to cross the energy barrier from the pore wall and passes over the pores, which is another justification based on the adsorption of water on hydrophilic pore walls. The actual pore diameter might get reduced by the water adsorbed on pore walls. In such a case, the adsorbed water layer can be thinner at higher temperatures resulting in the effective pore diameter becoming more extensive [41].

The separation of metal ions by NF is attained by size exclusion, and electrical interactions between the ions in the feed aqueous solution and the charged NF membranes. The degree of ionization of these functional groups is a function of the solution pH, which influences the membrane charge and, therefore, the rejection properties of the membrane [41]. The rejection of Cu, Ni, and Zn ions increased with the increase in the solution pH (Figure 3a–c). The feed solution pH determines the ion charge in the solution and the surface charge density of the membranes. The more the pH increases, the more the membrane charge becomes positive, leading to a stronger electrostatic repulsion between the membrane and the metal ions [35]. Copper hydroxide precipitation starts at pH 5.24, and the precipitation of the other metals (Zn and Ni) at a still higher pH. At the different pH values studied, the rejection of copper was higher than for Zn and Ni ions, as reported earlier [42]. The maximum rejection of Cu, Zn, and Ni was demonstrated as 99.99%, 99.96% and 99.63%, respectively, in the experiments where concentrations ranged between 10–25 ppm and pressure between 10–25 bar. It was observed that the rejection of metal ions decreased when the concentration of feed solution increased, a common phenomenon for NF membranes [37]. The increase in concentration in the feed solution apparently generates a screen formation of cation adjacent to the membrane on the high-pressure side, which neutralizes the negative charges of the NF membrane. Thus, the total negative charge of the membrane decreases, and the repulsion between membrane and anion decreases. As a result, the co-ions (ions with the same charge as the membrane) quickly escape through the membrane and due to electro-neutrality, the rejection of counter-ions is reduced [8,43]. Figure 3d–f shows a slight decrease in the rejection of Ni and Zn ions with an increase in feed solution concentration, whereas the rejection of the Cu ions was not much affected. Temperature and pressure also have not shown much influence upon the rejection of the metal ions. Overall, the findings of this study are quite in line with some other relevant studies, as displayed in Table 3.

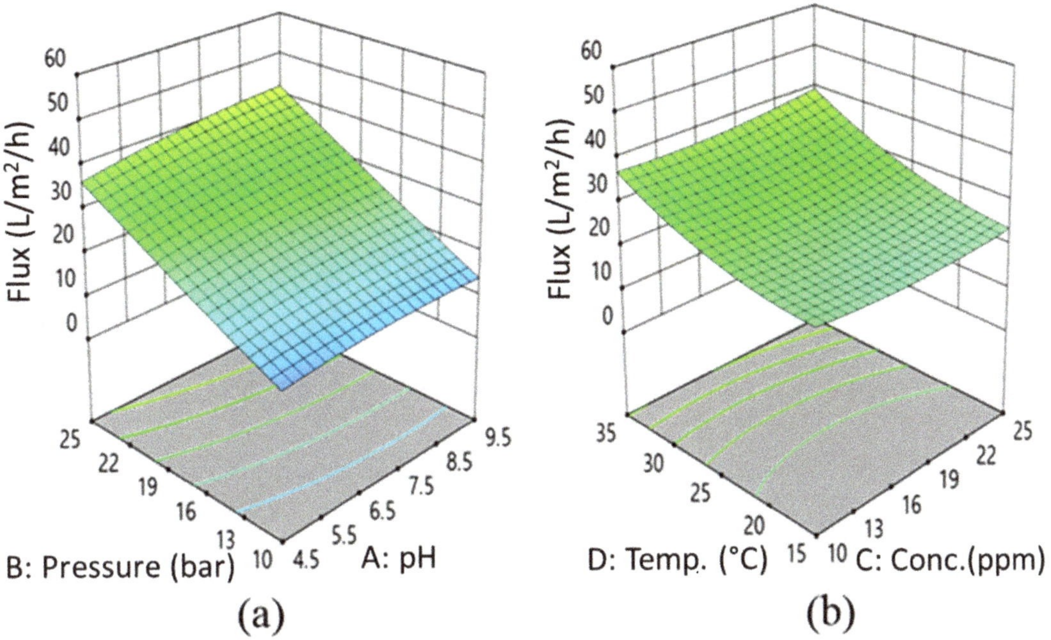

Figure 2. RSM plot for permeate flux. (**a**) Effect of pressure and pH. (**b**) Effect of temperature and concentration.

Table 3. Comparison of rejection efficiency of metals ions by NF membrane between this and other studies in the literature.

Pollutants/Wastewater Type	Membrane	Metal Rejection/Removal	References
Synthetic metal ion solution	NF	Zn (98.64%), Ni (90.54%), and Cu (99.79%)	Current study
Synthetic metal ion solution	NF	Zn (96.7%), Mg (95.01%), Cd (92.4%), Cu (91.9%), Ca (91.3%), Ni (90.7%), and Pb (90.5%).	[44]
Synthetic metal ion solution	NF and FO	Cu-(II) 95% and 99% in NF and F.O., respectively.	[45]
Synthetic metal ion solution	NF	Cu ion rejection was >90%	[46]
Synthetic metal ion solution	NF	Cu (99.82%), As (V) (96.75%), and Cr (97.22%).	[47]
Electroplating wastewater	NF	Ni (88.093%) and Cr (80.271%)	[48]
Synthetic metal ion solution	NF	Ni (93%)	[8]
Synthetic metal ion solution	NF	Cu (100%), Cd (99%), Mn (89%), and Pb (74%)	[37]
Synthetic metal ion solution	NF	Cd (97.26%) and Ni (98.90%)	[49]
Synthetic metal ion solution	NF	Ni (98.94%) and Cd (82.69%)	[50]

Note: NF—nanofiltration membrane; FO—Forward Osmosis.

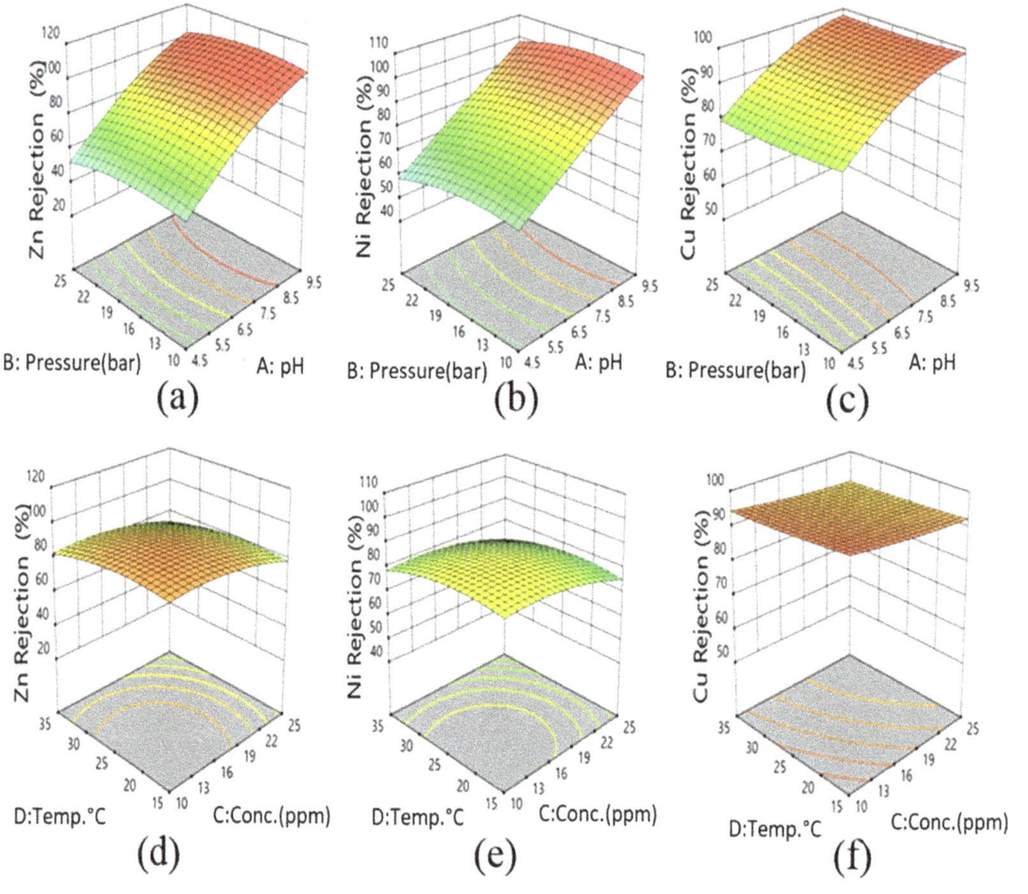

Figure 3. NF membrane RSM plots for pH and pressure effects on metals ions rejection% efficiency—(**a**) Zn ions, (**b**) Ni ions, and (**c**) Cu ions; and effect of temperature and concentration on metal ions rejection% efficiency—(**d**) Zn ions, (**e**) Ni ions, and (**f**) Cu ions.

3.1.3. Multi Response Optimization

The optimization of all input variables was done using the desirability function approach to arrive at the best response values of the factors Y_1, Y_2, Y_3, Y_4. As depicted in Table 4, at the optimal condition, the predicted response values of factors (Y_1, Y_2, Y_3, Y_4) were observed as 36.9 (L/m^2·h), 94.77%, 88.67%, 95.89%, respectively. The average values of factor responses after three runs, were found to be 41.93 (L/m^2·h) for Y_1, 98.64% for Y_2, 90.54% for Y_3, and 99.79% for Y_4. All the experimentally derived values are close to the predicted response values, showing a good correlation (Table 4).

Table 4. NF Optimization of response through RSM.

Optimum Input Variables (Feed Solution)		Predicted Responses by RSM		Experimental Validation	Difference%
Concentration (ppm)	14.24	Flux (L/m^2·h)	36.92	41.93	13.58
Temperature (°C)	25	Zn Removal%	94.77	98.64	4.08
pH	7.84	Ni Removal%	88.67	90.54	2.11
Pressure (Bar)	22.79	Cu Removal%	95.89	99.79	4.07

3.2. Experimental Performance of EC

The details of the coded variables (X_1, X_2, X_3 and X_4), and their response values are presented in Supplementary Table S2.

3.2.1. Statistical Analysis and Modelling by RSM

The findings of the experimental studies were analyzed statistically through analysis of variance (ANOVA). Table 5 shows the ANOVA models.

Table 5. ANOVA analysis for the EC process.

Source	Sum of Squares	Degree of Freedom	Mean Square	F-Value	p-Value	R^2	R^2_{adj}	Responses
Model	10,750.36	14	767.88	6.03	0.0007	0.8492	0.7085	Zn Removal%
Model	13,024.85	14	930.35	9.39	<0.0001	0.8976	0.8020	Ni Removal%
Model	940.48	14	67.18	2.78	0.0293	0.7220	0.4625	Cu Removal%

The F and p values presented in Table 5 indicate that the fitted models are significant. The values of R^2 and adjusted R^2 in the Table 5 also reveal the high significance levels of the models.

The quadratic regression model Equations (6)–(8) for Zn removal (Y_1), Ni removal (Y_2) and Cu removal (Y_3) in terms of coded factors are given below.

$$Y_1 = +93.51 + 7.05X_1 + 10.35X_2 - 5.22X_3 + 10.69X_4 - 0.030X_1^2 - 7.46X_2^2 + 0.58X_3^2 - 6.74X_4^2 + 0.039X_1X_2 + 4.07X_1X_3 - 4.70X_1X_4 + 0.81X_2X_3 - 0.60X_2X_4 + 4.38X_3X_4 \quad (6)$$

$$Y_2 = +90.88 + 9.06X_1 + 12.42X_2 - 5.53X_3 + 10.75X_4 - 2.14X_1^2 - 8.60X_2^2 + 0.069X_3^2 - 7.71X_4^2 - 1.75X_1X_2 + 1.35X_1X_3 - 4.46X_1X_4 + 1.01X_2X_3 + 0.93X_2X_4 + 2.24X_3X_4 \quad (7)$$

$$Y_3 = +97.05 + 2.00X_1 + 3.69X_2 - 1.00X_3 + 3.06X_4 + 0.10X_1^2 - 2.19X_2^2 + 0.79X_3^2 - 1.55X_4^2 + 0.081X_1X_2 - 0.57X_1X_3 - 1.25X_1X_4 + 0.91X_2X_3 - 0.16X_2X_4 + 0.32X_3X_4 \quad (8)$$

3.2.2. Response Surface Plots

It is well documented in the literature that initial pH is an essential operating parameter that strongly affects the EC process performance. The pH effect on the removal efficiencies of metal ions after EC treatment was validated in the experimental observations. Maximum removal efficiency for Zn (99.46%), Ni (98.14%), and Cu (99.96%) has been observed at pH 6. Figure 4a–c demonstrates an increase in the removal efficiency with an increase in pH. This indicates that metal ion elimination decreases in an acidic medium [50]. As reported, in an intensely acidic medium, the protons in the solution get reduced to H_2 gas at the cathode, and a sufficient number of hydroxyl ions are not generated. The pH of the initial solution affects the EC process by changing the solution's physico-chemical properties, such as solubility of metal hydroxides, electric conductivity, and size of colloidal particles of iron (III) complexes, which are most reactive agents for metal ions [51]. A slight reduction in the removal efficiency with the rise in the initial concentration of the metals in solution, as shown in Figure 4, is attributed to the fact that the amount of dissolved iron from the electrode may not have been enough to treat the metal ions present in the wastewater. The higher initial concentration in the feed solution was also reportedly found to significantly affect the EC process [52].

It was observed (Figure 4d–f) that increasing the constant current substantially reduces metal ions. The constant current emerged as a crucial parameter in improving metal ion removal, which may have contributed to the direct current field, and potential electrolysis, resulting in more release of ferric ions and generating more iron hydroxides, further forming coagulants for metal removal.

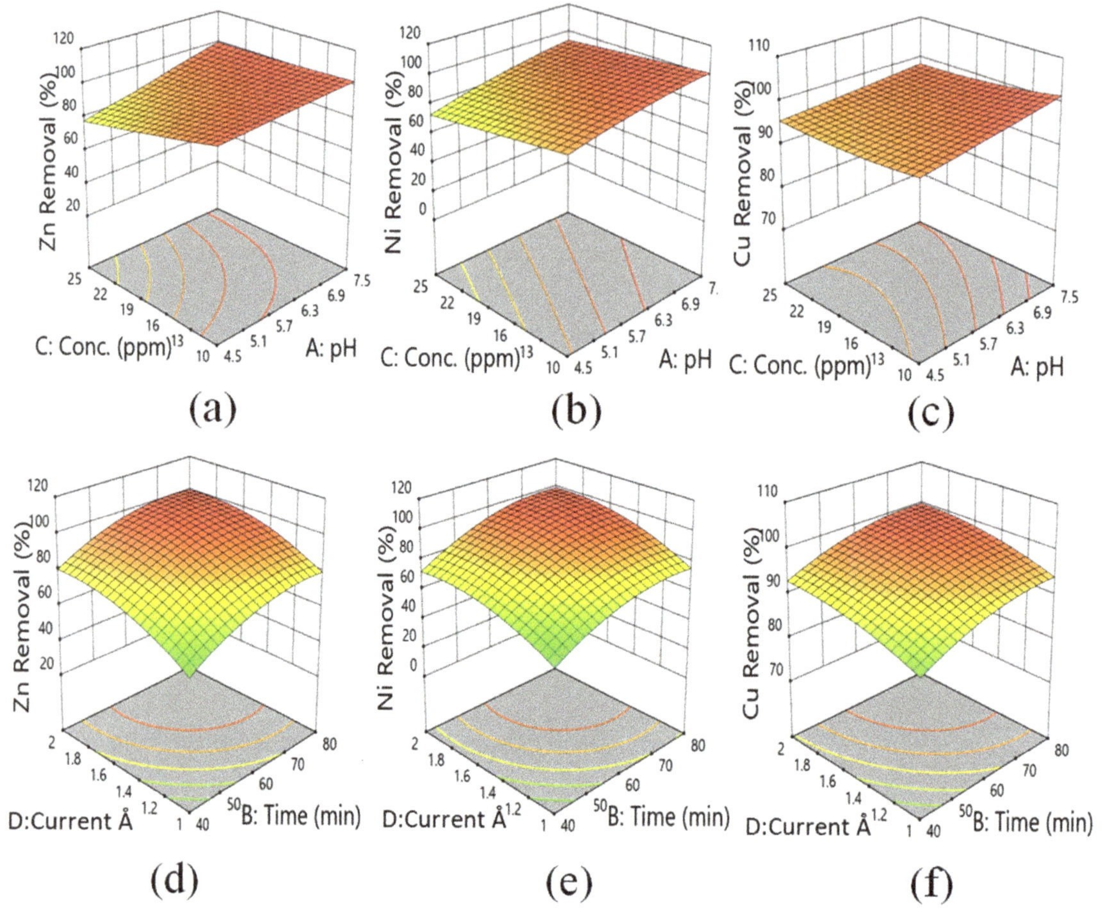

Figure 4. Electrocoagulation RSM graphs: effects of concentration and pH on removal% efficiency of metal ions: (**a**) Zn ions, (**b**) Ni ions and (**c**) Cu ions. Effect of current and reaction time on removal% efficiency of metal ions: (**d**) Zn ions, (**e**) Ni ions and (**f**) Cu ions.

Electrolysis time period plays a vital role in metal ion removal studies along with the constant current, pH and concentration. The concentration of Zn, Ni, and Cu has been observed to decrease with an increase in the electrolysis time. The complete reduction of metal ions was possible at a lower constant current by extending the electrolysis time. It may be stated that higher metal ion concentration consumes the adsorption ability of flocs formed, with fewer flocs being accessible for adsorption. Moreover, removal was limited by the formation rate of flocs of iron hydroxide complexes at the anode surface. It has been shown (Figure 4) that the minimum reduction was observed at lower electrolysis reaction times. The present study highlights that both the current and reaction time play a vital role in the removal efficiency of the EC process. Table 6 reports the results of this study vis-à-vis other studies reported in the literature on metal removal through EC processes.

Table 6. Overview of metals ion removal efficiency by EC processes described in the literature.

Pollutants/Wastewater	E.C. Process	Metal Removal Efficiency	References
Synthetic wastewater	SS electrode	Zn (99.81%), Ni (99.99%), and Cu (99.98%)	Current study
Synthetic wastewater	Aluminum Electrode	Ni^{2+} and Cu^{2+} ions% removal efficiency ranged from 34.56–100%	[53]
Electroplating Industry wastewater	Aluminium plates	Ni (99.75%)	[54]
Mine wastewater	Iron Electrodes	Co (99%) Ni (97%), Zn (99%), Cu (97%), and Cd (99%).	[55]
Smelting wastewater	Iron Plates	Zn^{2+}, Cd^{2+}, and Mn^{2+} removal efficiency 99.93%, 97.15%, and 85.46% respectivilty	[40]
Synthetic Smelting Wastewater	Fe electrode	Zn^{2+}, Cd^{2+}, and Mn^{2+} removal efficiency 99.5%, 99.9%, and 55% respectively.	[56]
Metal Plating Wastewater	Iron (carbon steel) plates	Cu^{2+}, Cr^{3+}, Ni^{2+}, and Zn^{2+} over 97%	[51]
Waste Fountain Solution	Aluminium and Iron electrode	Ni (>95%), Cu (>95%), and Zn (>80%)	[57]
Metal Plating Effluent	SS Electrode	Ni (100%) and Zn (100%)	[58]

3.2.3. Multi Response Optimization

For the EC System, the predicted response values of the factors (Y_1, Y_2, and Y_3) were obtained as 101.50%, 94.452%, and 98.866% under optimal operating conditions. Input variables of current and time are the dominant factor in reaction conditions, so the predicted response value shows a higher value. After three experimental runs, the average response values of Y_1, Y_2, and Y_3 were 99.81%, 99.99%, and 99.98%, respectively. All the experimentally attained values are quite close to the predicted response values and show a good correlation (Table 7).

Table 7. EC Optimization of response through RSM.

Optimum Input Variables (Feed Solution)		Predicted Responses by RSM		Experimental Validation	Difference%
pH	7	Zn Removal%	101.50	99.81	1.67
Time	60	Ni Removal%	94.45	99.99	5.86
Initial Conc. (ppm)	24.923	Cu Removal%	98.86	99.98	1.13
Current (Å)	1.912	–	–	–	–

3.3. *Comparison with Chemical Precipitation and Adsorption Processes*

Chemical precipitation is a commonly used treatment process for the removal of heavy metals from industrial wastewater because it is relatively inexpensive and easy to operate. This process involves the precipitation of heavy metals in the form of hydroxide and sulfide. Hydroxide precipitation depends on pH adjustment (9–11) to basic conditions [58]. The metal ions dissolved in the solution are precipitated into the insoluble solid phase as metal hydroxide through the chemical reaction when quicklime (CaO) is employed as a precipitant. Yet another common treatment process viz., adsorption, on the other hand, is a mass transfer process involving the migration of the metal ions (adsorbate) from the wastewater to a solid surface (adsorbent, commonly $CaCO_3$) and binding through physical (weak Van der Waals force) and chemical (strong covalent bond) adsorption mechanisms [59,60]. With an idea to compare the performance of NF and EC processes with the above-stated routine ones, the present study employed concentration, contact time, and dosing amount as operational variables for the conventional chemical treatment process employing CaO and $CaCO_3$. Twenty experiments were conducted for each process and Table 8 provides details of the experimental design.

Table 8. Factors and Range of Design for CaO and $CaCO_3$.

Factor	Name	Units	Low Actual	High Actual	Low Coded	High Coded
A	Conc.	ppm	10.00	25.00	−1.000	1.000
B	Contact Time	Minute	30.00	90.00	−1.000	1.000
C	Dosing Amount	g/L	4.00	8.00	−1.000	1.000

In CaO precipitation, the removal efficiency of the process was quite high for Zn, Ni, and Cu ions, as expected and as indicated by the results in Table 9 However, understandably, there are also many demerits in this process. It requires a large amount of chemical precipitant and produces a considerable amount of low-density sludge due to the poor settling properties, duly followed by further dewatering and disposal issues [60,61]. Aggregation of metal precipitates also has long term environmental impacts. Treated water also has a very high pH (10–12), so it cannot be further used in industrial process and requires treatment.

Table 9. Removal% Efficiency of Conventional (CaO and $CaCO_3$) vs. EC and NF processes.

S.N.	Elements Name	Unit	CaO (Maximum)	$CaCO_3$ (Maximum)	EC (Maximum)	NF (Maximum)
1	Zn Removal	%	99.72	99.82	99.46	99.96
2	Ni Removal	%	99.96	43.46	98.14	99.63
3	Cu Removal	%	99.99	99.99	99.96	99.99

Regarding the $CaCO_3$ adsorption process, the results presented in Table 9 indicate that the removal of Ni ions is not as efficient as for Cu and Zn ions. In the adsorption process, generated sludge needs to be separated from the solution and requires regeneration or labelling as a hazardous waste due to the strong possibility of leaching out of metals ions in the environment, while needing post-treatment sludge management. Van der Waal forces are very weak to strong for different adsorbents, due to which the process is unable to deliver promising results [62–64].

NF process lies between ultrafiltration (UF) and reverse osmosis (RO). Designed to separate contaminants smaller than 10 nm, it emerges as one of the exemplary processes for eliminating dissolved metals ions from wastewater. The leading gains of this process are higher removal efficiency, reliability and easy operation, lesser space requirement, and relatively lower energy requirement [60,63,65]. Table 9 shows an outstanding rejection rate for metals ions from this study.

The EC process is also widely recognized as an effective treatment method for eliminating heavy metal ions from industrial wastewater. It does not require any additional chemicals because the electron is a crucial reagent in the process. EC is considered a rapid and well-controlled technique, provides good reduction yield, produces less sludge, has the potential of metal recovery, requires less labor, can save significant energy, and is eco-friendly [57,59]. Table 9 shows an excellent metal ion reduction in the present experimental work.

4. Conclusions

The present study examines the removal efficiency of heavy metals (Cu, Ni, and Zn) in a mixed aqueous solution in a batch mode through a nanofiltration (NF) membrane and an electrocoagulation (EC) process and compares it with conventional chemical treatment processes. Solution pH is seen to significantly affect the removal efficiency in both the NF and EC processes. The highest permeate flux of 59.34 $L/m^2 \cdot h$ was observed at the experimental condition of pH 9.5, pressure 25 Bar, concentration 25 ppm and temperature 35 °C in NF process. The rejection rate of Zn, Ni and Cu was demonstrated as 95.32%, 94.98% and 96.93%, respectively. A marked synergistic effect of temperature and pressure has been observed, which increased the flux to a high value. The EC process has shown a maximum removal of Zn (99.46%), Ni (98.14%), and Cu (99.87%) at the operational conditions viz., pH 6, time 60 (min), concentration 2.5 ppm, and current 1.5 Å. The results for the EC process indicated that a lower concentration and approximately neutral pH helped the system to reach its full potential. Overall, both NF and EC processes have shown excellent removal for all the studied metal ions and the outcome of the experiments described above projects them as promising solutions in comparison to conventional chemical treatment approaches.

Supplementary Materials: The following supporting information can be downloaded at: https://www.mdpi.com/article/10.3390/app12010280/s1, Table S1: Experimental design and responses of the NF process; Table S2: Experimental design and responses of the EC process.

Author Contributions: Conceptualization, J.K. and H.J.; methodology, J.K. and H.J.; software, J.K.; formal analysis, J.K. and S.K.M.; investigation, J.K.; resources, H.J. and S.K.M.; writing—original draft preparation, J.K.; writing, review and editing, H.J. and S.K.M. All authors have read and agreed to the published version of the manuscript.

Funding: This research received no external funding.

Data Availability Statement: All relevant data are within the manuscript and available from the corresponding author on request.

Acknowledgments: This work has been carried out with the Junior/Senior Research Fellowship of the University Grant Commission (UGC), New Delhi, India, under the Ph.D. programme. Authors are highly thankful to the Indian Institute of Technology Roorkee, Roorkee for providing the necessary support and encouragement for bringing out this publication.

Conflicts of Interest: The authors declare no conflict of interest or state.

References

1. Tchounwou, P.B.; Yedjou, C.G.; Patlolla, A.K.; Sutton, D.J. Molecular, clinical and environmental toxicicology Volume 3: Environmental Toxicology. *Mol. Clin. Environ. Toxicol.* **2012**, *101*, 133–164. [CrossRef]
2. Vareda, J.P.; Valente, A.J.M.; Durães, L. Assessment of heavy metal pollution from anthropogenic activities and remediation strategies: A review. *J. Environ. Manag.* **2019**, *246*, 101–118. [CrossRef] [PubMed]
3. Hussain, J.; Husain, I.; Arif, M.; Gupta, N. Studies on heavy metal contamination in Godavari river basin. *Appl. Water Sci.* **2017**, *7*, 4539–4548. [CrossRef]
4. Malyan, S.K.; Singh, R.; Rawat, M.; Kumar, M.; Pugazhendhi, A.; Kumar, A.; Kumar, V.; Kumar, S.S. Biocatalysis and Agricultural Biotechnology An overview of carcinogenic pollutants in groundwater of India. *Biocatal. Agric. Biotechnol.* **2019**, *21*, 101288. [CrossRef]
5. Kumar, A.; Malyan, S.K.; Kumar, S.S.; Dutt, D.; Kumar, V. Biocatalysis and Agricultural Biotechnology An assessment of trace element contamination in groundwater aquifers of. *Biocatal. Agric. Biotechnol.* **2019**, *20*, 101213. [CrossRef]
6. Kumar, A.; Jigyasu, D.K.; Subrahmanyam, G.; Mondal, R.; Shabnam, A.A.; Cabral-Pinto, M.M.S.; Malyan, S.K.; Chaturvedi, A.K.; Gupta, D.K.; Fagodiya, R.K.; et al. Nickel in terrestrial biota: Comprehensive review on contamination, toxicity, tolerance and its remediation approaches. *Chemosphere* **2021**, *275*, 129996. [CrossRef]
7. Kumar, A.; Pinto, M.C.; Candeias, C.; Dinis, P.A. Baseline maps of potentially toxic elements in the soils of Garhwal Himalayas, India: Assessment of their eco-environmental and human health risks. *Land Degrad. Dev.* **2021**, *32*, 3856–3869. [CrossRef]
8. Maher, A.; Sadeghi, M.; Moheb, A. Heavy metal elimination from drinking water using nanofiltration membrane technology and process optimization using response surface methodology. *Desalination* **2014**, *352*, 166–173. [CrossRef]
9. Tonhá, M.S.; Garnier, J.; Araújo, D.F.; Cunha, B.C.A.; Machado, W.; Dantas, E.; Araújo, R.; Kutter, V.T.; Bonnet, M.P.; Seyler, P. Behavior of metallurgical zinc contamination in coastal environments: A survey of Zn from electroplating wastes and partitioning in sediments. *Sci. Total Environ.* **2020**, *743*, 140610. [CrossRef] [PubMed]
10. Rodrigues, S.K.; Machado, W.; Guerra, J.V.; Geraldes, M.; Morales, S.; Vinzón, S.B. Changes in Cd and Zn distribution in sediments after closure of an electroplating industry, Sepetiba bay, Brazil. *Mar. Pollut. Bull.* **2020**, *161*, 111758. [CrossRef] [PubMed]
11. Nriagu, J. Zinc Toxicity in Humans. *Encycl. Environ. Health* **2011**, 801–807. [CrossRef]
12. Qu, L.; Huang, H.; Xia, F.; Liu, Y.; Dahlgren, R.A.; Zhang, M.; Mei, K. Risk analysis of heavy metal concentration in surface waters across the rural-urban interface of the Wen-Rui Tang River, China. *Environ. Pollut.* **2018**, *237*, 639–649. [CrossRef] [PubMed]
13. Kumar, A.; Subrahmanyam, G.; Mondal, R.; Cabral-pinto, M.M.S.; Shabnam, A.A.; Jigyasu, D.K.; Malyan, S.K.; Kishor, R.; Khan, S.A.; Kumar, A.; et al. Chemosphere Bio-remediation approaches for alleviation of cadmium contamination in natural resources. *Chemosphere* **2021**, *268*, 128855. [CrossRef] [PubMed]

14. Tatiana, A.; Manceau, A.; Spadini, L.; Panfili, F.; Marcus, M.A.; Jacquet, T. Speciation and solubility of heavy metals in contaminated soil using X-ray microfluorescence, EXAFS spectroscopy, chemical extraction, and thermodynamic modeling. *Geochim. Cosmochim. Acta* **2006**, *70*, 2163–2190. [CrossRef]
15. Li, Y.; Tsend, N.; Li, T.K.; Liu, H.; Yang, R.; Gai, X.; Wang, H.; Shan, S. Microwave assisted hydrothermal preparation of rice straw hydrochars for adsorption of organics and heavy metals. *Bioresour. Technol.* **2019**, *273*, 136–143. [CrossRef] [PubMed]
16. Sadeek, A.; Negm, N.A.; Hefni, H.H.H.; Wahab, M.M.A. Metal adsorption by agricultural biosorbents: Adsorption isotherm, kinetic and biosorbents chemical structures. *Int. J. Biol. Macromol.* **2015**, *81*, 400–409. [CrossRef]
17. WHO. *2017 World Health Statistics 2017: Monitoring Health for the SDGs, Sustainable Development Goals*; World Health Organization: Geneva, Switzerland, 2017.
18. UN-Water. *World Water Assessment Programme (Nations Unies), the United Nations World Water Development Report 2018*; United Nations Educational, Scientific and Cultural Organization: New York, NY, USA, 2018. Available online: https://www.unwater.org/publications/world-water-development-report-2018/ (accessed on 14 October 2021).
19. Pandit, A.B.; Kumar, J.K. Clean Water for Developing Countries. *Annu. Rev. Chem. Biomol. Eng.* **2015**, *6*, 217–246. [CrossRef] [PubMed]
20. Price, J.R.; Ledford, S.H.; Ryan, M.O.; Toran, L.; Sales, C.M. Science of the Total Environment Wastewater treatment plant effluent introduces recoverable shifts in microbial community composition in receiving streams. *Sci. Total Environ.* **2018**, *613–614*, 1104–1116. [CrossRef] [PubMed]
21. Kumar, S.S.; Kadier, A.; Malyan, S.K.; Ahmad, A. Phytoremediation and Rhizoremediation: Uptake, Mobilization and Sequestration of Heavy Metals by Plants. In *Plant-Microbe Interactions in Agro-Ecological Perspectives*; Singh, D., Singh, H., Prabha, R., Eds.; Springer: Singapore, 2017. [CrossRef]
22. Malyan, S.K.; Yadav, S.; Sonkar, V.; Goyal, V.C.; Singh, O.; Singh, R. Mechanistic understanding of the pollutant removal and transformation processes in the constructed wetland system. *Water Environ. Res.* **2021**, *93*, 1882–1909. [CrossRef] [PubMed]
23. Wu, J.; Huang, R.; Zhou, Q.; Lu, H.; Li, F.; Wu, K.; Li, Z. Science of the Total Environment Magnetic biochar reduces phosphorus uptake by Phragmites australis during heavy metal remediation. *Sci. Total Environ.* **2021**, *758*, 143643. [CrossRef] [PubMed]
24. Malyan, S.K.; Kumar, S.S.; Fagodiya, R.K.; Ghosh, P.; Kumar, A.; Singh, R.; Singh, L. Biochar for environmental sustainability in the energy-water-agroecosystem nexus. *Renew. Sustain. Energy Rev.* **2021**, *149*, 111379. [CrossRef]
25. Kumar, D.; Sharma, C. Remediation of Pulp and Paper Industry Effluent Using Electrocoagulation Process. *J. Water Resour. Prot.* **2019**, *11*, 296–310. [CrossRef]
26. Kurniawan, T.A.; Chan, G.Y.S.; Lo, W.H.; Babel, S. Physico-chemical treatment techniques for wastewater laden with heavy metals. *Chem. Eng. J.* **2006**, *118*, 83–98. [CrossRef]
27. Singh, V.; Ram, C. Physico-Chemical Characterization of Electroplating Industrial Effluents of Chandigarh and Haryana Region. *J. Civ. Environ. Eng.* **2016**, *6*, 4. [CrossRef]
28. Sivakumar, D.; Nouri, J.; Modhini, T.M.; Deepalakshmi, K. Nickel removal from electroplating industry wastewater: A bamboo activated carbon. *Glob. J. Environ. Sci. Manag.* **2018**, *4*, 325–338. [CrossRef]
29. Dasgupta, J.; Mondal, D.; Chakraborty, S.; Sikder, J.; Curcio, S.; Arafat, H.A. Nano filtration based water reclamation from tannery effluent following coagulation pretreatment. *Ecotoxicol. Environ. Saf.* **2015**, *121*, 22–30. [CrossRef] [PubMed]
30. Almeida, M.; Erthal, R.; Padua, E.; Silveira, L.; Am, L. Response surface methodology (RSM) as a tool for optimization in analytical chemistry. *Talanta* **2008**, *76*, 965–977. [CrossRef]
31. Srivastava, A.; Aghilesh, K.; Nair, A.; Ram, S.; Agarwal, S.; Ali, J.; Singh, R.; Chandra, M. Response surface methodology and artificial neural network modelling for the performance evaluation of pilot-scale hybrid nanofiltration (NF) & reverse osmosis (RO) membrane system for the treatment of brackish ground water. *J. Environ. Manag.* **2021**, *278*, 111497.
32. Sinha, A.; Biswas, P.; Sarkar, S.; Bora, U.; Purkait, M.K. Separation of chloride and sulphate ions from nano fi ltration rejected wastewater of steel industry. *J. Water Process Eng.* **2020**, *33*, 101108. [CrossRef]
33. Madaeni, S.S.; Arast, N.; Rahimpour, F.; Arast, Y. Fabrication optimization of acrylonitrile butadiene styrene (ABS)/polyvinylpyrrolidone (PVP) nano fi ltration membrane using response surface methodology. *Desalination* **2011**, *280*, 305–312. [CrossRef]
34. Patidar, R.; Srivastava, V.C. Chemosphere Mechanistic insight into ultrasound-induced enhancement of electrochemical oxidation of o fl oxacin: Multi-response optimization and cost analysis. *Chemosphere* **2020**, *257*, 127121. [CrossRef] [PubMed]
35. Roy, Y.; Warsinger, D.M.; Lienhard, J.H. Effect of temperature on ion transport in nanofiltration membranes: Diffusion, convection and electromigration. *Desalination* **2017**, *420*, 241–257. [CrossRef]
36. Figoli, A.; Cassano, A.; Criscuoli, A.; Mozumder, M.S.I.; Uddin, M.T.; Islam, M.A.; Drioli, E. Influence of operating parameters on the arsenic removal by nanofiltration. *Water Res.* **2010**, *44*, 97–104. [CrossRef] [PubMed]
37. Al-Rashdi, B.A.M.; Johnson, D.J.; Hilal, N. Removal of heavy metal ions by nanofiltration. *Desalination* **2013**, *315*, 2–17. [CrossRef]
38. Gherasim, C.V.; Mikulášek, P. Influence of operating variables on the removal of heavy metal ions from aqueous solutions by nanofiltration. *Desalination* **2014**, *343*, 67–74. [CrossRef]
39. Zhou, R.; Liu, F.; Wei, N.; Yang, C.; Yang, J.; Wu, Y.; Li, Y. Comparison of Cr (VI) removal by direct and pulse current electrocoagulation: Implications for energy consumption optimization, sludge reduction and floc magnetism. *J. Water Process Eng.* **2020**, *37*, 101387. [CrossRef]
40. Xu, L.; Xu, X.; Cao, G.; Liu, S.; Duan, Z.; Song, S.; Song, M.; Zhang, M. Optimization and assessment of Fe e electrocoagulation for the removal of potentially toxic metals from real smelting wastewater. *J. Environ. Manag.* **2018**, *218*, 129–138. [CrossRef] [PubMed]

41. Pino, L.; Vargas, C.; Schwarz, A.; Borquez, R. Influence of operating conditions on the removal of metals and sulfate from copper acid mine drainage by nanofiltration. *Chem. Eng. J.* **2018**, *345*, 114–125. [CrossRef]
42. Tsuru, T.; Izumi, S.; Yoshioka, T.; Asaeda, M. Temperature effect on transport performance by inorganic nanofiltration membranes. *AIChE J.* **2000**, *46*, 565–574. [CrossRef]
43. Micari, M.; Diamantidou, D.; Heijman, B.; Moser, M.; Haidari, A.; Spanjers, H.; Bertsch, V. Experimental and theoretical characterization of commercial nanofiltration membranes for the treatment of ion exchange spent regenerant. *J. Memb. Sci.* **2020**, *606*, 118117. [CrossRef]
44. Peydayesh, M.; Mohammadi, T.; Nikouzad, S.K. A positively charged composite loose nanofiltration membrane for water purification from heavy metals. *J. Memb. Sci.* **2020**, *611*, 118205. [CrossRef]
45. Hamid, M.F.; Abdullah, N.; Yusof, N.; Ismail, N.M.; Ismail, A.F.; Salleh, W.N.W.; Jaafar, J.; Aziz, F.; Lau, W.J. Effects of surface charge of thin-film composite membrane on copper (II) ion removal by using nanofiltration and forward osmosis process. *J. Water Process Eng.* **2020**, *33*, 101032. [CrossRef]
46. Wang, J.; Yu, W.; Graham, N.J.D.; Jiang, L. Evaluation of a novel polyamide-polyethylenimine nanofiltration membrane for wastewater treatment: Removal of Cu^{2+} ions. *Chem. Eng. J.* **2020**, *392*, 123769. [CrossRef]
47. Roy, S.; Majumdar, S.; Sahoo, G.C.; Bhowmick, S.; Kundu, A.K.; Mondal, P. Removal of As(V), Cr(VI) and Cu(II) using novel amine functionalized composite nanofiltration membranes fabricated on ceramic tubular substrate. *J. Hazard. Mater.* **2020**, *399*, 122841. [CrossRef] [PubMed]
48. Hosseini, S.S.; Nazif, A.; Shahmirzadi, M.A.A.; Ortiz, I. Fabrication, tuning and optimization of poly (acrilonitryle) nanofiltration membranes for effective nickel and chromium removal from electroplating wastewater. *Sep. Purif. Technol.* **2017**, *187*, 46–59. [CrossRef]
49. Chaudhari, L.B.; Murthy, Z.V.P. Separation of Cd and Ni from multicomponent aqueous solutions by nanofiltration and characterization of membrane using IT model. *J. Hazard. Mater.* **2010**, *180*, 309–315. [CrossRef] [PubMed]
50. Murthy, Z.V.P.; Chaudhari, L.B. Separation of binary heavy metals from aqueous solutions by nanofiltration and characterization of the membrane using Spiegler-Kedem model. *Chem. Eng. J.* **2009**, *150*, 181–187. [CrossRef]
51. Al-shannag, M.; Al-qodah, Z.; Bani-melhem, K.; Rasool, M. Heavy metal ions removal from metal plating wastewater using electrocoagulation: Kinetic study and process performance. *Chem. Eng. J.* **2015**, *260*, 749–756. [CrossRef]
52. Mollah, M.Y.A.; Schennach, R.; Parga, J.R.; Cocke, D.L. Electrocoagulation (EC)—Science and applications. *J. Hazard. Mater.* **2001**, *84*, 29–41. [CrossRef]
53. El-Ashtoukhy, E.S.Z.; Amin, N.K.; Fouad, Y.O.; Hamad, H.A. Intensification of a new electrocoagulation system characterized by minimum energy consumption and maximum removal efficiency of heavy metals from simulated wastewater. *Chem. Eng. Process.-Process Intensif.* **2020**, *154*, 108026. [CrossRef]
54. Sarwanto, S.; Nugroho, R.; Handayani, M.; Pratama, M.A. Heliyon Optimization and reaction kinetics on the removal of Nickel and COD from wastewater from electroplating industry using Electrocoagulation and Advanced Oxidation Processes. *Heliyon* **2020**, *6*, e03319. [CrossRef]
55. Mamelkina, M.A.; Vasilyev, F.; Tuunila, R.; Sillanpää, M.; Häkkinen, A. Investigation of the parameters affecting the treatment of mining waters by electrocoagulation. *J. Water Process Eng.* **2019**, *32*, 100929. [CrossRef]
56. Xu, L.; Cao, G.; Xu, X.; Liu, S.; Duan, Z.; He, C.; Wang, Y.; Huang, Q. Simultaneous removal of cadmium, zinc and manganese using electrocoagulation: Influence of operating parameters and electrolyte nature. *J. Environ. Manag.* **2017**, *204*, 394–403. [CrossRef]
57. Prica, M.; Adamovic, S.; Dalmacija, B.; Rajic, L. The electrocoagulation/flotation study: The removal of heavy metals from the waste fountain. *Process Saf. Environ. Prot.* **2014**, *94*, 262–273. [CrossRef]
58. Arslan, T.; Arslan-alaton, I.; Tünay, O. Complexing agent and heavy metal removals from metal plating effluent by electrocoagulation with stainless steel electrodes. *J. Hazard. Mater.* **2009**, *165*, 838–845. [CrossRef]
59. Malik, L.A.; Bashir, A.; Qureashi, A.; Pandith, A.H. Detection and removal of heavy metal ions: A review. *Environ. Chem. Lett.* **2019**, *17*, 1495–1521. [CrossRef]
60. Bolisetty, S.; Peydayesh, M.; Mezzenga, R. Sustainable technologies for water purification from heavy metals: Review and analysis. *Chem. Soc. Rev.* **2019**, *48*, 463–487. [CrossRef]
61. Fu, F.; Wang, Q. Removal of heavy metal ions from wastewaters: A review. *J. Environ. Manag.* **2011**, *92*, 407–418. [CrossRef]
62. Vunain, E.; Mishra, A.K.; Mamba, B.B. Dendrimers, mesoporous silicas and chitosan-based nanosorbents for the removal of heavy-metal ions: A review. *Int. J. Biol. Macromol.* **2016**, *86*, 570–586. [CrossRef] [PubMed]
63. Abdullah, N.; Yusof, N.; Lau, W.J.; Jaafar, J.; Ismail, A.F. Recent trends of heavy metal removal from water/wastewater by membrane technologies. *J. Ind. Eng. Chem.* **2019**, *76*, 17–38. [CrossRef]
64. Zhao, M.; Xu, Y.; Zhang, C.; Rong, H.; Zeng, G. New trends in removing heavy metals from wastewater. *Appl. Microbiol. Biotechnol.* **2016**, *100*, 6509–6518. [CrossRef]
65. Vardhan, K.H.; Kumar, P.S.; Panda, R.C. A review on heavy metal pollution, toxicity and remedial measures: Current trends and future perspectives. *J. Mol. Liq.* **2019**, *290*, 111197. [CrossRef]

applied sciences

MDPI

Article

Expanded S-Curve Model of Relationship between Domestic Water Usage and Economic Development: A Case Study of Typical Countries

Xiaoqian Guo [1,2,*], Anjian Wang [1,2], Guwang Liu [1,2] and Boyu Du [1,2]

1 MLR Key Laboratory of Metallogeny and Mineral Assessment, Institute of Mineral Resources, China Academy of Geological Science, Beijing 100037, China; ajwang@cgs.ac.cn (A.W.); lgw110@126.com (G.L.); dby970313@163.com (B.D.)

2 Research Center for Strategy of Global Mineral Resources, Institute of Mineral Resources, China Academy of Geological Science, Beijing 100037, China

* Correspondence: xiaoqianguo88@163.com; Tel.: +86-010-6899-9656

Abstract: Domestic water plays a growing role with the unprecedented economic development and rising urbanization. The lack of long-term evaluation of domestic water usage trends limits our understanding of the relationship between domestic water usage and economics. Here, we present a pragmatic approach to assess the long-term relationship between domestic water usage and economics through historical data of the last 100 years from 10 typical countries to establish an evaluation method for different economics. The relationship between domestic water usage and GDP per capita was described as an expanded S-curve model and the mathematical modeling was derived to simulate this relationship for four typical countries as case studies. The simulation results show that the expanded S-curve of different countries can be calibrated with three key points: takeoff point, turning point, and zero-growth point, and four transitional sections: slow growth, accelerated growth, decelerated growth, and zero/negative growth, corresponding to the same economic development level. In addition, other factors influencing domestic water usage are also discussed in this research, including urbanization, industrial structure, and technical progress. We hope to provide a case study of an expanded S-curve as a foundation for forecasting domestic water usage in different countries or in the same economy at different developmental stages.

Keywords: expanded S-curve model; domestic water usage; economic development; mathematical model

Citation: Guo, X.; Wang, A.; Liu, G.; Du, B. Expanded S-Curve Model of Relationship between Domestic Water Usage and Economic Development: A Case Study of Typical Countries. *Appl. Sci.* **2022**, *12*, 6090. https://doi.org/10.3390/app12126090

Academic Editors: Amit Kumar, Santosh Subhash Palmate and Rituraj Shukla

Received: 29 April 2022
Accepted: 11 June 2022
Published: 15 June 2022

1. Introduction

As an essential resource for human development, water is required throughout the life-cycle processes of all of society. In the context of unprecedented economic development and rising urbanization, water usage (i.e., withdrawal) by humans has increased from 500 km^3yr^{-1} to nearly 4000 km^3yr^{-1} over the last century, with an annual increase rate of 1.5% between 1960 and 2010 [1,2]. This increasing water usage has aggravated water scarcity, affecting more than 2 million people globally. In addition, it is predicted that more than half of the global population will live in regions suffering from at least moderate water shortage by 2050 [3,4].

Among global water usage, the principal user of water is in the agriculture sector, accounting for 70% of total water usage, with the remaining part being attributable to the industrial sector and domestic sectors [5]. The global domestic water demand is projected to see a 130% increase by 2050, which is much faster compared to other water sectors [6]. Therefore, with economic development, domestic water will play a major role in total water usage in the near future. A long-term evaluation of domestic water usage trends could provide references for policymakers, and is emerging as a paramount issue for efficient and sustainable management of water resources [7].

A series of studies has been conducted on the evaluation of water usage trends, based on different theories and methodologies. These published studies were performed based on two integrated criteria: drivers and approaches. The first one investigated water-usage drivers in a short timeframe. Domestic water usage is considered to be related to a series of drivers, including economic, climate, population, water price, and policies. Zhou et al. quantified socioeconomic drivers, such as urban population and service GVA, to investigate the key drivers of water changes [7,8]. Manouseli et al. considered climate change as a factor affecting domestic water [9,10]. Meng et al. proved the significant relationship between regional GDP, population, and water consumption [11]. Suarez-Varela modeled the linear relationship between water usage and water price [12]. Bijl et al. described the GDP per capita, population, and water withdrawal efficient as synthetic factors for domestic water change [13]. Among these factors, the investigation of long-term drivers was constrained by the lack of continuous data, except for the economic drivers, which could be traced back to 1900 by the World Bank [14]. The second one is the investigation of approaches. These approaches can be divided into two types, namely, single-equation models and hybrid models. Single-equation models include the linear regression model [10], whale optimization algorithm [15], artificial neural network [16], pseudo-panel approach [8], and so on. Hybrid models are driven by macroscale socioeconomic activity to simulate water use in specific regions. Hybrid models include IMAGE (Integrated Model to Assess the Global Environment) in IMAGE regions [17], QUAIDS (Quadratic Almost Ideal Demand System) models in Spanish [12], and IUWM (Integrated Urban Water Management) in Australia [18]. Therefore, a universal and long-term evaluation of water-usage trends is still missing, and a new approach should be introduced.

In this research, we adopted the expanded S-curve model to assess the long-term relationship between water usage and economics, which are presented by domestic water usage per capita (DWPC) and GDP per capita, respectively. The S-curve model has been widely used in mineral and energy resource evaluation. The S-curve pattern was first proposed by French mathematician Verhulst in 1838 for the description of biological population [19], and was employed by Wang et al. [20] and Gao et al. [21] to quantify the relationships between economics and energy and steel, respectively. Besides, in the water-usage studies by Zuo [22] and Florke et al. [23], the S-curve pattern was also used as country curve qualitatively. Here, water-usage data from 10 typical countries and regions in the last 100 years are described through the relationship between domestic water usage and GDP per capita, and the key points and transitional sections are identified in the expanded S-curve model. Other factors affecting domestic water usage are also discussed.

2. Material and Methods

2.1. Data and Key Drivers

This study first collected a vast amount of data about domestic water resources from 17 typical countries from 1950 to 2020 according to the publicly available data. The water data in this research are mainly from three parts. Firstly, the global water-related database with free access was established in AQUASTAT by the Food and Agriculture Organization of the United Nations (FAO) and the water databases in the World Bank Open Data [5,14]. It should be noted that national water-use records were conducted every 5 years or longer, and most of the water records could only be traced to 1960 or later. Secondly, a few of the detailed water-use categories were collected from national statistical offices, such as the U.S. Geological Survey (USGS) and the Eurostat and German Association of Energy and Water Industries [24–26]. Thirdly, a series of published literature reviews and statical surveys about water use were consulted. Gleick [27] and Shiklomanov [28] tried to conduct an adequate data survey for the World's Water Report and USA water data. Florke et al. used the WaterGAP 3 model for back-calculating water-use data on a global scale [23]. However, many historical records on domestic water use were incomplete or discontinuous, as shown Figure 1, for the primary selection of 17 countries. The relationship between water usage and economic development is shown as the domestic water per capita (DWPC) and GDP per capita to offset the regional disparity.

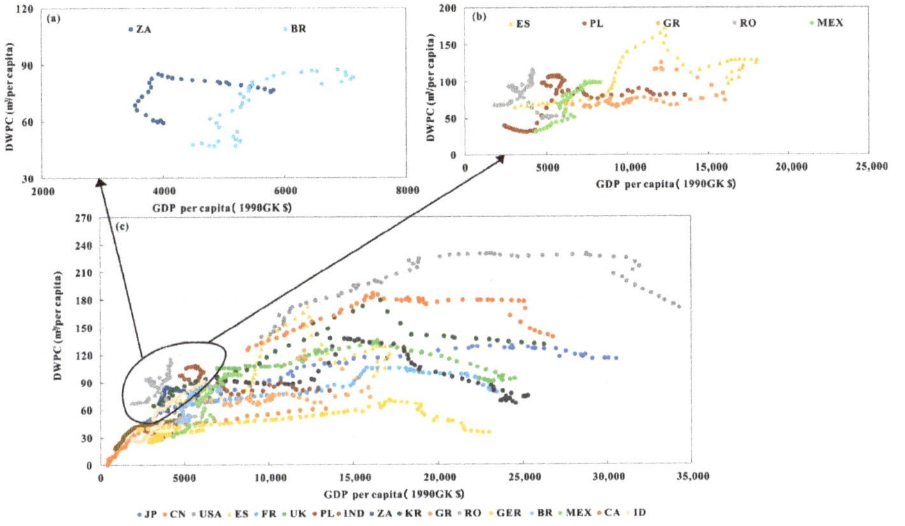

Figure 1. Relationship between GDP per capita and DWPC for (**a**) South Africa and Brazil; (**b**) Spain, Poland, Greece, Romania, and Mexico; and (**c**) all 17 countries.

Figure 1 shows the collected domestic water data from 17 countries, including Japan (JP), China (CN), the United States (USA), Spain (ES), France (FR), the United Kingdom (UK), Poland (PL), India (IND), Indonesia (ID), South Africa (ZA), South Korea (KR), Greece (GR), Romania (RO), Germany (GER), Brazil (BR), Mexico (MX), and Canada (CA). However, several countries showed poor data availability. South Africa and Brazil, as shown in Figure 1a, showed a C-type curve, meaning that the water usage was reduced during economic recession and increased during economic recovery. Spain, Poland, Greece, Romania, and Mexico showed an entangled type, showing that the water usage drastically changed during economic transition. These 7 countries with cluttered data were excluded and the remaining 10 countries were collected as our research objects.

To get consistent and long-term DWPC data, the relationship between urbanization rate and DWPC was derived first, and changes in water-usage intensity were expressed as urbanization rate change due to the observation that as urbanization rate increases, water users in a more urban population trend toward a more water-intensive lifestyle. After the maximum level was reached, DWPC was either stable or declined with the increasing urbanization rate. The relationship between urbanization rate and DWPC is shown in Figure 2. Instead of using solely regional curves to estimate past DWPC, the current data model version was derived for 9 countries. Germany was an exception due to the lower DWPC, so the DWPC with urbanization between 20% and 60% in Germany was derived from its original data. Where data availability was missing with an urbanization of between 20% and 60%, the available information was combined in order to allow for the fitting of a simulating curve to the historical data, as shown in Figure 2.

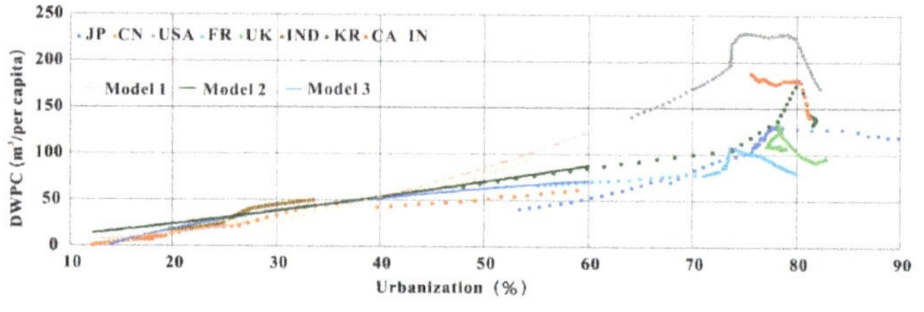

Figure 2. Relationship between urbanization and DWPC.

The urbanization data could be traced back to 1900 and the relationship between urbanization and DWPC was simulated as three different models in Table 1, with the R-squared above 0.85. In addition, the corresponding models for applicable countries were listed according to the trends of the curves after urbanization above 60%, as shown in Table 1. When the urbanization was acquired, the DWPC could be derived with corresponding models.

Table 1. Simulated models between urbanization and DWPC.

Model	Fitting Curve	(R-Squared) R^2	Applicable Countries
Model 1	$y = 0.0208*x^{2.1216}$	0.8908	USA, CA
Model 2	$y = 0.49617*x^{1.22812}$	0.8689	KR, IND, IN, UK
Model 3	$y = 48.314\ln(x) - 128.18$	0.8743	JP, FR, CN

Therefore, the long-term trends between DWPC and GDP per capita during 1900 to 2020 are shown in Figure 3 for 10 countries. It should be noted that the DWPC shown in Figure 3 is adjusted data, which were derived from the urbanization. Consequently, the DWPC is somewhat higher or lower than the observed data.

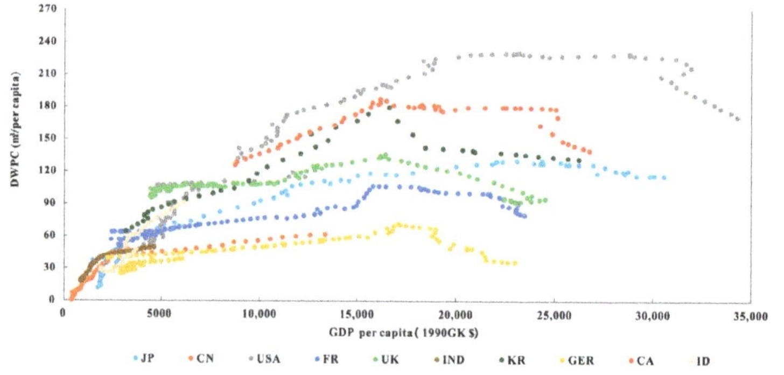

Figure 3. Relationship between GDP per capita and DWPC for 10 typical countries.

2.2. Mathematical Modeling of the Expanded S-Curve

The expanded S-curve model illustrating the relationship between DWPC and GDP per capita offers a tool to identify critical transitions from one stable state to another during economic development. A mathematical technique is employed to describe the expanded S-curve model. According to the expanded S-curve in previous studies [20,21,29], the relationship between DWPC (W) and GDP per capita (G) can be expressed as follows:

$$W - W_i = A\frac{\exp[\alpha_1(G - G_i)] - \exp(-\alpha_3(G - G_i))]}{2\cos h[\alpha_2(G - G_i)]} \quad (1)$$

where α_1, α_2, and α_3 are the exponential constraints, and A is the amplitude of the equation. W_i and G_i are the corresponding turning points on the expanded S-curve for DWPC and GDP per capita, respectively. Equation (1) is expressed as a hyperbolic tangent function.

Then, the linearity changes before the takeoff point, around the turning point, and after the zero-growth point are derived from Equation (1) as Equations (2)–(4).

$$W - W_i = A + A(\alpha_2 - \alpha_3)(G - G_i) = A + \rho_l(G - G_i) \tag{2}$$

$$W - W_i = 0.5A(\alpha_1 + \alpha_3)(G - G_i) = \rho_i(G - G_i) \tag{3}$$

$$W - W_i = A + A(\alpha_1 - \alpha_2)(G - G_i) = A + \rho_v(G - G_i) \tag{4}$$

where ρ_l, ρ_i, and ρ_v are the slopes of the curve before the takeoff point, around the turning point, and after the zero-growth point, respectively. They can be calculated from the systems of Equations (5)–(7):

$$\alpha_1 = \frac{\rho_l + 2\rho_i + \rho_v}{2A} \tag{5}$$

$$\alpha_2 = \frac{\rho_l + 2\rho_i - \rho_v}{2A} \tag{6}$$

$$\alpha_3 = \frac{-\rho_{l+}2\rho_i - \rho_v}{2A} \tag{7}$$

Equation (1) has a first-order partial derived from 0 at the zero-point of the S-curve:

$$\tanh[\alpha_1(G_v - G_i)]\tan\mathrm{h}[\alpha_2(G_V - G_i)] = \alpha_1\alpha_2^{-1}, \frac{dW}{dG} = 0 \tag{8}$$

By substituting Equations (5) and (6) into Equation (8), Equation (9) can be obtained

$$\tanh\left(\varphi_1 A^{-1}\right)\tanh\left(\varphi_2 A^{-1}\right) = \varphi_3 \tag{9}$$

where

$$\varphi_1 = 0.5(\rho_l + \rho_i + \rho_v)(G_v - G_i) \tag{10}$$

$$\varphi_2 = 0.5(\rho_l + 2\rho_i - \rho_v)(G_v - G_i) \tag{11}$$

$$\varphi_3 = \frac{\rho_l + \rho_i + \rho_v}{\rho_l + 2\rho_i - \rho_v} \tag{12}$$

In summary, the W_i, G_i, ρ_l, ρ_i, and ρ_v were from research data, and the A, α_1, α_2, and α_3 were from the equations.

3. Results

3.1. Expanded S-Curve in Typical Developed Countries

According to the universal equation, the expanded S-curve equations of the DWPC were established for four typical countries as Equations (13)–(16), which were the US, Japan, UK, and France, respectively.

(1) US

$$W = 170 + 181 \times \frac{\exp[0.000311 \times (G - 11{,}500)] - \exp[-0.0000872(G - 11{,}500)]}{2\cosh[0.000088(G - 11{,}500)]} \tag{13}$$

(2) Japan

$$W = 95 + 90 \times \frac{\exp[0.0000338 \times (G - 11{,}000)] - \exp[-0.0000668(G - 11{,}000)]}{2\cosh[0.0000673(G - 11{,}000)]} \tag{14}$$

(3) UK

$$W = 120 + 30 \times \frac{\exp[0.000395 \times (G - 13{,}000)] - \exp[-0.000311(G - 13{,}000)]}{2\cosh[0.000395(G - 13{,}000)]} \tag{15}$$

(4) France

$$W = 90 + 35 \times \frac{\exp[0.00000763 \times (G - 12{,}000)] - \exp[-0.000825(G - 12{,}000)]}{2\cosh[0.000178(G - 12{,}000)]} \quad (16)$$

Figure 4 gives the expanded S-curve simulation for these four typical countries.

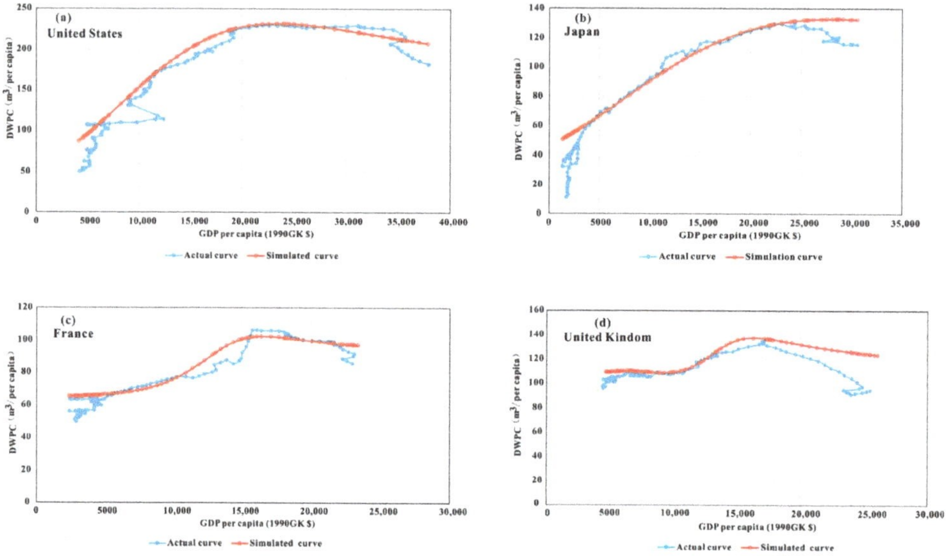

Figure 4. Expanded S-curve simulation of DWPC and GDP for (**a**) the United States (**b**) Japan, (**c**) France, and (**d**) the United Kingdom.

The changing trajectories of DWPC with GDP per capita in these countries generally experienced three stages.

For the US in Figure 4a, the first stage was before the GDP per capita of USD 6500 in 1930 during Great Depression, and its economy sequence entered a special period with a winding curve until World War II. The DWPC maintained a flat trend during the first stage. In the second stage, the DWPC in the US kept growing to a high level of 230 m^3 in 1990, with a GDP per capita of USD 20,000. After the GDP per capita of USD 35,000 in 2000, the DWPC started to decrease due to technology improvement, with the wide use of dishwashers and water-saving toilets. The efficiency improvements dramatically reduced the water usage.

Japan, in Figure 4b, showed a similar evolution pattern. The DWPC decreased distinctly during World War II in 1940s with a GDP per capita of around USD 2500, and it dropped to 32 m^3 with an annual decreasing rate of 7%. Then, with the post-war construction in 1950 with GDP per capita around USD 3500, the economic development model enabled Japan to enter a rapid development process of urbanization. From 1990 to 2000, after 40 years of linear growth, the DWPC in Japan peaked at approximately 120 m^3 with a GDP per capita of USD 20,000. After 2000, more efficient appliances and fixtures contributed to significant reductions in DWPC in Japan, with the same reduction trends in US.

France's and Britain's economic development was similar to that of US and Japan in the first stage, as shown in Figure 4c, d. They were stagnant for a long time after World War I and World War II before the 1950s. Meanwhile, the DWPC showed stationary trends until 1950. Then the DWPC in these two countries developed differently. For France, the DWPC showed no variation around 65 m^3 until 1970, with a GDP per capita of USD 11,000. Then it started to experience a slight increase due to the soaring urbanization rate. In the third stage, the DWPC in France diminished from 106 m^3 after 2000 with a GDP per capita of USD 20,000 due to the application of water-saving machines on a large scale. The DWPC variations in UK were directly related to the evolution of water bureau management, which was further propelled by urbanization development. In the late 1960s with a GDP per capita above USD 10,000, the DWPC increased from 113 m^3 with the increasing urbanization rate and population, and the management framework of water in the UK was optimized to improve the water efficiency. So, from 1970 to 1990, with a GDP per capita of between USD 11,000 and USD 18,000, an increase in the DWPC of between 115 m^3 to 138 m^3 was evident in the UK. In 1980s, with the stagflation in economics, the UK government published the Government White Paper on Privatization of Water Industry in 1986 and the top 10 water industries in the UK completed the privatization in 1989, which led to an increase in water prices and a decrease in DWPC in the 1990s with a GDP per capita of above USD 18,000 [30].

3.2. Implication of the Expanded S-Curve Model

According to the correlation analysis of the increase in DWPC and GDP per capita in Sections 2.2 and 3.1, we can conclude that the expanded S-curve can be calibrated with three key points, which are the takeoff point, the turning point, and the zero-growth point. Meanwhile, the long-term DWPC trends with GDP per capital were also divided into four stages according to the growth rate transition, including slow growth, accelerated growth, decelerated growth, and zero/negative growth. Figure 5 shows the key points and stages of the S-curve, and the points for each country are summarized in Table 2.

Table 2. Key points of expanded S-curve for each country (1990 GK in USD).

Country	Takeoff Points GDP per Capita	Turning Points GDP per Capita	Zero-Growth Points GDP per Capita
US	4000–4200	10,100–11,500	20,000–22,000
UK	4500–4800	12,000–13,000	17,000–18,000
France	3500–3800	12,000–13,000	19,000–20,500
Japan	3000–3500	10,000–11,000	22,000–23,000
China	2000–2500	11,000–12,000	–
Germany	3200–3800	11,000–12,000	18,000–19,000
India	1500–1800	–	–
Indonesia	2300–2500	–	–
Canada	–	–	19,000–21,000
South Korea	3000–4000	10,000–12,000	18,000–20,000

The takeoff point is the starting point for the accelerated growth in DWPC in the range of USD $1500–5000, implying an adjustment of agriculture society to industrial society with the economic boom. Before this takeoff point, the DWPC was in the slow-growth section. The takeoff points for developed economics, such as the UK, the USA, France, and South Korea, occurred after USD 3000, whereas for developing economies, such as India and China, it occurred between USD 1500 and USD 2500. The turning point made an adjustment period of an industrial structure in the process of industrialization, with a GDP of USD 10,000–USD 13,000 for the researched countries without diversity. After the turning point, the growth rate in the DWPC transited from accelerated growth to decelerated growth until the zero-growth point. The zero-growth points were concentrated around USD 17,000–USD 22,000, which indicates that the DWPC entered a zero-growth or slow-decline stage. This is also consistent with the post-industrial stage [20], when the living standards were improved substantially to promote the technical progress of water-saving facilities.

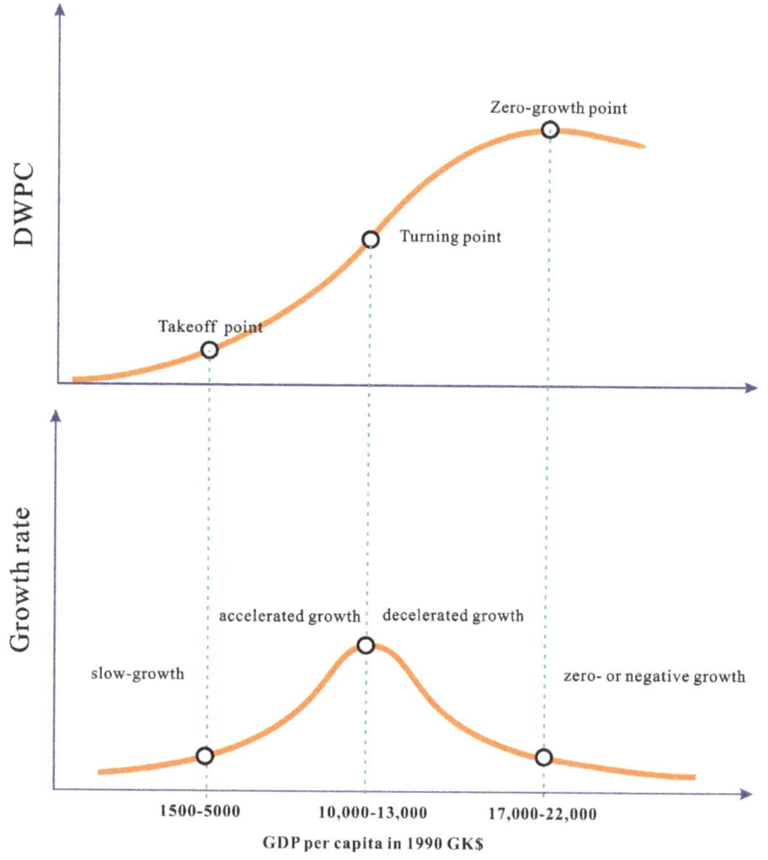

Figure 5. Three key points and four stages with different growth rates of the expanded S-curve.

4. Causes for the Changes in Domestic Water Usage

The expanded S-curve model describes the effect of economics on domestic water usage; however, domestic water-usage changes also have a close connection with industrial structure, urbanization, and scientific–technical progress, which were also promoted by economics. The relationship between tertiary industry proportion, urbanization, DWPC, and GDP per capita for these 10 typical countries is summarized in Figure 6. The tertiary industry proportion in Figure 6a and urbanization in Figure 6b both present a similar regularity with GDP per capita, with accelerate growth before a GDP per capita of USD 10,000, decelerated growth between a GDP per capita of USD 10,000 and USD 13,000, and peak value around a GDP per capita of USD 20,000, which is consistent with the DWPC in Figure 6c.

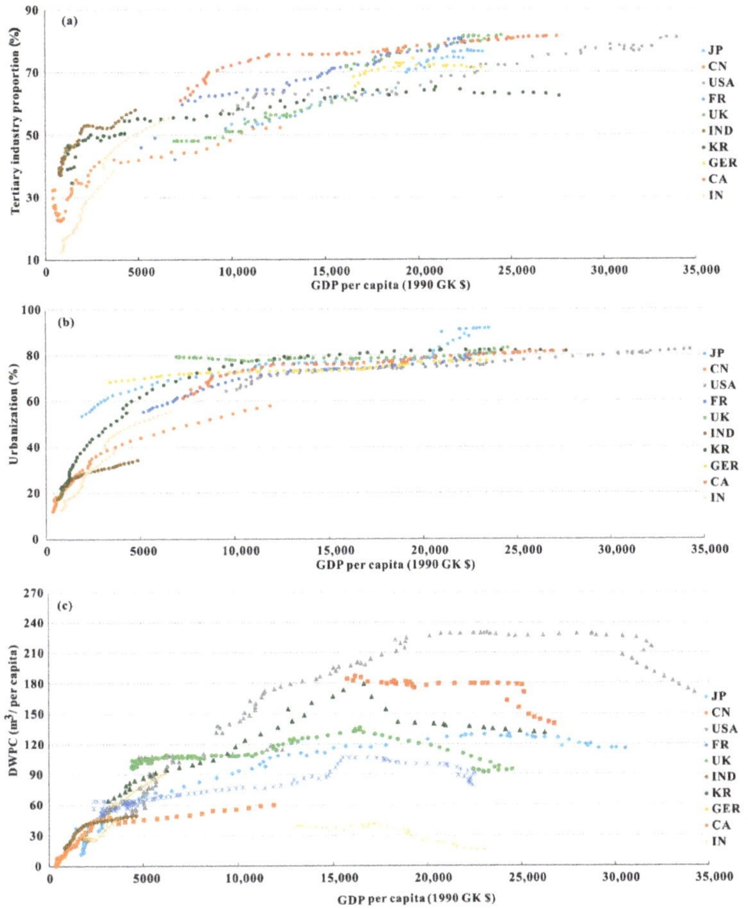

Figure 6. Relationships between GDP per capita and (**a**) tertiary proportion, (**b**) urbanization, and (**c**) tertiary proportion of GDP.

In order to clearly characterize the relationship between these factors, a schematic diagram was established in Figure 7. The three key points of the expanded S-curve were annotated, and the corresponding turning points of tertiary industry proportion and urbanization were also calibrated to compare the corresponding relations.

4.1. Urbanization

The urbanization rate represents the population structure, which has a significant effect on the DWPC. According to previous study, water users in a more urban population in the first trended toward a more water-intensive lifestyle with increasing urbanization rate [13,31,32]. Chen et al. tested the urbanization factor in promoting the DWPC by LMDI [33,34]. The urbanization rate and DWPC showed a positive relationship with conformal key points, as seen in Figures 1 and 7. The urbanization rate can be divided into three stages: It rapidly grew from an urbanization of 20% to 70% before GDP per capita reaches USD 10,000, and then the growth rate decelerated until urbanization reached 80% and the GDP per capita reaches USD 20,000. After this stage, the urbanization was generally saturated, with a stable trend. The turning point for DWPC also occurred between USD

10,000 and USD 13,000, as seen in Figure 7, which is consistent with urbanization. Therefore, we can arrive at the conclusion that urbanization could accelerate domestic water usage.

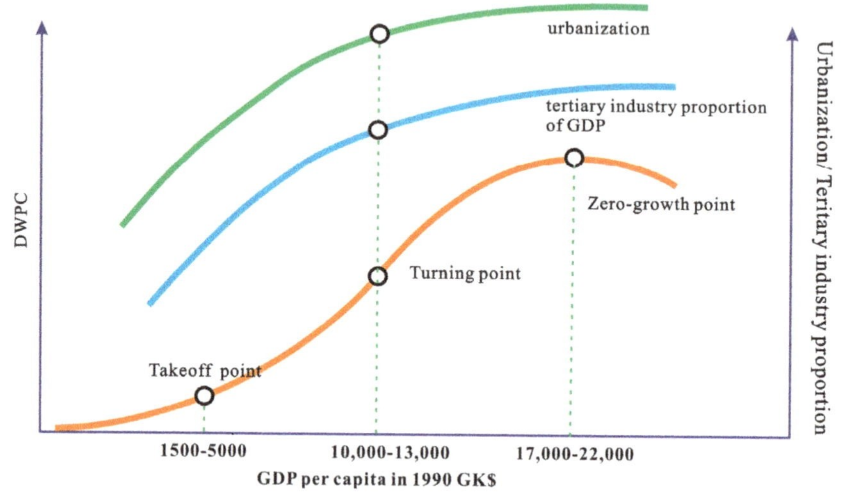

Figure 7. Relationship between key points of expanded S-curve and important indicators of economic and social development.

4.2. Industrial Structure

Domestic water is used in the tertiary industry for urban households, rural households, and commercial service [5,14,24]. As the tertiary industry proportion increases, the urban infrastructure and commercial service would be enhanced with economic development, resulting in an increasing trend in the DWPC. During the industrialization and post-industrialized era, the tertiary industry proportion showed a linear increasing shape. This rapid increasing trend discontinued until a GDP per capita of USD 10,000. Then, it entered a slowly increasing stage between a GDP per capita of USD 10,000 and USD 15,000. The peak of the tertiary industry proportion was around USD 20,000, coinciding with the zero-growth point of the expanded S-curve of DWPC in Figures 6 and 7.

4.3. Technical Progress

The technical progress was usually promoted by the economic development, and it can be considered a main driver for water-usage change [35]. Figure 8 shows the indoor DWPC subsectors for the US and Japan. The indoor DWPC can be divided into five parts, and showers accounted for the greatest share for the US in Figure 8a and toilets account for the greatest share for Japan in Figure 8b. Besides, the shares for kitchen (mainly for dishwashers), clothes washers, and others were also substantial. Between 1990 and 2016, there was a statistically significant reduction in DWPC for toilets in both the US and Japan. According to previous studies, the declines are easy to understand due to the wide use of water-saving toilets. In addition, the increasing the efficiency of fixtures and appliances significantly reduced the amounts of clothes washers and kitchen use. Therefore, the reduction of DWPC after 2000 for most causes could be partly attributed to technological progress.

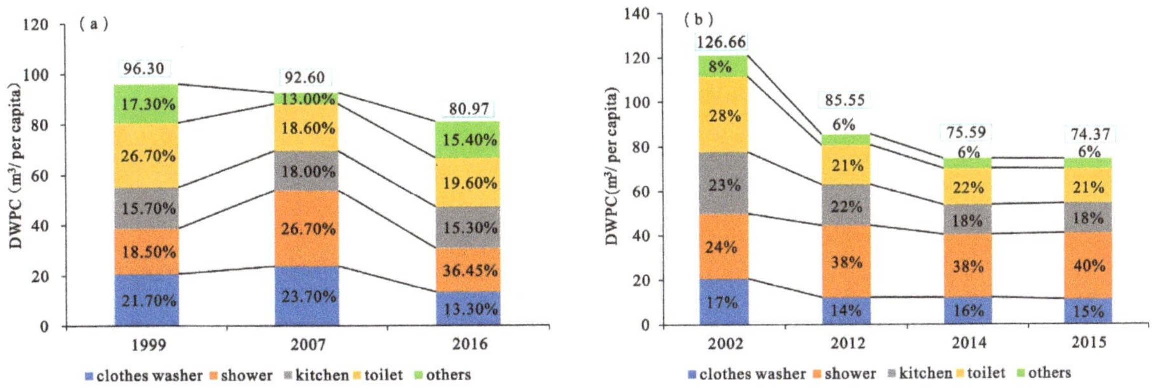

Figure 8. DWPC subsector proportions in (**a**) the US and (**b**) Japan.

5. Conclusions

In this paper, we focused on the evaluation of long-term domestic water-usage trends with economic development from typical countries. The relationship between domestic water usage and socio-economic developments on a country scale for the time period from 1900 to 2020 in 10 typical countries was demonstrated. The simulation results show that with the growth in GDP per capita, domestic water per capita showed an expanded S-curve of 'slow growth–rapid growth–zero growth, or even negative growth, with three key points, which were the takeoff point, turning point, and zero-growth point, respectively. The takeoff point of the expanded S-curve was located at a GDP per capita of USD 1500–USD 5000, according to the different development levels. The turning point was located at a GDP per capita of USD 10,000–USD 13,000, and the zero-growth point was concentrated around a GDP per capita of USD 17,000–USD 22,000, consistent with the post-industrial stage. Besides, the urbanization was proven to accelerate domestic water usage, and the higher tertiary industry proportion of GDP enhanced the domestic water-usage trends. The decreased water usage was attributed to technological progress, with widely used water-saving appliances.

The results of this research show that the expanded S-curve is applicable to the relationship between domestic water usage and economic development on a country scale. We hope this conclusion can contribute to the development of future solutions and strategies for domestic water prediction in different economies or similar economies under different development stages. However, there is still a lot of uncharted territory of the application of the expanded S-curve models. In this paper, only four of 10 typical countries were simulated in detail, and studies on a series of countries with few water usage data are still deficient. We will apply this expanded S-curve model to other countries in our future research, and hope that this model will encourage the efficient and sustainable management of water resources.

Author Contributions: Conceptualization, X.G. and A.W.; methodology X.G. and G.L.; validation and data curation, X.G. and B.D.; writing—original draft preparation, X.G. All authors have read and agreed to the published version of the manuscript.

Funding: This research was funded by the National Natural Science Foundation of China, grant numbers 72088101, 71991485, and 71991480. The APC was funded by grant number 72088101.

Institutional Review Board Statement: Not applicable.

Informed Consent Statement: Not applicable.

Data Availability Statement: Not applicable.

Conflicts of Interest: The authors declare no conflict of interest.

References

1. Wada, Y.; de Graaf, I.E.M.; van Beek, L.P.H. High-resolution modeling of human and climate impacts on global water resources. *J. Adv. Model. Earth Syst.* **2016**, *8*, 735–763. [CrossRef]
2. Haddeland, I.; Heinke, J.; Biemans, H.; Eisner, S.; Flörke, M.; Hanasaki, N.; Konzmann, M.; Ludwig, F.; Masaki, Y.; Schewe, J.; et al. Global water resources affected by human interventions and climate change. *Proc. Natl. Acad. Sci. USA* **2014**, *111*, 3251–3256. [CrossRef] [PubMed]
3. Owen, A.; Scott, K.; Barrett, J. Identifying critical supply chains and final products: An input-output approach to exploring the energy-water-food nexus. *Appl. Energy* **2018**, *210*, 632–642. [CrossRef]
4. Xiao, Z.; Yao, M.; Tang, X.; Sun, L. Identifying critical supply chains: An input-output analysis for Food-Energy Water Nexus in China. *Ecol. Model.* **2019**, *392*, 31–37. [CrossRef]
5. FAO (Food and Agriculture Organization of the United Nations). AQUASTAT. 2020. Available online: https://www.fao.org/statistics/zh (accessed on 20 December 2021).
6. The OECD Environmental Outlook to 2050: The Consequences of Inaction. Available online: https://www.oecd.org/g20/topics/energy-environment-green-growth/oecdenvironmentaloutlookto2050theconsequencesofinaction.htm (accessed on 25 February 2022).
7. Zhou, F.; Bo, Y.; Ciais, P.; Dumas, P.; Tang, Q.; Wang, X.; Liu, J.; Zheng, C.; Polcher, J.; Yin, Z.; et al. Deceleration of China's human water use and its key drivers. *Proc. Natl. Acad. Sci. USA* **2020**, *117*, 7702–7711. [CrossRef] [PubMed]
8. Gam, I.; Rejeb, B.J. Micro-economic analysis of domestic water demand: Application of the pseudo-panel approach. *Environ. Chall.* **2021**, *4*, 110118. [CrossRef]
9. Manouseli, D.; Anderson, B.; Nagarajan, M. Domestic Water Demand During Droughts in Temperate Climates: Synthesising Evidence for an Integrated Framework. *Water Resour. Manag.* **2018**, *32*, 433–447. [CrossRef]
10. Wang, X.J.; Zhang, J.Y.; Shahid, S.; Xie, W.; Du, C.Y. Modeling domestic water demand in Huaihe River Basin of China under climate change and population dynamics. *Environ. Dev. Sustain.* **2018**, *20*, 911–924. [CrossRef]
11. Meng, X.M.; Tu, L.P.; Yan, C.; Wu, L.F. Forecast of annual water consumption in 31 regions of China considering GDP and population. *Sustain. Pro. Consum.* **2021**, *21*, 713–736.
12. Marta, S.V. Modeling residential water demand: An approach based on household demand systems. *J. Environ. Manag.* **2020**, *261*, 109921.
13. David, B.L.; Bogaart, P.W.; Kram, T.; Vries, B. Long-term water demand for electricity, industry and households. *Environ. Sci. Policy* **2016**, *55*, 75–86.
14. World Bank Group. 2022. Available online: https://data.worldbank.org.cn/indicator/NY.GDP.MKTP.KD (accessed on 24 February 2022).
15. Guo, W.Y.; Liu, T.; Dai, F.; Xu, P. An improved whale optimization algorithm for forecasting water resources demand. *Appl. Soft Comput.* **2020**, *86*, 105925. [CrossRef]
16. Zhang, W.S.; Zhou, L.F.; Cheng, Q. Eco-environmental Demand for Water in Wetlands at Liaohe Estuary. *J. Irrig. Drain.* **2017**, *36*, 122–130.
17. Integrated Assessment of Global Environmental Change with IMAGE 3.0: Model Description and Policy Applications. Available online: https://www.pbl.nl/en/publications/integrated-assessment-of-global-environmental-change-with-IMAGE-3.0PBL (accessed on 11 December 2021).
18. Rathnayaka, K.; Malano, H.; Arora, M.; Georgeb, B.; Maheepalac, S. Prediction of urban residential end-use water demands by integrating known and unknown water demand drivers at multiple scales. *Resour. Conserv. Recycl.* **2017**, *118*, 1–12. [CrossRef]
19. Verhulst, P.F. Notice sur la loi que la population suit dansson accroissement: Correspondence. *Math. Phys.* **1983**, *10*, 113–121.
20. Wang, A.J.; Wang, G.S.; Zhou, F.Y. The Limits and Cycles of the Growth of Energy and Mineral Resources Consumption. *Acta Geosic. Sin.* **2017**, *38*, 1–10.
21. Gao, X.R.; Wang, A.J.; Liu, G.W.; Liu, C.H.; Yan, K. Expanded S-Curve Model of a Relationship Between Crude Steel Consumption and Economic Development: Empiricism from Case Studies of Developed Economies. *Nat. Resour. Res.* **2019**, *28*, 547–562. [CrossRef]
22. Zuo, Q.T. Interval S-model for forecasting per capita domestic water consumption. *J. Hydrau. Eng.* **2008**, *39*, 351–354.
23. Florke, M.; Kynast, E.; Barlund, I.; Eisner, S.; Wimmer, F. Domestic and industrial water uses of the past 60 years as a mirror of socio-economic development: A global simulation study. *Global Environ. Change* **2013**, *23*, 144–156. [CrossRef]
24. USGS (United States Geological Survey). 2021. Available online: https://www.usgs.gov/mission-areas/water-resources/science/water-use-united-states?qt-science--center--objects=0#qt-science--center--objects (accessed on 22 February 2022).
25. European Statics. 2022. Available online: https://ec.europa.eu/eurostat (accessed on 21 February 2022).
26. German Association of Energy and Water Industries. Available online: https://www.eureau.org/about/members/germany-bdew (accessed on 18 December 2021).
27. Gleick, P.H.; Palaniappan, M. Peak water limits to freshwater withdrawals and use. *Proc. Natl. Acad. Sci. USA* **2010**, *107*, 11155–11162. [CrossRef]

28. Shiklomanov, I.A.; Rodda, J.C. *World Water Resources at the Beginning of the 21st Century;* International Hydrology Series; Cambridge University Press: Cambridge, UK, 2003.
29. Wang, A.; Wang, G.; Chen, Q.; Yu, W.; Yan, K.; Yang, H. S-curve model of relationship between energy consumption and economic development. *Nat. Resour. Res.* **2015**, *24*, 53–64. [CrossRef]
30. *The Development of the Water Industry in England and Wales;* Ofwat: Birmingham, UK, 2006.
31. He, W.; Wang, Y.L. Calculation of urban water resources utilization efficiency in the Yellow River basin and analysis of its influencing factors. *Acta Sci. Circ.* **2021**, *41*, 4760–4770.
32. Vieira, P.; Jorge, C.; Cova, D. Assessment of household water use efficiency using performance. *Resour. Conserv. Recycl.* **2017**, *116*, 94–106. [CrossRef]
33. Chen, L.; Qiao, C.H.; Xia, L.L.; Cai, Y.P.; Yang, Z.F. Changes of domestic water consumption and its driving mechanism in the period of new urbanization in Guangdong. *South-North Water Transf. Water Sci. Technol.* **2021**, *19*, 273–280.
34. Zhang, Z.; Liu, J.; Cai, B.; Shan, Y.; Zheng, H.; Li, X.; Li, X.; Guan, D. City-level water withdrawal in China: Accounting methodology and applications. *J. Ind. Ecol.* **2020**, *24*, 951–964. [CrossRef]
35. Residential End Uses of Water, Version 2. Available online: https://www.waterrf.org/resource/residential-end-uses-water-version-2 (accessed on 22 February 2022).

applied sciences

MDPI

Article

Prediction of the Long-Term Performance Based on the Seepage-Stress-Damage Coupling Theory: A Case in South-to-North Water Diversion Project in China

Xinyong Xu [1,2,*], Wenjie Xu [1], Chenlong Xie [3] and Mohd Yawar Ali Khan [4]

1 School of Water Conservancy, North China University of Water Resources and Electric Power, Zhengzhou 450045, China; hsxuwenjie@163.com
2 Collaborative Innovation Center of Water Resources Efficient Utilization and Protection Engineering, Zhengzhou 450045, China
3 Shanghai Investigation, Design and Research Institute Co., Ltd., Shanghai 200335, China; xiechenlong@sidri.com
4 Department of Hydrogeology, Faculty of Earth Sciences, King Abdulaziz University, Jeddah 21589, Saudi Arabia; makhan7@kau.edu.sa
* Correspondence: xuxinyong@ncwu.edu.cn

Abstract: The South-to-North Water Diversion Project has been in operation since 2014, directly benefiting more than 79 million people in China. Thus, its service life and long-term performance have gained much attention from scholars. To predict its life and performance, this study used the seepage/stress-damage coupling method. In addition, a seepage/stress-damage coupling theory was proposed and a finite element model of a deep excavated canal in the Xichuan Section of the South-to-North Water Diversion Project was established. The results showed that this canal subsided greatly in the first two years of operation, which can be confirmed by the monitoring data. It is predicted that, after 50 years of normal operation, the canal damage may start and spread from the water level, and reach 37.6%, but such damage will not affect its normal water delivery function. The purpose of this study is to provide guidance for the safe operation of the project.

Keywords: settlement; damage evolution; seepage/stress-damage method; data monitoring

Citation: Xu, X.; Xu, W.; Xie, C.; Khan, M.Y.A. Prediction of the Long-Term Performance Based on the Seepage-Stress-Damage Coupling Theory: A Case in South-to-North Water Diversion Project in China. *Appl. Sci.* **2021**, *11*, 11413. https://doi.org/10.3390/app112311413

Academic Editor: Amit Kumar

Received: 9 October 2021
Accepted: 29 November 2021
Published: 2 December 2021

1. Introduction

The South-to-North Water Diversion Project (SNWDP) aims to optimize the temporal and spatial allocation of water resources in China. As a national strategic project, it safeguards China's land management and sustainable development. Canal engineering is an integral part of SNWDP, and its seepage failure involves complicated hydraulic problems, particularly in some deep excavated sections, due to the high groundwater level, complex geological conditions, soil consolidation and deformation, and rainfall or channel infiltration [1]. The seepage–stress coupling may occur between the concrete lining and the foundation, damaging the lining plate. If the damage persists, the water from the canal will seep into the soil of the canal more quickly, altering the seepage field and causing structural damage between the soil of the canal, the concrete lining, and the seepage field [2]. Therefore, scholars at home and abroad are all concerned about the SNWDP's service life and performance evolution in long-term operation, for it matters to water delivery safety and further affects the people's living conditions, social and economic development, and environmental protection [3]. The Xichuan Section is the first section of the main channel of the SNWDP Middle Route, classified into Class I project according to the engineering grade, so its safe running is of great significance.

A seepage/stress-damage (SSD) coupling theory was proposed, and a finite element model of a deep excavated canal in the SNWDP Xichuan Section was established in the same scale as its actual design drawings. Even the materials used and the surrounding environment were the same as the actual situation. First, the SSD coupling theory proposed in this paper needs a stable seepage field, on which a lot of research has been performed. For example, Cai et al. [4] used the strength reduction method to establish a seepage–stress coupled numerical model. They discussed the impact of various factors on the slope stability under different working conditions and concluded that the groundwater and rainfall infiltration have the most considerable effect on slope stability. In southern Jiangxi province, Pan et al. [5] investigated the failure modes of granite residual soil slopes and employed normal soil material parameters to analyze precipitation infiltration under three operating situations. They discovered that the wet front's depth and slope coefficient have varied over time. Zhou et al. [6] studied the spatial–temporal characteristics of water movement on fractured soil slopes under rainfall conditions. They investigated the mechanism of fractured slope instability as a function of soil saturation variations and discovered that matrix suction is the primary driver of overall instability. Huang et al. [7] analyzed the stability of hydraulic landslides with different permeability coefficients under fluctuating reservoir water levels and rainstorm conditions. They found that due to heavy rains, the stability of the landslide was considerably reduced, and the coefficient of stability increased with increased permeability. Kim et al. [8] analyzed and compared water pressure and pore pressure data from hydraulic wells to observe the influence of seepage changes. Luo et al. [9]. analyze the sudden pipeline crash of a specific project and derive its evolution process. Zhao et al. [10] proposed an innovative permeability evolution equation. They found that the seepage pressure will continuously intensify fracture propagation and penetration in the rock mass due to the time effect of permeability and failure. Nian et al. [11] used pore pressure as a controlled condition to analyze the rainfall infiltration and seepage on slopes under different rainfall intensities. They obtained the relationship between the rainfall intensity and the actual infiltration rate. The above research results showed that the permeability coefficient will change with time, rainfall intensity and other external factors, which can be verified by the SSD theory (taking the permeability coefficient as a variable) in this study.

Second, the SSD coupling method used in this paper needs the coupling of the seepage field and the stress field, which is mainly achieved by solving the seepage field and converting it into an equivalent load acting on the model nodes. Many research on the seepage–stress coupling has been conducted at national and global scales. For example, Wang et al. [12] established a theoretical model of micro-fracture grouting seepage based on the fluid–solid coupling between grout seepage and micro-fractures. They studied the fracturing conditions, fractures' spatial distribution, and the variation law of mud seepage distance. Through analysis and comparison, Ma et al. [13] obtained the failure mode and seepage characteristics of unloaded rock with and without water pressure. Ma et al. [14] used finite-difference to analyze the influence of saturated or unsaturated seepage on the slope stability. They obtained the influence of the flow rate on the stability of the slope. Liu et al. [15] analyzed that the external water pressure of the tunnel is related to the basement and seepage of the basement and is the main influencing factor related to safety. Xiao [16] used the neural network method to develop the program to combine the seismic load effect and the fluid–solid coupling effect. He analyzed the seepage stability of the earth dam and determined the dangerous sliding surface of the dam slope. Chen [17] gave a method of calculating the safety factor of slope stability considering seepage conditions based on the law of seepage–stress coupling evolution. He estimated that the effect of seepage has a significant impact on slope stability. Cai et al. [18] evaluated the slope stability under rainfall infiltration conditions based on the shear strength reduction technology. They considered the non-coupled conditions of seepage and deformation, combined with statistics and observation methods. Rahardjo et al. [19] studied the factors influencing slope stability under rainfall infiltration conditions. They found that slope instability mainly depends on the rainfall intensity and the nature of the soil, as well as slope type and groundwater level. Baum et al. [20] established saturated and unsaturated transient rainfall infiltration models. Based on correlated groundwater transients, unsaturated infiltration analysis, and groundwater pressure diffusion, the models predicted the time and main source areas of landslides caused by rainfall. Rahardjo et al. [21] considered different groundwater levels, rainfall intensity, and soil properties to analyze the stability of the residual soil slope under rainfall infiltration conditions. The results are in good agreement with the research trend in the parameter study. Muntohar et al. [22] analyzed the failure laws of shallow slopes under rainfall infiltration conditions based on the Green–Ampt infiltration model and the infinite slope stability model. The proposed model can be used to estimate the first-order approximation of the time when a rainfall-induced shallow landslide occurs and its sliding depth. Tsai et al. [23] compared the design plan with actual case data. They investigated the influence of unit weight and the function of unsaturated shear strength and saturation on shallow landslides triggered by rainfall infiltration. Borja et al. [24] established a finite element model that couples solid deformation with fluid pressure in unsaturated soil to evaluate slope stability. However, most of the above research was conducted by using the seepage method or the seepage–stress coupling method. Based on these results, the SSD coupling method was adopted in this paper.

Third, the SSD coupling method used in this paper correlates seepage–stress with damage to reflect the impact of damage on the seepage of concrete linings. The relationship between the seepage coefficient and damage was used as a bridge connecting the seepage field, stress field and damage field. Some scholars have also explored such methods. For example, Zhou [25] derived the permeability coefficient conversion equation taking into account the damage to the tunnel rock and depicted the SSD multi-field coupling model of the surrounding rock. He analyzed the stability of the surrounding rock excavated in the tunnel construction based on the fluid–solid coupling theory. Zhou et al. [26] established an SSD coupling algorithm based on the permeable lining theory and applied it to high-pressure hydraulic tunnels. Their results are consistent with the general engineering laws and provide a reference for solving practical engineering problems. Sheng et al. [27] believe that the influence of groundwater on slope stability cannot be replaced by pore water, and the synergy of the seepage field and the stress field must be considered in foundation pit slope engineering. Xu et al. [28] established the equation of relation between rock failure and coefficient of permeability based on damage variables and seepage–stress coupling. They described the evolutions of the rock-failure-based permeability and groundwater seepage field. They studied the evolutionary relationship between rock mass stress and strain, permeability and strain, strain and failure, as well as permeability and failure. Zhu et al. [29] coupled failure and fluid flow to the Mohr–Coulomb failure criterion, based on the dynamic evolution of damage, porosity and permeability, and proposed SSD models under the effect of hydraulic fracturing and natural fracturing based on the dynamic evolution of damage, porosity, and permeability. The results are very close to the engineering practices. The above research findings and the SSD coupling method used in this paper make the predictions more realistic.

Aiming at the above problems, this study used the SSD coupling method to predict the SNWDP's service life and performance evolution after long-term operation. Taking into account the change in the permeability coefficient induced by soil consolidation over time and the evolution of the infiltration field and its performance after long-term operation, it is necessary to discover the internal mechanism of the seepage failure and further explore the long-term changes in the performance of typical deep excavated sections of the SNWDP. The ultimate goal of this study is to predict the performance evolution of the deep excavated canal in the SNWDP Xichuan Section after long-term operation, providing a theoretical basis for the actual operation of the project.

2. Basic Theory and Realization Method of SSD Coupling

2.1. Basic Theory of SSD Coupling

Changes in the seepage–stress coupling environment can cause changes in the internal microstructure (i.e., meso-damage) [30], macro-mechanical properties, and permeability of concrete. Changes in permeability and mechanical properties can affect the concrete's stress state, the distribution of soil pore pressure and worsen the meso-damage of concrete. This phenomenon is called SSD coupling.

Conventional seepage–stress coupled governing equations include solid-based geometric equations and equilibrium equations, fluid-based mass conservation equations and flow equations, and seepage–stress coupled constitutive equations [31]. The SSD coupled governing equations can be obtained by introducing concrete damage variables into the conventional seepage–stress coupled equations. The following derivation techniques are explained by using direct tensor notation to simplify the theoretical formulations mathematically [32].

Assuming that the seepage process follows the nonlinear Darcy's law in the entire section, water and materials are incompressible, and the volumetric deformation of the saturated porous solid framework is equal to the deformation of the pores, then, the seepage field conforms to the continuity equation of three-dimensional single-phase porous fluid [33,34].

$$\frac{\partial}{\partial x}\left[k_x\frac{\partial H}{\partial x}\right]+\frac{\partial}{\partial y}\left[k_y\frac{\partial H}{\partial y}\right]+\frac{\partial}{\partial z}\left[k_z\frac{\partial H}{\partial z}\right]+Q=0 \tag{1}$$

where k_x, k_y and k_z are the permeability coefficients in the x, y, and z directions, respectively; hydraulic potential $H=\frac{p}{\gamma}$, in which p is the pore water pressure and γ is the water unit weight; z is the elevation head; and Q is the source sink term.

Assuming that concrete and rock masses are equivalent continuum models, then, after finite element discretization, interpolation, and integration, the matrix equation for solving the seepage field can be obtained as follows [35]:

$$[A]\{H\}=\{F\} \tag{2}$$

where $[A]$ is the total permeability matrix, $\{H\}$ is the column vector of the node head, and $\{F\}$ is the nodal load obtained by integrating the seepage boundary. After the seepage field is calculated, the water load generated by the hydraulic gradient acts on the inside of the structure in the form of seepage force. In the equivalent continuum model, the seepage gradient acts on the node in the form of seepage force. After the node head is obtained through the seepage field calculation, the seepage load acting on the element node is calculated as follows:

$$\{F_p\}=-\int\int_{\Omega}\int\gamma[N]^T\left\{\frac{\partial H}{\partial x},\frac{\partial H}{\partial y},\frac{\partial H}{\partial z}-1\right\}^T d\Omega \tag{3}$$

where $[N]$ is the interpolation function, Ω is the integral domain of the seepage force node and T is the transposition of a matrix.

The computational space domain is discretized to obtain the seepage–stress coupling equation:

$$[K][U]=\{F_V\}+\{F_s\}+\{F_p\}+\{F_{\sigma_0}\} \tag{4}$$

where $[K]$ is the structural stiffness matrix; $[U]$ is the nodal displacement matrix; F_V and F_s are the body and surface loads, respectively; F_p is the equivalent load formed by pore pressure; and F_{σ_0} is the initial stress load.

According to the incremental theory of plasticity in the plastic damage model, the total strain tensor, ε, is composed of the elastic strain rate, ε^{el}, and the equivalent plastic strain rate, ε^{pl}:

$$\varepsilon=\varepsilon^{el}+\varepsilon^{pl} \tag{5}$$

When there is no damage to the concrete, the stress–strain relationship of the concrete is as follows:

$$\sigma=D^{el}\left(\varepsilon-\varepsilon^{pl}\right) \tag{6}$$

where σ is the total stress, and D^{el} is the elastic stiffness matrix.

When the concrete material is damaged, according to the theory of continuum damage mechanics, the internal micro-cracks, micro-pores, and other micro-defects under the action of external loads can be described by the damage factor, d. The damage factor is mainly used to reflect the concrete stiffness degradation under uniaxial or multiaxial loads. Assuming that the damage is isotropic, then, the relationship between the damage and stress of the concrete under the three-dimensional multiaxial state can be expressed by the damage elasticity equation, and the concrete stress, σ, is calculated as follows [36]:

$$\sigma = (1-d)\overline{\sigma} = (1-d)D^{el}\left(\varepsilon - \varepsilon^{pl}\right) \tag{7}$$

where $\overline{\sigma}$ is the effective stress, which represents the stress on the net section of the concrete material.

The element damage, d, is expressed by the equivalent plastic strain $\widetilde{\varepsilon}^{pl}$:

$$\begin{cases} d_t = d_t\left(\widetilde{\varepsilon}_t^{pl}\right), 0 \le d_t \le 1 \\ d_c = d_c\left(\widetilde{\varepsilon}_c^{pl}\right), 0 \le d_c \le 1 \end{cases} \tag{8}$$

where d_t is the tensile damage factor, d_c is the compressive damage factor, t is the tensile state, and c is the compressive state.

The equivalent plastic strain $\widetilde{\varepsilon}^{pl}$ is calculated as follows:

$$\begin{cases} \widetilde{\varepsilon}_t^{pl} = \int_0^t \dot{\widetilde{\varepsilon}}_t^{pl} dt \\ \widetilde{\varepsilon}_c^{pl} = \int_0^t \dot{\widetilde{\varepsilon}}_c^{pl} dt \end{cases} \tag{9}$$

$$\begin{cases} \dot{\widetilde{\varepsilon}}_t^{pl} = r(\hat{\overline{\sigma}})\hat{\dot{\varepsilon}}_{max}^{pl} \\ \dot{\widetilde{\varepsilon}}_c^{pl} = -(1 - r(\hat{\overline{\sigma}}))\hat{\dot{\varepsilon}}_{min}^{pl} \end{cases} \tag{10}$$

where $\hat{\dot{\varepsilon}}_{max}^{pl}$ is the maximum value the plastic strain rate tensor, $\hat{\dot{\varepsilon}}_{min}^{pl}$ is the minimum value of the plastic strain rate tensor, $\dot{\widetilde{\varepsilon}}_t^{pl}$ is the equivalent plastic strain rate in tension, and $\dot{\widetilde{\varepsilon}}_c^{pl}$ is the equivalent plastic strain rate in compression. The multiaxial stress weighting factor $r(\hat{\overline{\sigma}})$ can be defined as follows:

$$r(\hat{\overline{\sigma}}) = \frac{\sum_{i=1}^{3}\langle\hat{\overline{\sigma}}_i\rangle}{\sum_{i=1}^{3}|\hat{\overline{\sigma}}_i|}, 0 \le r(\hat{\overline{\sigma}}) \le 1 \tag{11}$$

where $\hat{\overline{\sigma}}_i$ (i = 1, 2, 3) are the principal stress components, respectively, $\langle\cdot\rangle$ is defined as $\langle x\rangle = (|x| + x)/2$, and $|x|$ is the absolute value of x.

Under periodic alternating loads, the complex concrete damage mechanism is related to the cracking and merging of the initial cracks and their interrelation during changes. When the concrete is subject to compression after tension, its stiffness will be partially restored, that is, the unilateral effect is more significant. To reflect this effect, the relationship between tensile and compressive damage variables, d_t and d_c, is:

$$(1-d) = (1 - s_t d_c)(1 - s_c d_t) \tag{12}$$

where $0 \le s_t,\ s_c \le 1$, s_t, and s_c are the relational expressions after stiffness recovery.

$$\begin{cases} s_t = 1 - \omega_t r(\hat{\overline{\sigma}}), 0 \le \omega_t \le 1 \\ s_c = 1 - \omega_c[1 - r(\hat{\overline{\sigma}})], 0 \le \omega_c \le 1 \end{cases} \tag{13}$$

where ω_t and ω_c are the weighting factors of stiffness recovery related to the material properties. Figure 1 shows the stiffness recovery curve of the concrete damage model when the weighting factors are $\omega_t = 0$ (compression → tension) and $\omega_c = 1$ (tension → compression) under uniaxial alternating loads.

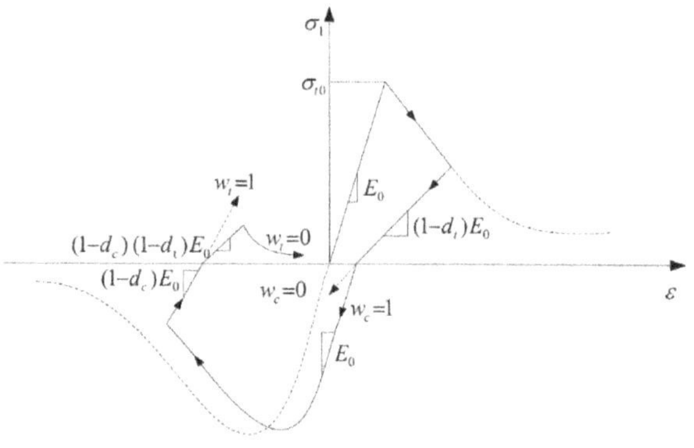

Figure 1. Stress–strain relation under the uniaxial alternating load.

When the plastic damage model is used, the damage may cause the degradation of the concrete structure's stiffness. Based on it, the influence of concrete damage and cracking on the stress state of the structure can be simulated. At the same time, the concrete damage and cracking have a significant impact on the permeability characteristics of the structure. The material element is composed of a damaged phase and an undamaged phase. The element permeability coefficient is calculated as follows [37]:

$$k = (1-d)k_m + dk_d\left(1 + \varepsilon_v{}^{pf}\right)^3 \tag{14}$$

where k_m is the permeability coefficient of the undamaged phase, and k_d is the permeability coefficient of the damaged phase. Assuming that no damage occurs in the case of elastic deformation, while plastic deformation and damage occur simultaneously, then, the plastic volumetric strain of the damaged phase is $\varepsilon_v{}^{pf} = d\varepsilon_v^p$, in which ε_v^p is the plastic volumetric strain of the element.

Once macroscopic cracks appear, brittle material's permeability will suddenly increase; thus, the sudden jump factor, ξ, is introduced to calculate the permeability coefficient of the damage phase [25]:

$$k_d = \xi k_m \tag{15}$$

where, for compression–shear damage, $\xi = 100$, and for tensile damage,

$$\xi = \begin{cases} 10, 0 < d \leq 0.1 \\ \frac{1000-10}{0.9-0.1}d + 10, 0.1 < d < 0.9 \\ 1000, 0.9 \leq d \leq 1 \end{cases}$$

Compared with conventional seepage–stress coupling models, the model in this paper couples the effect of damage and extends the study of the seepage–stress coupling problem from the simple stress state analysis to the damage process analysis, which lays a theoretical basis for further studying the concrete failure process and seepage evolution under seepage–stress coupling conditions.

2.2. SSD Coupling Method

The lining supports most of the water pressure from the deep excavated canal of the SNWDP [38]. In this paper, the constitutive elastoplastic relationship is used to simulate the canal lining, and it is based on the Mohr–Coulomb criterion; and the SSD coupling method is used to analyze and predict the long-term functioning of the lining of the canal.

The SSD coupling method used in this paper is based on the ABAQUS finite element software for secondary development [39]. The SSD coupling method used in this paper is based on the ABAQUS finite element software for secondary development. The damage was obtained via the FORTRAN language program, namely GETBRM. According to the damage curve and the permeability coefficient, the subprogram USDFLD (ABAQUS subprogram, which can define the constant variable on the material point as a time function) is used to update the canal permeability coefficient with damage changes. The element's permeability coefficient is defined as a field variable, and the subprogram was utilized in each incremental calculation to obtain the maximum principal strain and the equivalent plastic strain of the material integration point, thereby determining the element's stress state. The element's damage variable was solved. The lining's permeability coefficient was revised based on relevant information of the element and node to predict the long-term operation of the SNWDP more accurately. The established SSD coupling analysis process is shown in Figure 2.

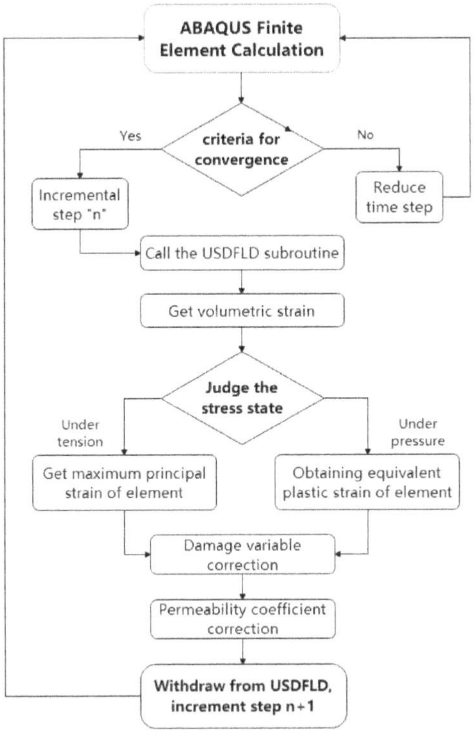

Figure 2. Flowchart of seepage/stress-damage coupling analysis.

3. Model Parameters and Boundary Conditions

3.1. Project Overview

There is a canal excavated with a depth of 36–47 m in the SNWDP Xichuan Section. The canal is located on the edge of the northern subtropical zone and also in the humid area. Affected by the monsoon climate all year round, it has four distinct seasons and abundant rainfall, with an average annual rainfall of more than 730 mm. In addition, this canal has a high groundwater level. Given that canals with high fills are more likely to undergo slope instability, this paper selects a typical section of the said deep canal in Xichuan Section for research. Its slope is reinforced by the combination of large-section excavation and water collection well. The canal consists of concrete lining plates, a geomembrane, sand–gravel cushions, and foundations. There are three-level bridleways on both banks, which can withstand vehicle loads, canal water pressure, groundwater pressure, and gravity.

As illustrated in Figure 3, a three-dimensional finite element numerical simulation model of a representative portion of a deep dug channel was created. It was accomplished by considering the central point of the canal bottom as the origin, the X-axis as the horizontal direction of the vertical water flow, the Y-axis as the direction parallel to the flow direction, and the Z-axis as the vertical direction the vertical water flow. The canal structure in the numerical model was discrete based on the C3D8RP (hexahedral reduced-integration) element. It had 185,000 elements and 208,098 nodes in total. The bottom of the finite element model was constrained fully, and the surrounding was constrained normally. The boundary conditions of the total head and the free seepage section were set. The monitored seepage flow of the canal was converted into the seepage velocity and set as the seepage velocity boundary condition.

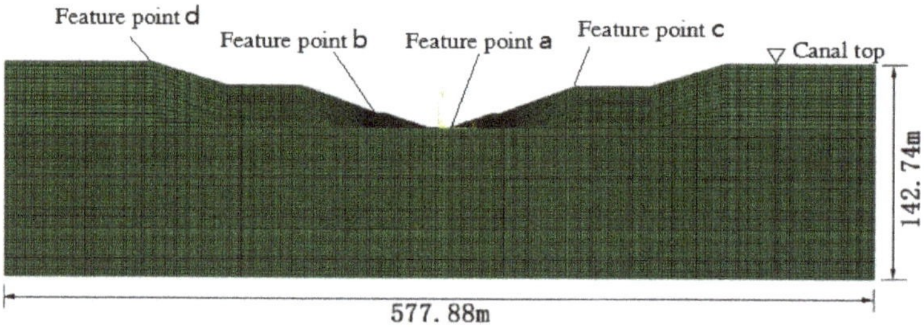

Figure 3. Three-dimensional finite element model of a typical section of the deep excavated canal.

3.2. Finite Element Model and Material Properties

This study chose a specific section of a high groundwater level for the numerical simulation analysis of the infiltration–stress coupling. A three-dimensional finite element numerical simulation model based on the drawings of the typical section design was established. Figure 4 depicts the canal's general details and the distribution of structural materials in each portion. Part of the model parameters are as follows: the canal bottom width is 13.5 m; the digging depth is 46 m; an 8 cm–thick C25 concrete slab is used as the lining plate, under which there is a composite geomembrane and then a 25 cm–thick coarse sand cushion, with the foundation at the bottommost; the designed water level is 8 m; the increased water level is 8.77 m; and the underground water level is 41.28 m.

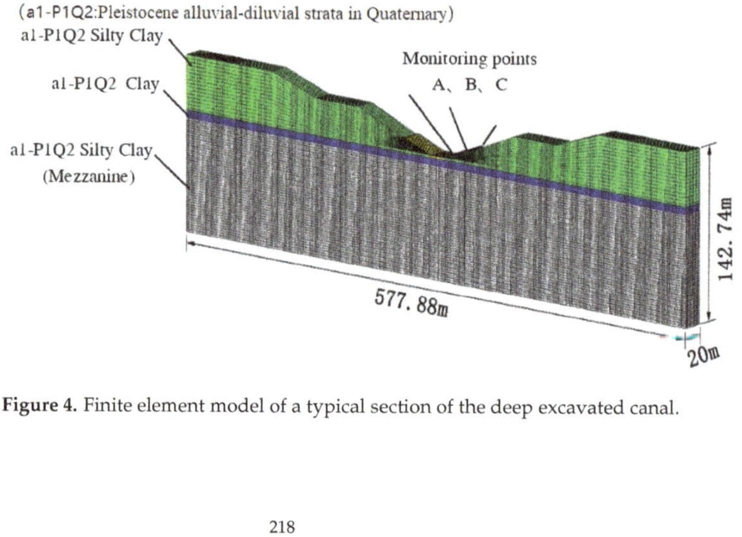

Figure 4. Finite element model of a typical section of the deep excavated canal.

Three monitoring points, A, B, and C, were selected in the deep excavated canal to monitor the canal's displacement and settlement. The clay materials are mainly used for the foundation and slope of the deep excavated canal section of the SNWDP. The anti-seepage system is mainly achieved by a concrete lining board, geomembrane, coarse sand cushion, and polysulfide sealant, as shown in Figure 5. During the actual operation of the SNWDP, the canal's infiltrated surface is subject to the continuous change of the permeability coefficient. Therefore, in this study, the characteristics of the water section were constantly assumed to ensure the continuity and accuracy of the results.

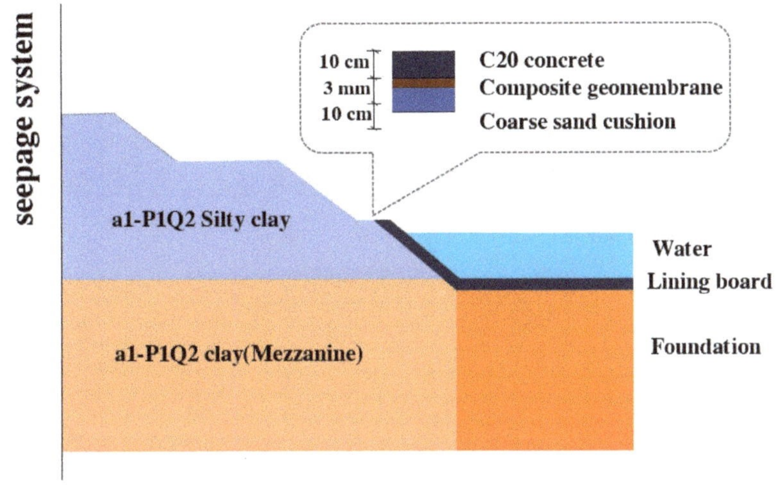

Figure 5. Canal seepage system.

4. Comparison between Monitored Data and Numerical Simulation

This section provides a concise and precise description of the experimental results, their interpretation, and the experimental conclusions that can be drawn.

Three monitoring points, A, B, and C, were selected on the canal, bottom, and slope, respectively (as shown in Figure 6a). Their safety-monitoring data from January 2014 to January 2018 were calculated and analyzed. Settlement monitoring points were used to conduct on-site surveys of the settlement displacement of the deep excavated section of the SNWDP (Figure 6b,c). The calculation results were compared and analyzed. Figure 7 shows the correlation curve between the simulated settlement and the monitored settlement of the monitoring points.

It can be seen from Figure 7 that, (1) during the 5-year operation of the deep excavated canal, the displacement gradually increased from the bottom to the top of the canal, reaching the highest at the top. The maximum difference between the monitored displacement and settlement value and the calculated value is 0.559 mm, and the minimum is 0.02 mm; (2) the canal subsided rapidly during the first two years, and then the settlement slowed down and got close to final settlement; and (3) compared to the displacement curve of the monitored section, the settlement trend is similar. According to the Adj.R-square coefficient and Pearson's correlation coefficient, the calculated data curve is highly fitted to the actually monitored data curve. After five years of operation, the settlements are the same, suggesting that the numerical simulation of the canal's long-term settlements is consistent with the actual project operation.

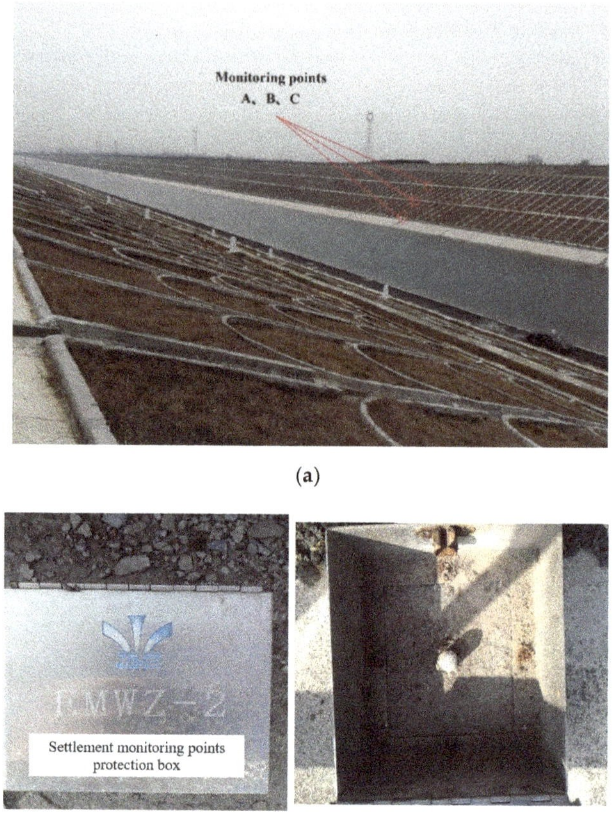

(a)

(b) (c)

Figure 6. Canal of the South-to-North Water Diversion Project and settlement monitoring points. (**a**) A section of the South-to-North Water Diversion Project, (**b**) Settlement monitoring points protection box, (**c**) Settlement monitoring points.

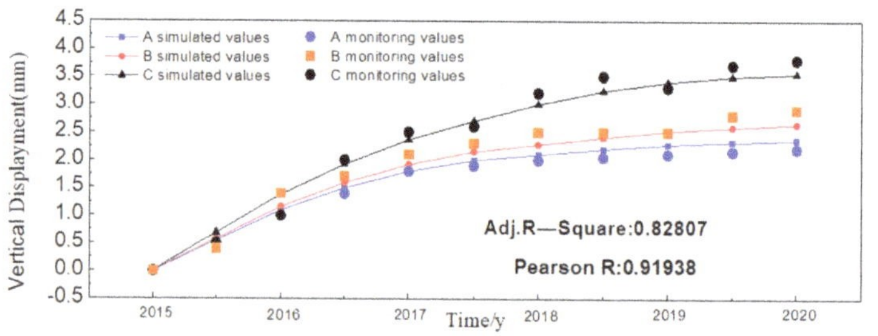

Figure 7. Correlation curve between the estimated settlement and the monitored settlement of the monitoring points.

5. Evolution of the Canal's Long-Term Behavior Based on SSD

5.1. Evolution of Canal Pore Pressure

Long-term seepage failure has a significant impact on the safe operation of the canal. Taking as an example the deep excavated canal in the section of the SNWDP, the long-term settlement of the canal and seepage field changes under the effect of coupling of seepage and stresses are calculated. Based on the above parameters and conditions, the ABAQUS software is used to estimate the deep excavated canal's long-term settlements and seepage field changes.

Figures 8 and 9 respectively show the saturation contours of the deep excavated canal after 10 and 20 years of operation. Here, the saturation is used to describe the moisture content of the foundation soil under the concrete lining plate. It can be seen from the figures that in the deep excavated canal, most of the soil is in a saturated state, and some are unsaturated, and the seepage effect is relatively large. By comparing the two contours, it can be seen that under given conditions, as the operating time of the canal increases, the saturated zone decreases and the unsaturated zone increases. It may be due to the fluctuation in the void ratio and the permeability coefficient produced by canal settlement. With the continuous settlement of the canal, the void ratio decreases, and the coefficient of permeability also decreases. In the macroscopic view, a change in the seepage field generates a change in the stress field, which shows a change in the settlement.

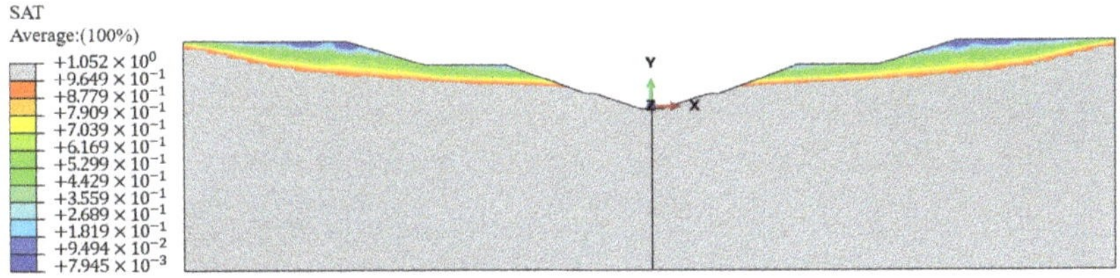

Figure 8. Canal saturation contour when t = 10 years.

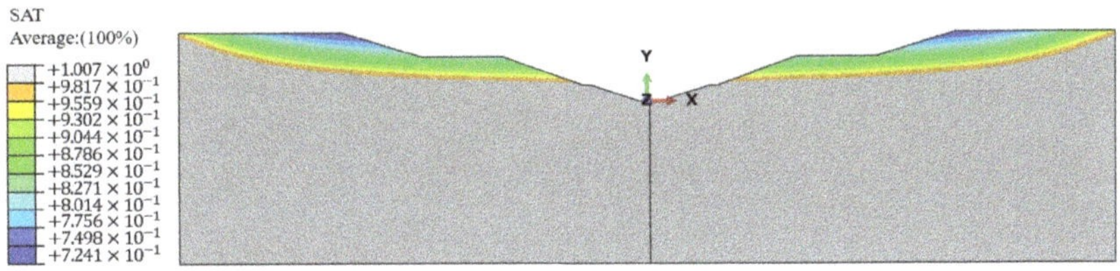

Figure 9. Canal saturation contour when t = 20 years.

Figures 10 and 11 respectively show the pore pressure contours of the deep excavated canal after ten years and 20 years of operation. It can be seen from the figures that there is negative pore pressure at the top of the canal, indicating the existence of unsaturated zones. The foundation exhibits both saturated seepage and unsaturated seepage, which is consistent with the numerical results of saturation. Pore pressure decreases on both sides towards the middle. Due to the high groundwater level on both sides of the deep excavated canal, the groundwater flows into the canal center under the action of gravity. Therefore, drainage measures should be taken on the slope of the canal to reduce the groundwater seepage and improve the security of the canal water supply.

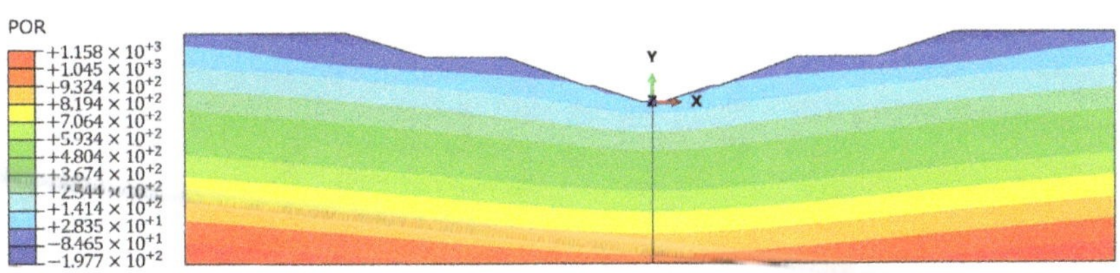

Figure 10. Canal pore pressure contour when t = 10 years.

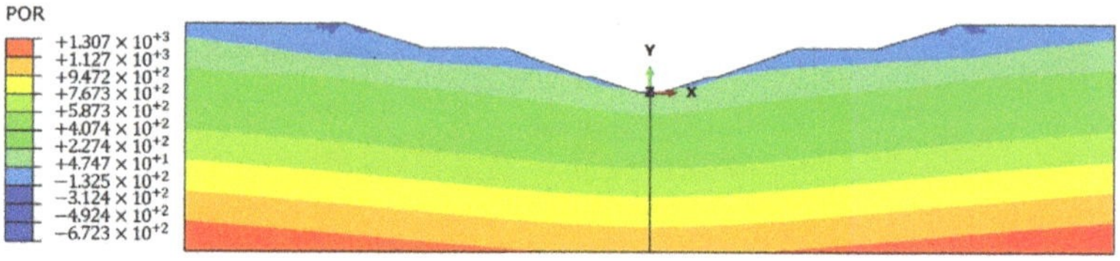

Figure 11. Canal pore pressure contour when t = 20 years.

5.2. Evolution of the Canal's Long-Term Settlements

In the calculation process, the design water level of the canal, the highest groundwater level of the slope and underground drainage measures are not applicable. The anti-seepage system is damaged, the seepage of the slope is stable, and the interior slope has no boundary flow. Based on the finite element model of the specific cross-section of the deep excavated canal, the characteristic points a, b, c, and d respectively were selected for settlement analysis at the canal bottom, embankment, slope, and top. The water level adopted the canal's design water level, the vehicle load on the first-level bridleway adopted the car-10 level load, the crowd load was 0.3 t/m^2, and the highest groundwater level was 180.218 m.

The SSD coupling method was used to compute the canal settlements after five years of operation based on the above working conditions. The nephogram of the canal's vertical displacements after a 5-year operation is shown in Figure 12, and the settlement of each distinctive point is shown in Figure 13. The highest vertical displacement of the canal was 4.151 mm, which was within the authorized settlement range, as shown in the figures. The U2 is the vertical component of the total settlement U.

Figure 12. Canal settlement nephogram when t = 5 years.

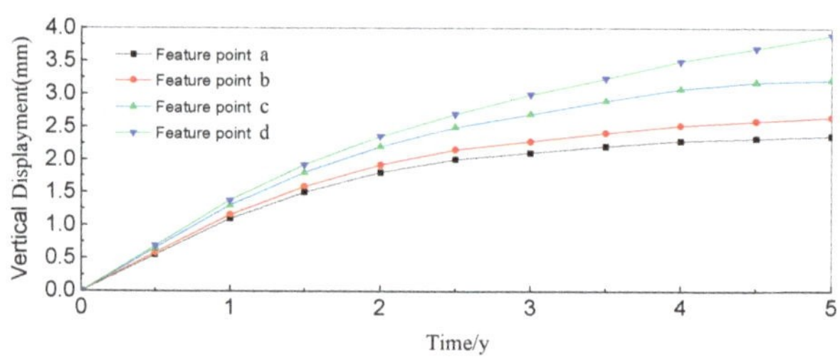

Figure 13. Settlements of each characteristic point when t = 5 years.

Based on the foregoing conclusions, a numerical simulation of the deep excavated canal is carried out, and a nephogram of the canal's settling clouds after ten years of operation is obtained (see Figure 14). According to the nephogram of canal settlements, the settlement curve of each characteristic point was obtained (as shown in Figure 15). It can be seen from the figures that the maximum settlement is 5.128 mm, which represents an increase of 0.977 mm compared to the five years of operation. This change is not significant, so it is believed that the settlement has reached the final settlement. Table 1 shows the settlement rate of the canal top's characteristic point at different times after the 10-year operation. The canal subsides at a decreasing rate after its operation. After 10 years of operation, the canal settlement rate is 0.16 mm/year. Although its absolute number is minimal and may not have much impact, the settlement continues.

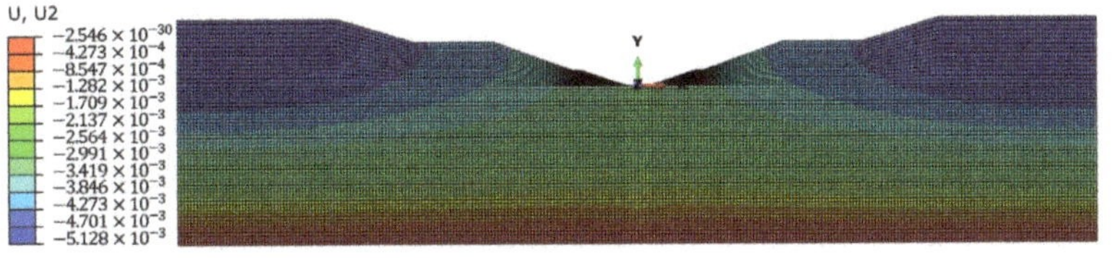

Figure 14. Canal settlement nephogram when t = 10 years.

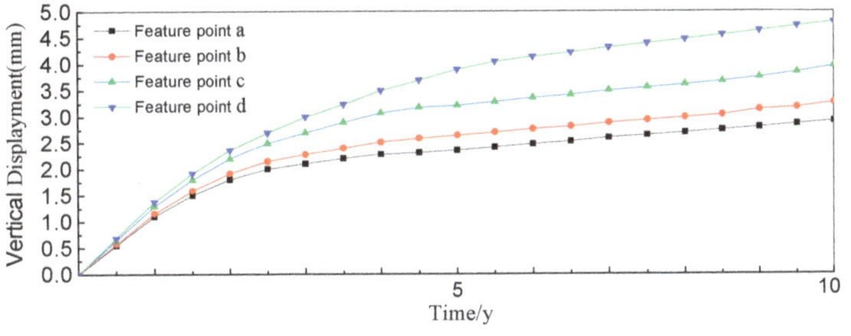

Figure 15. Settlement of each characteristic point when t = 10 years.

Table 1. Canal top's settlement rate.

Time/Year	1	2	3	4	5
Rate (mm/year)	1.38	0.98	0.64	0.5	0.4
Time/year	6	7	8	9	10
Rate (mm/year)	0.24	0.18	0.17	0.165	0.16

The numerical simulation is repeated to investigate the deep excavated canal's long-term settlement and deformation to get the nephogram of canal settlements after 20 years of operation, as shown in Figure 16. It is evident from the figure that the maximum settlement is very close to that calculated after 10 years of operation, indicating the deep excavated canal has reached a stable state, as shown in Figure 17. However, in actual operation, the stability of the deep excavated canal is affected by many factors such as complex water distribution conditions and varying environments, and significant settlements can occur locally. As the operating time increases, the canal's settlement becomes smaller, not affecting its operating safety.

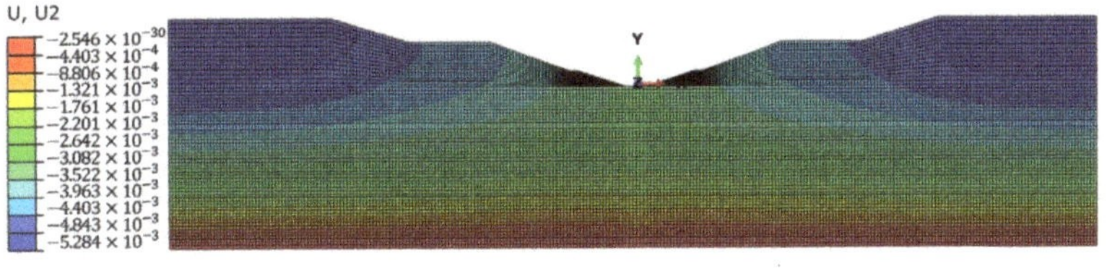

Figure 16. Canal settlement nephogram when t = 20 years.

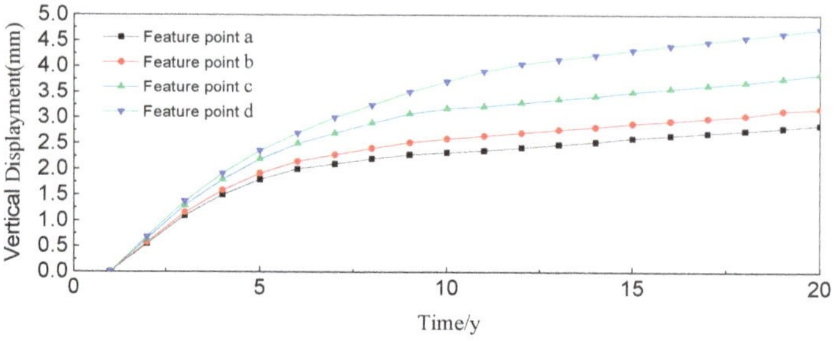

Figure 17. Settlement of each characteristic point when t = 20 years.

In this paper, the hypothetical SSD coupling method was used to predict the life and performance of the deep excavated canal in the SNWDP Xichuan Section. The numerical results showed that under normal operating conditions, the canal may only subside a little after 20 years of operation. The numerical simulation in this paper is based on the coupling of the seepage field and the stress field. It is expected that the canal operates will work stably for some time in the future.

5.3. Canal Lining Damage and Crack after Long-Term Operation

A numerical simulation of the evolution of the SNWDP's deep excavated canal after 50 years of regular operation was conducted by using the SSD coupling. The canal settlement nephogram was obtained (Figure 18). The displacement of each characteristic point is shown in Figure 19. Compared with the settlement nephogram after 20 years of operation, the settlement is insignificant and remains stable.

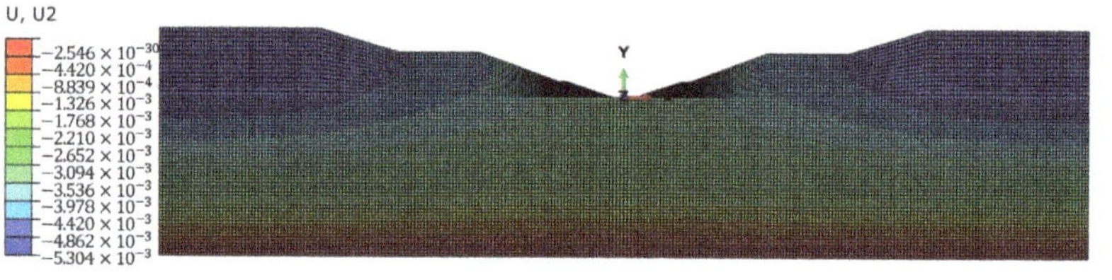

Figure 18. Canal settlement nephogram when t = 50 years.

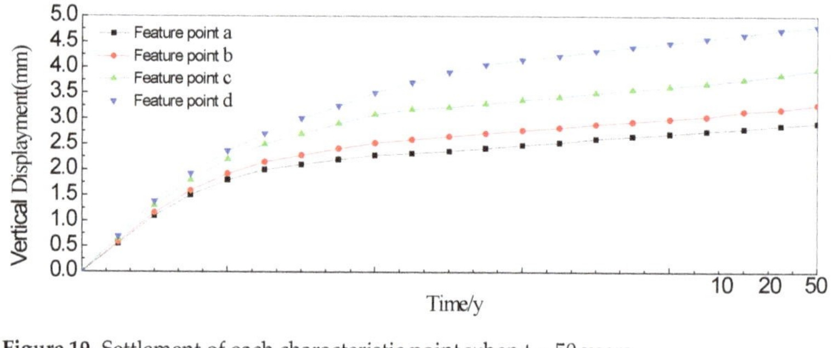

Figure 19. Settlement of each characteristic point when t = 50 years.

The seepage effect is larger in the deep excavated canal because much of the soil is saturated. As the settlement increases, both the void ratio and permeability coefficient decrease. The change in the seepage field affects the stress field and settlement at the macro level.

After 50 years of operation, the canal was discovered to be slightly damaged when investigating the evolution of its long-term behavior (Figure 20). In terms of overall damage, the maximum damage to the canal lining after the long-term operation is 37.6%. It is predicted that the deep excavated canal in the SNWDP Xichuan Section can still be used under normal operating conditions, for there is no large-area damage, except the concrete damage near the water surface. The damage first appeared near the water level of the left lining plate because the elevation of the top of the canal on the left bank was higher than that on the right bank, and the seepage–stress coupling effect was greater. The damage then occurred on the right lining plate, particularly at the point of the overflow of the water table in the canal, which was symmetrical to the point of damage on the left bank. The high level of the water table and its increasing water pressure with depth are the main causes of damage to the canal lining board. The water pressure differential between the inside and outside of the canal lining board is significant. The water pressure difference between the top and lower positions of the water surface inside the canal is minor, resulting in damage. The canal lining board is generally in a relatively safe condition, and there is no large-scale damage. Over time, the lining board may eventually be damaged in the form of uplifting after 50 years of operation, which may provide theoretical indications for the actual operation of the project. This study used the SSD coupling method to predict the service life and running status of the deep excavated canal in the SNWDP Xichuan Section under normal conditions, but it was limited to such external factors as environment, climate, and rainfall intensity [40]. More other influencing factors need to be studied further.

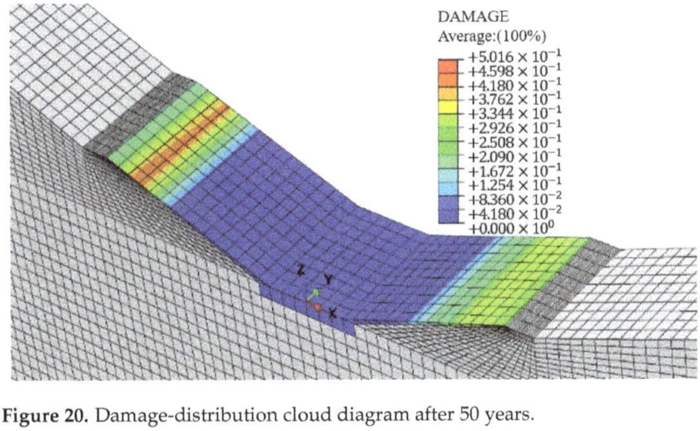

Figure 20. Damage-distribution cloud diagram after 50 years.

6. Conclusions

In this study, the SSD coupling method was adopted. The canal "lining-foundation" is considered a whole coupled system. The concrete lining's damage and the foundation's seepage damage were linked together, and long-term effects were introduced for numerical simulation and analysis of the canal lining structure. The conclusion is summarized as follows:

(1) The applicability of the SSD coupling method to the SNWDP in China: The long-term settlement of the canal and the modification of the seepage field were investigated by using a constitutive coupled plastic damage model of nonlinear dynamic damage of concrete, which is used in the numerical simulation of the coupled seepage–stress model. After five years of operation through modeling, the maximum settlement of the deep excavated canal in the SNWDP Xichuan Section is 4151 mm, which is consistent with the monitoring data, thus verifying the rationality of the numerical simulation method.

(2) Prediction on the settlement of the canal after long-term operation: Through the numerical simulation of the settlement after 10 years of operation, it is found that the maximum settlement is 5.128 mm, and the canal settlement mainly occurred in the first two years. Numerical simulations were used to obtain the settlement

nephograms after 20 and 50 years of operation. It is evident from the nephograms that the settlement finally remains the same.

(3) The damage evolution process of the canal after long-term operation: Judging by the overall damage of the lining plates on the canal of the SNWDP, the lining plate on the left bank was damaged first, and the damage was concentrated near the water level. After a long-term operation, the lining structure of the deep excavated canal in the SNWDP Xichuan Section is still safe, and there is no apparent damage.

(4) Causes of the damage and limitations of this study: The deep excavated canal in the SNWDP Xichuan Section is destroyed from the water surface, mainly due to the seepage of the high underground water. The damage spreads gradually from the local area, eventually leading to failure of the concrete lining plate. However, this study was limited by the environment, climate, and rainfall intensity. Therefore, more other influencing factors need to be studied further.

Author Contributions: Conceptualization, X.X. and W.X.; methodology, C.X.; software, C.X.; validation, W.X. and C.X.; formal analysis, W.X.; investigation, W.X.; resources, X.X.; data curation, X.X.; writing—original draft preparation, W.X.; writing—review and editing, W.X. and M.Y.A.K.; visualization, W.X.; supervision, X.X.; project administration, X.X.; funding acquisition, X.X. All authors have read and agreed to the published version of the manuscript.

Funding: This study was supported by the National Key Research and Development Program of China (Grant No. 2018YFC0406901), the National Natural Science Foundation of China (Grant No. 51979109), Henan Science and Technology Innovation Talent Program (Grant No. 174200510020), and Henan Province University Science and Technology Innovation Team Support Plan (Grant No. 19IRTSTHN030). These supports are gratefully acknowledged and greatly appreciated.

Institutional Review Board Statement: Not applicable.

Informed Consent Statement: Not applicable.

Data Availability Statement: Not applicable.

Conflicts of Interest: The authors declare no conflict of interest.

References

1. Kumar, A.; Taxak, A.K.; Mishra, S.; Pandey, R. Long term trend analysis and suitability of water quality of River Ganga at Himalayan hills of Uttarakhand, India. *Environ. Technol. Innov.* **2021**, *22*, 101405. [CrossRef]
2. Cui, H.D.; Zhang, J.F.; Zhang, W.; Wang, J.L. Numerical analysis of seepage field in typical confined water strata in the middle route of the South-to-North Water Diversion Project. *Rock Soil Mech.* **2010**, *31*, 447–451.
3. Kumar, A.; Cabral-Pinto, M.; Kumar, A.; Kumar, M.; Dinis, P.A. Estimation of Risk to the Eco-Environment and Human Health of Using Heavy Metals in the Uttarakhand Himalaya, India. *Appl. Sci.* **2020**, *10*, 7078. [CrossRef]
4. Cai, Y.F.; Qi, C.M. Slope Stability under Seepage-stress Coupling and Rainfall Infiltration. *J. Univ. South China (Sci. Technol.)* **2019**, *33*, 33–39.
5. Pan, Y.L.; Jian, W.X.; Li, L.J.; Lin, Y.Q.; Tian, P.F. A study on the Rainfall Infiltration of Granite Residual Soil Slope with An Improved Green-Ampt model. *Rock Soil Mech.* **2020**, *41*, 1–8.
6. Zhou, Z.; Zhang, J.M.; Ning, F.L.; Luo, Y.; Wang, J.L. Temporal and Spatial Characteristics of Moisture Migration and Instability Mechanism of Cracked Soil Slope under Rainfall Infiltration. *J. Traffic Transp. Eng.* **2020**, *8*, 1–8.
7. Huang, F.; Luo, X.; Liu, W. Stability Analysis of Hydrodynamic Pressure Landslides with Different Permeability Coefficients Affected by Reservoir Water Level Fluctuations and Rainstorms. *Water* **2017**, *9*, 450. [CrossRef]
8. Kim, J.; Han, H.; Jin, Y. Analysis of Pore Water Pressure and Piping of Hydraulic Well. *Water* **2021**, *13*, 502. [CrossRef]
9. Luo, B.; Sun, Y.; Xu, Z.; Chen, G.; Zhang, L.; Lu, W.; Zhao, X.; Yuan, K. Damage Characteristics and Mechanism of the 2017 Groundwater Inrush Accident That Occurred at Dongyu Coalmine in Taiyuan, Shanxi, China. *Water* **2021**, *13*, 368. [CrossRef]
10. Zhao, E.; Jiang, Y. Seepage Evolution Model of the Fractured Rock Mass under High Seepage Pressure in Dam Foundation. *Adv. Civ. Eng.* **2021**, *2021*, 8832774. [CrossRef]
11. Nian, G.Q.; Chen, Z.H.; Zhang, L.F.; Bao, M. Treatment of Two Boundary Conditions for Rainfall Infiltration in Slope and Its Application. *Rock Soil Mech.* **2020**, *12*, 1–11.
12. Wang, K.; Wang, L.; Ren, B.; Fan, H. Study on Seepage Simulation of High Pressure Grouting in Microfractured Rock Mass. *Geofluids* **2021**, *2021*, 6696882. [CrossRef]
13. Ma, Z.T.; Cui, Y.Q.; Lu, K.Y.; Song, D.F.; Yang, Y.C. Study on Fracture and Permeability Characteristics of Unloading Rock with High Water Pressure. *Math. Probl. Eng.* **2021**, *2021*, 6685405. [CrossRef]

14. Ma, Z.; Zhu, C.; Yao, X.; Dang, F. Slope Stability Analysis under Complex Stress State with Saturated and Unsaturated Seepage Flow. *Geofluids* **2021**, *2021*, 6637098. [CrossRef]
15. Liu, Z.J.; Huang, Y.; Zhou, D.; Ge, H. Analysis of External Water Pressure for a Tunnel in Fractured Rocks. *Geofluids* **2017**, *2017*, 8618613. [CrossRef]
16. Xiao, X.L. Analysis of Slope Stability on Account of Fluid-Structure Coupling Seepage Effect. Ph.D. Thesis, Huazhong University of Science and Technology, Wuhan, China, 2011.
17. Chen, L.G. The Slope Stability Analysis of Taking into Account the Seepage and Stress Coupling Effect Based on ABAQUS. Ph.D. Thesis, Zhengzhou University, Zhengzhou, China, 1 May 2010.
18. Cai, F.; Ugai, K. Numerical Analysis of Rainfall Effects on Slope Stability. *Int. J. Geomech.* **2004**, *4*, 69–78. [CrossRef]
19. Rahardjo, H.; Ong, T.H.; Rezaur, R.B.; Leong, E.C. Factors Controlling Instability of Homogeneous Soil Slopes under Rainfall. *J. Geotech. Geoenviron. Eng.* **2007**, *133*, 1532–1543. [CrossRef]
20. Baum, R.L.; Godt, J.W.; Savage, W.Z. Estimating the timing and location of shallow rainfall-induced landslides using a model for transient, unsaturated infiltration. *J. Geophys. Res. Earth Surf.* **2010**, *115*, 26. [CrossRef]
21. Rahardjo, H.; Nio, A.S.; Leong, E.C.; Song, N.Y. Effects of Groundwater Table Position and Soil Properties on Stability of Slope during Rainfall. *J. Geotech. Geoenviron. Eng.* **2010**, *136*, 1555–1564. [CrossRef]
22. Muntohar, A.S.; Liao, H.J. Rainfall infiltration: Infinite slope model for landslides triggering by rainstorm. *Nat. Hazards* **2010**, *54*, 967–984. [CrossRef]
23. Tsai, T.L.; Chen, H.F. Effects of degree of saturation on shallow landslides triggered by rainfall. *Environ. Earth Sci.* **2010**, *59*, 1285–1295. [CrossRef]
24. Borja, R.I.; White, J.A. Continuum deformation and stability analyses of a steep hillside slope under rainfall infiltration. *Acta Geotech.* **2010**, *5*, 1–14. [CrossRef]
25. Zhou, Y.F. Study on Seepage-stress-damage/crack Coupling Theory and Method for Hydraulic Tunnel. Ph.D. Thesis, Wuhan University, Wuhan, China, 1 May 2015.
26. Li, Z.; Kai, S.; Yafen, Z. Hydro-mechanical Coupling Analysis of Pervious Lining in High Pressure Hydraulic Tunnel. *J. Hydraul. Eng.* **2018**, *49*, 313–322.
27. Sheng, J.L.; Ye, J. Slope Stability of Foundation Pit Considering the Seepage-stress Coupling Action. *J. Wuhan Univ. Sci. Technol.* **2015**, *38*, 391–395.
28. Xu, B.; Chen, B.; Huang, H.; Qi, Y.; Xinhua, X.; Xu, H. Underground water migration studies based on the damage variable coupling with seepage and rock stress. *J. Food Agri. Environ.* **2013**, *11*, 786–790.
29. Zhu, H.Y.; Jin, X.C.; Guo, J.C.; An, F.C.; Wang, Y.H.; Lai, X.D. Coupled flow, stress and damage modelling of interactions between hydraulic fractures and natural fractures in shale gas reservoirs. *Int. J. Oil Gas Coal Technol.* **2016**, *13*, 359–390. [CrossRef]
30. Kumar, A.; Sharma, M.P.; Taxak, A.K. Analysis of water environment changing trend in Bhagirathi tributary of Ganges in India. *Desalin Water Treat* **2017**, *63*, 55–62. [CrossRef]
31. Chen, H.F. *Constitutive Equation of Concrete and Soil [M]*; China Construction Industry Press: Beijing, China, 2004; p. 258.
32. Kumar, A.; Mishra, S.; Taxak, A.; Pandey, R.; Yu, Z.-G. Nature rejuvenation: Long-term (1989–2016) vs. short-term memory approach based appraisal of water quality of the upper part of Ganga River, India. *Environ. Technol. Innov.* **2020**, *20*, 101164. [CrossRef]
33. Wu, M.X. Finite-element algorithm for Richards' equation for saturated-unsaturated seepage flow. *J. Hydraul. Eng.* **2009**, *40*, 1274–1279.
34. Ji, Y.J.; Liu, J.J.; Cheng, L.S. Numerical simulation of tunnel excavation considering fluid-solid coupling. *Rock Soil Mech.* **2011**, *32*, 1229–1233.
35. Kang, B.; Ming, X. Seepage-damage-stress coupling analysis of hydraulic tunnel lining during hydraulic fracturing. *Chin. J. Rock Mech. Eng.* **2010**, *29*, 3769–3776.
36. Lubliner, J.; Oliver, J.; Oller, S.; Onate, E. A plastic-damage model for concrete. *Int. J. Solids Struct.* **1989**, *25*, 299–326. [CrossRef]
37. Jia, S.P.; Chen, W.Z.; Yu, H.D. Study on the fully coupled damage model of seepage field and stress field during mudstone tunnel construction. *Rock Soil Mech.* **2009**, *30*, 19–26.
38. Lee, J.; Fenves, G.L. Plastic damage model for cyclic loading of concrete structure. *J. Eng. Mech.* **1998**, *124*, 892–900. [CrossRef]
39. Fei, K.; Zhang, J.W. *Application of ABAQUS in Geotechnical Engineering [M]*; China Water Conservancy and Hydropower Press: Beijing, China, 2010; p. 222.
40. Hu, J.; Zhang, J.K.; Yu, M.X.; Ma, F.H.; Xiao, W.S. Analysis and prediction of slope deformation characteristics of deep-excavated expansive soil channel. *Hydro-Sci. Eng.* **2021**, *4*, 1–9.

MDPI
St. Alban-Anlage 66
4052 Basel
Switzerland
Tel. +41 61 683 77 34
Fax +41 61 302 89 18
www.mdpi.com

Applied Sciences Editorial Office
E-mail: applsci@mdpi.com
www.mdpi.com/journal/applsci

www.ingramcontent.com/pod-product-compliance
Lightning Source LLC
Chambersburg PA
CBHW040827250926
51850CB00007BA/1282
9783036596433